Layered Materials for Energy Storage and Conversion

Smart Materials

Series editors:
Hans-Jörg Schneider, *Saarland University, Germany*
Mohsen Shahinpoor, *University of Maine, USA*

Titles in this series:

1: Janus Particle Synthesis, Self-Assembly and Applications
2: Smart Materials for Drug Delivery: Volume 1
3: Smart Materials for Drug Delivery: Volume 2
4: Materials Design Inspired by Nature
5: Responsive Photonic Nanostructures: Smart Nanoscale Optical Materials
6: Magnetorheology: Advances and Applications
7: Functional Nanometer-Sized Clusters of Transition Metals: Synthesis, Properties and Applications
8: Mechanochromic Fluorescent Materials: Phenomena, Materials and Applications
9: Cell Surface Engineering: Fabrication of Functional Nanoshells
10: Biointerfaces: Where Material Meets Biology
11: Semiconductor Nanowires: From Next-Generation Electronics to Sustainable Energy
12: Supramolecular Materials for Opto-Electronics
13: Photocured Materials
14: Chemoresponsive Materials: Stimulation by Chemical and Biological Signals
15: Functional Metallosupramolecular Materials
16: Bio-Synthetic Hybrid Materials and Bionanoparticles: A Biological Chemical Approach Towards Material Science
17: Ionic Polymer Metal Composites (IPMCs): Smart Multi-Functional Materials and Artificial Muscles Volume 1
18: Ionic Polymer Metal Composites (IPMCs): Smart Multi-Functional Materials and Artificial Muscles Volume 2
19: Conducting Polymers: Bioinspired Intelligent Materials and Devices
20: Smart Materials for Advanced Environmental Applications
21: Self-cleaning Coatings: Structure, Fabrication and Application
22: Functional Polymer Composites with Nanoclays
23: Bioactive Glasses: Fundamentals, Technology and Applications
24: Smart Materials for Tissue Engineering: Fundamental Principles
25: Smart Materials for Tissue Engineering: Applications
26: Magnetic Nanomaterials: Applications in Catalysis and Life Sciences
27: Biobased Smart Polyurethane Nanocomposites: From Synthesis to Applications
28: Inorganic Two-dimensional Nanomaterials: Fundamental Understanding, Characterizations and Energy Applications
29: Ionic Liquid Devices

30: Polymerized Ionic Liquids
31: Nanogels for Biomedical Applications
32: Reactive Inkjet Printing: A Chemical Synthesis Tool
33: Electrochromic Smart Materials: Fabrication and Applications
34: Layered Materials for Energy Storage and Conversion

How to obtain future titles on publication:
A standing order plan is available for this series. A standing order will bring delivery of each new volume immediately on publication.

For further information please contact:
Book Sales Department, Royal Society of Chemistry, Thomas Graham House, Science Park, Milton Road, Cambridge, CB4 0WF, UK
Telephone: +44 (0)1223 420066, Fax: +44 (0)1223 420247
Email: booksales@rsc.org
Visit our website at www.rsc.org/books

Layered Materials for Energy Storage and Conversion

Edited by

Dongsheng Geng
University of Science and Technology Beijing, China
Email: dgeng@ustb.edu.cn

Yuan Cheng
*Institute of High Performance Computing, A*STAR, Singapore*
Email: chengy@ihpc.a-star.edu.sg

and

Gang Zhang
*Institute of High Performance Computing, A*STAR, Singapore*
Email: zhangg@ihpc.a-star.edu.sg

ROYAL SOCIETY OF CHEMISTRY

Smart Materials No. 34

Print ISBN: 978-1-78801-426-7
PDF ISBN: 978-1-78801-619-3
EPUB ISBN: 978-1-78801-738-1
Print ISSN: 2046-0066
Electronic ISSN: 2046-0074

A catalogue record for this book is available from the British Library

The Royal Society of Chemistry is a charity, registered in England and Wales, Number 207890, and a company incorporated in England by Royal Charter (Registered No. RC000524), registered office: Burlington House, Piccadilly, London W1J 0BA, UK, Telephone: +44 (0) 20 7437 8656.

For further information see our web site at www.rsc.org

Printed in the United Kingdom by CPI Group (UK) Ltd, Croydon, CR0 4YY, UK

Contents

Smart Materials No. 34
Layered Materials for Energy Storage and Conversion
Edited by Dongsheng Geng, Yuan Cheng and Gang Zhang

Published by the Royal Society of Chemistry, www.rsc.org

CHAPTER 1

MOF-derived Materials for Extremely Efficient Electrocatalysis

ZHONGXIN SONG, LEI ZHANG, MATTHEW ZHENG AND XUELIANG SUN*

University of Western Ontario, Department of Mechanical and Materials Engineering, London N6A 5B9, Canada
*Email: xsun@eng.uwo.ca

1.1 Introduction

Nanostructured materials such as porous carbon, metal/metal oxides nanoparticles (NPs), and their composites have been widely investigated in the field of electrocatalysis.[1–5] Metal–organic frameworks (MOFs) as precursors and/or templates for the design of functional porous materials have become a rapidly expanding research area in recent years.[6] As their name suggests, MOFs are constructed by periodic coordination of metal ions and organic ligands to form one to three-dimensional (3D) networks. The organic linkers are enormously diverse and have a variety of configurations. The vast numbers of metal ions and organic linkers as well as their diverse assemblies thus have led to the reporting of more than 20 000 MOFs.[7] In addition to their adjustable compositions, another distinctive advantage of MOFs is their ordered pore structure, uniform pore size, and high specific surface area in contrast to traditional microporous and mesoporous

Smart Materials No. 34
Layered Materials for Energy Storage and Conversion
Edited by Dongsheng Geng, Yuan Cheng and Gang Zhang

Published by the Royal Society of Chemistry, www.rsc.org

materials. The pore size of MOFs can be adjusted from a few angstroms to nanometres,[8,9] which enables the access of reactants and removal of products within a specific size, ensuring rapid mass diffusion and transport.

The remarkable advantages of MOF-based materials in catalysis mainly arise from the inherent properties of MOF precursors: MOFs possess highly dense and uniformly dispersed active sites; the high surface area, porous structure, and open channels facilitate rapid mass transport and diffusion. These advantages enable MOF-based nanomaterials to be promising solid catalysts, especially in electrochemical catalysis. However, the coordinatively unsaturated metal sites and nonconductive organic ligands limit MOFs to certain electrocatalytic reactions. Fortunately, this challenge can be alleviated by two approaches: (i) functional modification: it is possible to convert the metal ions into metal/metal compounds while carbonizing the organic linker into a conductive carbon support by a precise post-treatment modification. (ii) Pore encapsulation: MOF structures can incorporate various catalytic active species into their pore space and behave as nanoreactors to host catalytic reactions.[10–12] Moreover, the obtained nanocomposites derived from MOF precursors display high surface area, porous structure, and uniformly dispersed active sites, which were found to be important properties in electrocatalysis. In this chapter, we describe several unique structures and compositions of MOF-derived materials, and then highlight the recent progress of MOF-derived nanocomposites for electrocatalysis. Finally, the major challenges of MOF-based materials and their research opportunities for further development in electrocatalysis are discussed.

1.2 MOF-derived Materials: Structures and Compositions

Recently, MOFs have gained increasing attention due to their ordered framework and porous structures.[13–15] Enabled by their diverse structures, high surface area, and permanent porosity, MOFs are considered as novel precursors to construct functional materials such as nanoporous carbon, nano-metal compounds (*e.g.* metal oxides/sulphides/carbides), and their composites.[11,16] Generally, MOF crystals with controllable size, shape, and compositions can be obtained by incorporating the desired metal ions and organic ligands during the MOFs' synthesis. Subsequently, post-synthetic modification can be used to treat MOF precursors and transfer the precursors into MOF-derived materials with diverse nanostructures and compositions. This section describes the multiple nanostructures (*e.g.* porous, core–shell, hollow structures, and 3D composites structure, as shown in Figure 1.1) and compositions designed from novel MOF precursors, with an emphasis on their attractive properties and unique functions for boosting electrocatalysis.

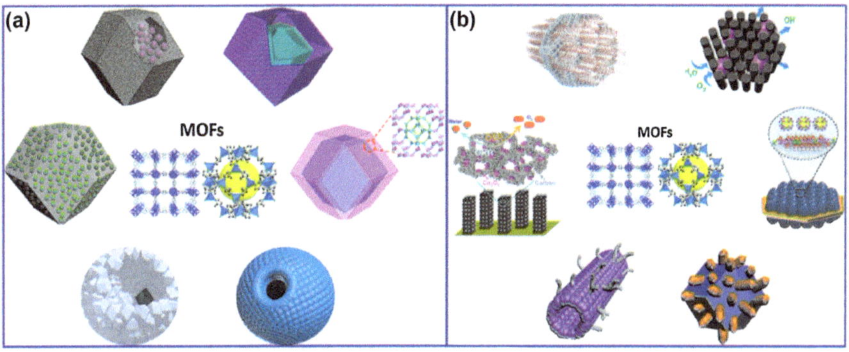

Figure 1.1 Schematic of MOF-derived nanostructures: (a) MOF-derived structures, (b) MOF-derived micro/nano-composites.

1.2.1 Structures of MOF-derived Materials

1.2.1.1 MOF-derived Porous Carbon

MOFs can be used as ideal sacrificial templates to construct diverse nanostructured materials such as porous carbon, metal compounds, and their composites. These MOF-derived materials can inherit the advantages of MOF precursors, especially their high surface area and tailorable porosity. As an example, Yamauchi *et al.*[10,17] used Zn-based MOFs (zeolitic imidazolate framework, ZIF-8) as self-sacrificing templates for the preparation of nitrogen-doped porous carbons (NPCs). The resulting NPCs not only maintained the polyhedral morphology of the ZIF-8 precursor, but also exhibited a large surface area (up to 1110 $m^2 g^{-1}$) and hierarchical porosity. The NPCs maintained an average pore size with a diameter of 1.06 nm, which was much closer to the pore size of the parent ZIF-8. By delicate design of the MOF precursor, for example, integrating the properties of ZIF-8 and ZIF-67 nanocrystals, the core–shell structured ZIF-8@ZIF-67 polyhedrons with ZIF-8 as the core and ZIF-67 as the shell can be well defined.[18] By choosing ZIF-8 seeds with different sizes, the core size of ZIF-8 can be tuned. Simultaneously, varying the feeding molar ratio of $Co^{2+}:Zn^{2+}$, the shell thicknesses of ZIF-67 can be adjusted. After thermal treatment of a ZIF-8@ZIF-67 precursor and then Co removal by acid etching, the functionalized nanoporous carbon hybrid with a core structure of nitrogen-doped porous carbon and shell of highly graphitic carbon (GC) was obtained. In particular, the resultant NPC@GC core–shell materials possessed an interconnected hierarchically micro/mesoporous structure due to the carbonization of organic linkers and removal of metal atoms from the parent ZIF-8@ZIF-67. These results suggest the possibility of designing MOF-derived nanocarbon with a porous character partially inherited from the parent MOFs.

1.2.1.2 MOF-derived Hollow Structures

Hollow structures with a high specific surface area (external and internal surface), enhanced material utilization and efficiency, as well as high stability, have been considered as one of the most attractive structures for energy storage and conversion application.[19–21] Construction of hollow structures with tuneable architectures can be enabled by using MOF strategies. Recently, Guan *et al.*[22] reported the ZIF-7 strategy for the fabrication of hollow structured cobalt/nitrogen-doped carbon (Co/NC) materials. In their work, ZIF-67 crystals were synthesized and self-assembled on the surface of polystyrene spheres (PS) to construct yolk–shell PS/ZIF-67 composite spheres. Subsequently, by a controlled pyrolysis, the PS/ZIF-67 composite was transformed into hollow particles with a single hole on the surface of the shell. During high temperature PS/ZIF-67 pyrolysis, the ZIF-67 micro-shells were transformed to Co/NC, while isolated Co nanocrystals were covered by a porous carbon layer. The PS sphere, a thermally degradable template, was decomposed and generated a strong gas outflux of hydrocarbon, leading to the formation of a single large-through hole on the shell of the resultant Co/NC spheres. Importantly, the authors found that the open size of the hole has a relationship with the heating rate of the pyrolysis. A higher heating rate decomposes PS more rapidly and results in a large-sized hole on the final hollow Co/NC shells.

By appropriate design of the composition of MOF precursors and by carefully controlling their subsequent post-treatment, hollow materials with unique architectures can be achieved. For instance, yolk–shell hollow structures and multi-shelled hollow polyhedrons have been fabricated by innovative MOFs' synthesis and post-synthetic modification. In 2012, Kuo and co-workers[23] coated pre-synthesized palladium (Pd) nanocrystals with a Cu_2O layer, then *in situ* synthesized an outer shell of polycrystalline ZIF-8. The clean Cu_2O layer with a capping-agent-free surface contributed to the growth of a ZIF-8 shell, which was simultaneously dissolved by the protons' environment during the ZIF-8 synthesis. The trace amount of Cu_2O residue can be removed by treatment with a solution of 3% NH_4OH in methanol, thus forming the obvious void between the metal Pd cores and the ZIF-8 shells. The morphology and property of the metal nanocrystals were well preserved during the coating of the ZIF-8 shell. Through this strategy, a series of metal nanocrystals@ZIF-8 yolk–shell composites can be constructed. The yolk–shell structures suitably incorporate the functions of metal cores, porous shells, and the cavity between the core and shell, which provides a typical example in the rational design MOF-derived materials with hollow architectures. Recently, Lou *et al.*[24] presented a two-step method to synthesize box-in-box double-shelled nanocages with different shell compositions. In their experiment, the uniform Co-based ZIF-67 NPs were synthesized in the first step. The ZIF-67 NPs were then dispersed in an ethanol solution of $Ni(NO_3)_2$ to generate ZIF-67/Ni-Co layered double hydroxides (LDH) yolk–shelled structures. After thermal calcination in air, the ZIF-67 cores and Ni-Co LDH shells can be further transformed into Co_3O_4 and $NiCo_2O_4$ nanocages, respectively. Both the Co_3O_4 inner shell and the $NiCo_2O_4$ outer shell can remain intact after thermal treatment, resulting in Co_3O_4-$NiCo_2O_4$ double-shelled nanocages. This two-step strategy can be applied to ZIF-67 NPs with different particle sizes, thus giving rise to the size-controlled synthesis of Co_3O_4-$NiCo_2O_4$ double-shelled nanocages. Several kinds of double-shelled nanocages with inner Co_3O_4 and outer metal oxide (*e.g.* Fe/Mg/CoO_x) nanocages can be developed *via* this highly versatile strategy.

1.2.2 Compositions of MOF-derived Materials

1.2.2.1 Metal-free Nanocarbon

Due to the presence of carbon-containing organic linkers in MOF crystals, nanocarbon materials are easily constructed through carbonization of MOF precursors and removal of metal species. Thus far, several Zn-based MOFs, such as ZIF-8, MOF-5, and MOF-74 have been demonstrated to be promising self-sacrificial precursors to produce metal-free nanocarbon.[25–27] Generally, The zeolitic imidazolate frameworks (ZIFs) have been adopted as both templates and precursors to develop nitrogen-doped porous nanocarbons, due to their highly ordered porosity and organic ligands with rich nitrogen.[28] In 2014, Zhang *et al.*[29] demonstrated the *in situ* synthesis of NPCs by high temperature pyrolysis of a ZIF-8 precursor. This approach enables the simultaneous incorporation of carbon and nitrogen species, resulting in the formation of NPCs after the carbonization of a ZIF-8 precursor. Additionally, the remarkable characteristics of ZIF-8 such as being rich in nitrogen, having hierarchical porosity, as well as ordered 3D networks, could be passed to the derivative material of NPCs (Figure 1.2a). Moreover, it was revealed that the total nitrogen content in NPCs could be controlled from 23.9 to 5.82 at% when the carbonization temperature was increased from 700 to 1000 °C. Considering that organic ligands are also composed of various heteroatoms (N, P, S, *etc.*) other than carbon, heteroatoms-doped nanocarbon with highly dispersed doping sites could also be designed from MOF precursors. To further tune the electronic properties and conductivity of nanocarbon, incorporating multi-heteroatoms into the MOF-derived nanocarbon has been increasingly investigated in recent years. For example, Sun *et al.* reported the facile synthesis of MOF-derived nanoporous carbon couple-doped with nitrogen and sulphur atoms.[30] Thiourea was used as a sulphur precursor, which was encapsulated into the pore structures of a ZIF-8 template (as a carbon and nitrogen source). Under a high-temperature carbonization process, nitrogen and sulphur couple-doped nanoporous carbon (N/S-NPC) was produced (Figure 1.2b). The physical characterizations indicated that N/S-NPC demonstrated a high specific surface area, graphitic and porous structure, and was rich in nitrogen and sulphur doping sites. The X-ray photoelectron spectroscopy indicated that 5.4 at% nitrogen and 0.3 at%

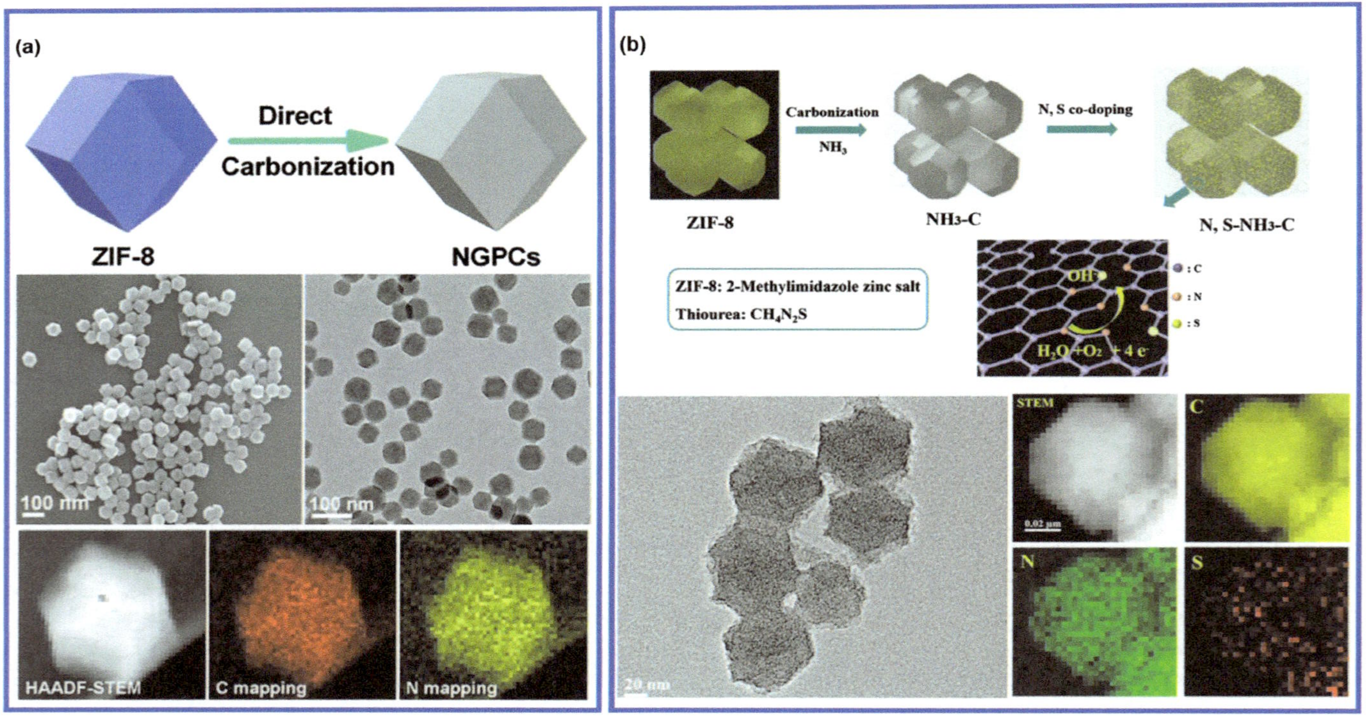

Figure 1.2 (a) Schematic illustration of the synthesis of nitrogen-doped porous carbon (NPC) from a ZIF-8 precursor; scanning electron microscope (SEM), transmission electron microscope (TEM) images and corresponding element maps of NPC nanopolyhedra. Reproduced from ref. 29 with permission from the Royal Society of Chemistry. (b) Schematic illustration of the fabrication of N, S-co-doped porous carbon from a ZIF-8 precursor; TEM images and element maps of the N/S-carbon polyhedral. Reproduced from ref. 30 with permission from the Royal Society of Chemistry.

sulphur were doped in the as-prepared N/S-NPC. Significantly, it was revealed that the MOF-derived N/S-co-doped nanocarbon exhibited excellent catalytic activity for the oxygen reduction reaction (ORR).

1.2.2.2 Transition Metal/Metal Compound-decorated Nanocarbon

MOFs are one class of compounds consisting of metal ions strongly coordinated to organic ligands to form one-, two- or three-dimensional structures. Based on such composition, MOFs can be converted into metal-decorated nanocarbon materials (*e.g.* metal/carbon, metal/metal-compound/carbon) by pyrolysis treatment at a range of temperatures between 400–1000 °C.[31–33] Yamauchi's group synthesized cobalt-decorated N-doped porous carbon (Co/NPC) by pyrolysis of a ZIF-67 precursor at 600–800 °C under N_2 atmosphere.[34] It was found that the rise in annealing temperature from 600 to 800 °C generated an increase in the crystallization of the prepared Co NPs. It was found that higher temperatures contribute to better graphitization of the porous carbon matrix and results in enhanced conductivity and stability for the as-prepared Co/NPC. This approach allowed the combination of the catalytic active sites of Co NPs with the conductive and porous N-doped nanocarbon matrix, which was considered as a potential material for electrocatalysis. Known from the metal–organic composition, the MOF precursors can be used to develop (i) metal NPs-doped carbon hybrids, or (ii) porous metal oxide materials. Li *et al.*[35] used a one-step air calcination strategy to directly convert a Co-MOF (ZIF-67) precursor into hierarchically porous Co_3O_4 with honeycomb-like architectures. The Co_3O_4 architectures displayed abundant porosity and oxygen vacancy, which exhibit distinct electrocatalytic activity for the oxygen evolution reaction (OER). As is commonly understood, the metal oxide NPs tend to agglomerate under electrochemical catalysis conditions, which is seriously detrimental to their activity and stability. Encapsulation of these metal oxide species into a porous stable substrate has been considered as a potential method to develop highly stable metal hybrids due to the pore entrapping effect. Hou *et al.* embedded Co_3O_4 NPs in N-doped porous carbon (Co_3O_4/NPC) by a two-step thermal decomposition of ZIF-67.[33] Firstly, the ZIF-67 precursor was heat-treated in an inert atmosphere to obtain Co/NPC. Then, by oxidation of Co_3O_4/NPC in air at 350 °C, Co_3O_4/NPC was finally achieved. The developed Co_3O_4/NPC product maintained the geometry of the parent ZIF-67 framework and generated highly stable Co_3O_4 NPs. The novel integration of Co_3O_4 NPs into N-doped carbon networks demonstrated a strong synergistic effect between Co_3O_4 and NPC, which enhanced the intrinsic properties of each individual component.

Apart from the metal oxide derivatives, metal phosphides (MPs) have gained great attention because of their excellent activity in electrocatalysis applications. In comparison to traditional synthetic methods of MPs, MOF-oriented approaches provide the opportunity to tailor MP materials with a high surface area and well-defined porous structure. In 2015, You *et al.* reported a ZIF-67-derived synthetic route to prepare porous CoPx/NC polyhedrons composed of CoP and Co_2P NPs embedded in a N-doped carbon matrix (as shown in Figure 1.3a).[36] In their experimental synthesis, the ZIF-67 precursors first underwent pyrolysis in an Ar atmosphere to obtain Co–N–C polyhedrons, which were subjected to phosphating at 300 °C with an extra phosphorous source (sodium hypophosphite monohydrate) to achieve the resultant CoPx/NC polyhedrons. The X-ray diffraction (XRD) patterns indicated that metallic Co NPs with a crystallite size of ~33 nm were transformed into CoPx NPs (composed of CoP and Co_2P) after the phosphating reaction. Energy dispersive X-ray spectroscopy (EDS) mapping confirmed the presence of Co, P, N, and C components in the resultant Co-Px/NC, and showed evidence that the P element is highly localized within the region of Co, indicating the formation of CoPx NPs.

The varied compositions and components of the hollow structured materials have also been investigated, both of which are of great importance for the development of MOF-derivates. Recently, Lou *et al.* developed a series of novel hollow structures ranging from iron-based nano-boxes, molybdenum carbides (MoCx) nano-octahedrons, nickel sulphide (NiSx) nano-frames, to complex double-shelled nanocages, derived from various MOF-based precursors.[24,37–39] A MOF-engaged strategy was applied to synthesize the Co NP embedded carbon@cobalt sulphide (Co_9S_8) double-shelled nanocages (Co-C@Co_9S_8 DSNCs) (as shown in Figure 1.3b).[37] A well-defined ZIF-67@CoS yolk–shell structure was initially synthesized, followed by thermal annealing of ZIF-67@CoS to convert ZIF-67 and CoS into Co-doped-carbon inner shells and Co_9S_8 outer shells, respectively. Interestingly, the Co-C@Co_9S_8 DSNCs with complex cores, robust shell structures and a cavity in-between can be designed. In the structure of Co-C@Co_9S_8 double-shelled nanocages, it was found that the inner Co–C shells as the active centres function by enhancing the catalytic activity, whereas the external Co_9S_8 shells construct the fence structure on the outside Co–C centres that benefit in preventing Co–C from aggregation and deactivation. The synergy between the Co–C and Co_9S_8 double shells endows the Co-C@Co_9S_8 DSNCs with excellent catalytic activity and superior stability.

Besides the examples of MOF-derived mono-metal-doped nanocarbons, multimetal-contained nanostructures can be prepared from MOFs or MOF composite precursors. As shown in the schematic illustration in Figure 1.3c, hierarchical nanotubes can be synthesized from the core-shelled Fe_2Ni-MIL-88@Fe-MIL-88 MOF precursors.[40] The MOFs of Fe_2Ni-MIL-88 nanorods were initially synthesized *via* a hydrothermal approach. Then a cover layer of Fe-MIL-88 was prepared on the outer surface of Fe_2Ni-MIL-88 nanorods. The core-shelled Fe_2Ni-MIL-88@Fe-MIL-88 architecture was thus fabricated *via* a consecutive seed-mediated growth technique. After thermal treatment

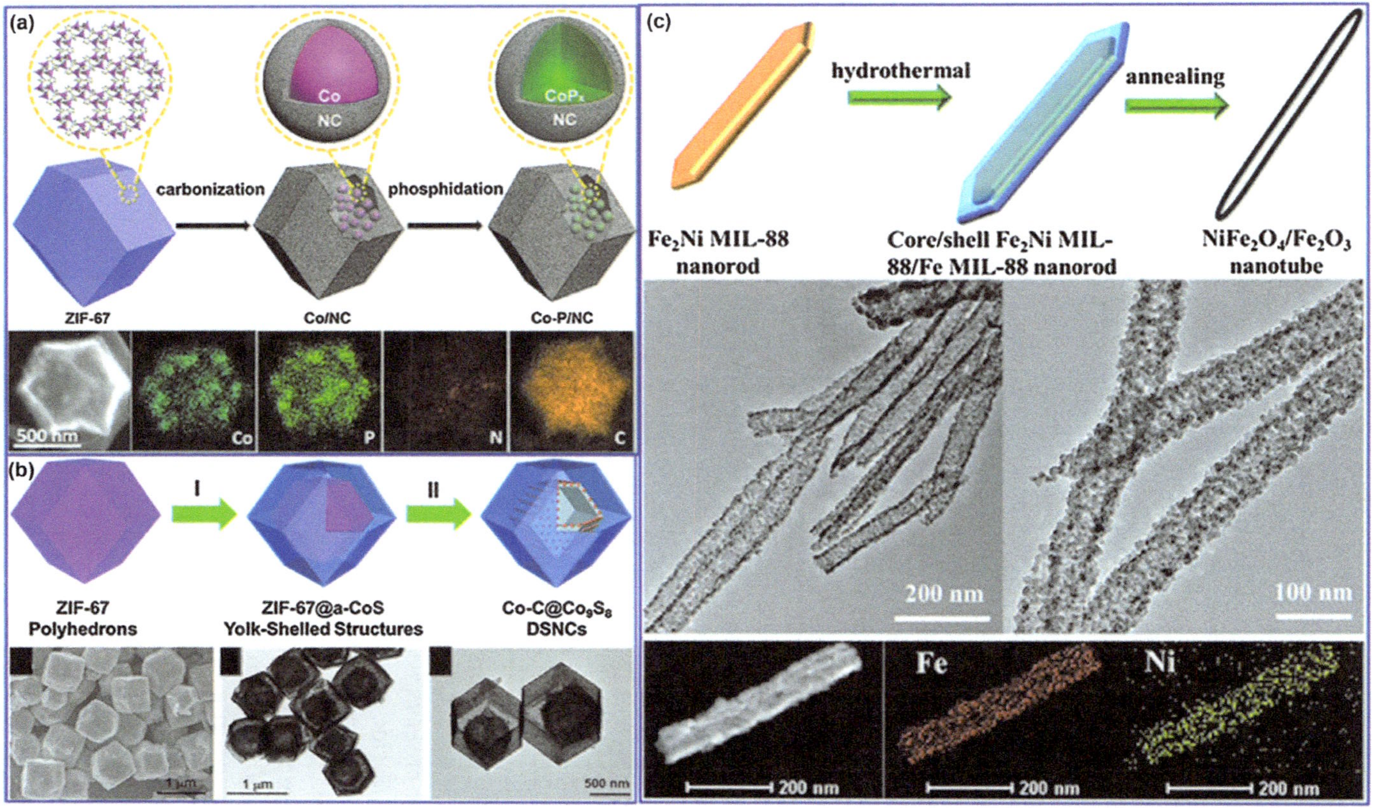

Figure 1.3 (a) Illustration of the two-step synthesis of CoP/NC nanopolyhedrons; TEM and elemental mapping images of CoP/NC. Reproduced from ref. 36 with permission from American Chemical Society, Copyright 2015. (b) Schematic illustration of the synthesis process of CoC@Co_9S_8 double-shelled nanocages; SEM and TEM revealing the formation of CoC@Co_9S_8 double-shelled nanocages. Reproduced from ref. 37 with permission from the Royal Society of Chemistry. (c) Schematic illustration of the synthesis of $NiFe_2O_4/Fe_2O_3$ nanotubes; TEM images and elemental mapping of the $NiFe_2O_4/Fe_2O_3$ nanotubes. Reproduced from ref. 40 with permission from the Royal Society of Chemistry.

of this core-shelled MOF composite in air, a new type of selectively functionalized $NiFe_2O_4/Fe_2O_3$ nanotubes was generated. Scanning electron microscope (SEM) and transmission electron microscope (TEM) images showed that the $NiFe_2O_4/Fe_2O_3$ composite exhibited the morphology of nanotubes with diameters of 78 nm and lengths of 1 μm. Element mapping in Figure 1.3c indicates that the Fe and Ni elements are uniformly distributed throughout the nanotubes matrix, suggesting a homogeneous mixture of $NiFe_2O_4$ and Fe_2O_3.

1.2.2.3 *Micro/Nano-structured MOF-derived Composites*

It has been observed that MOF crystals demonstrate poor stability and tend to randomly aggregate with neighbouring particles, which results in large particle formation during the post-treatment of MOF precursors. Although the MOF crystals' initial morphology and shape can be well maintained, the effective surface area would be reduced significantly under some high temperature post-treatment. Thus, it is desirable to effectively control the dimensions of MOF crystals by introducing an extra substrate. A combination of MOFs with functional substrates such as one-dimensional (1D) carbon nanotubes/nanofibers, 2D graphene sheets, and three-dimensional (3D) porous carbons, could enable the fabrication of supported composites with uniformly dispersed and highly stable MOF crystals. Much effort has been devoted to construct nanocarbon-supported MOF composites; their derivative materials exhibit a uniform distribution of MOFs crystals, desirable morphologies, and hierarchically porous structures, resulting in their reputation as promising materials for a wide range of applications in energy-related fields.[41,42]

As a typical example, N-doped porous carbon-coated carbon nanotubes (CNTs) with a 1D nanostructure were prepared by carbonization of ZIF-8 film coated-CNT precursors in an inert atmosphere.[43] The designed nanocarbon *via* integrating a CNT skeleton and N-decoration species, was effectively applied as an electrocatalyst with abundant active sites and a conductive substrate. Beside supporting MOFs on a CNT substrate, using a Co-based MOF precursor can enable CNTs to be grown *in situ* on the MOFs' surface with assistance from a Co catalyst and carbon source from self-sacrificial MOFs. Recently, Lou and co-workers[39,44] designed a MOF-strategy for the synthesis of a hierarchical tubular material consisting of Co_3O_4 hollow NPs dispersed in a CNT matrix. In the synthesis process, polyacrylonitrile (PAN)-cobalt acetate ($Co(Ac)_2$) nanofibers were chosen as both self-engaged templates and cobalt source for the ZIF-67 preparation. Due to the strong chemical bonding between the 2-methylimidazole organic linkers and the cobalt clusters within the nanofibers, a uniform shell of ZIF-67 nanocrystals was grown on the surface of the nanofibers. Then, the PAN support was removed and resulted in hollow structured ZIF-67 tubulars. After a two-step annealing treatment, the hierarchical $CNTs/Co_3O_4$ microtubes were obtained. For the annealing process, metallic Co NPs were yielded in the reduction atmosphere (H_2). Subsequently, the highly dispersed Co NPs catalyst boosted the *in situ* growth of the CNTs by consuming the carbon source from the ZIF-67, resulting in a hierarchical CNTs/Co/C composite. Finally, these Co NPs were oxidized to Co_3O_4 hollow NPs by thermal treatment in air, resulting in the CNTs/Co_3O_4 microtube composites as shown in Figure 1.4a. The well-distributed CNTs contributed to the further enhanced conductivity of the CNTs/Co_3O_4 microtubes, which were found to be of benefit for maintaining the high stability of Co_3O_4 NPs. In another study, Liu *et al.*[45] reported the fabrication of double-shelled carbon nanocages (DSNCs) derived from core-shell ZIF-8@ZIF-67 hybrid precursors. An inner shell of nitrogen-doped nanocarbon and an outer shell of Co-embedded N-doped carbon was fabricated by pyrolysis of the ZIF-8 core and ZIF-67 shell precursors, respectively. A cross-section SEM image of a DSNC's particle showed the presence of an inner cavity between the double nanocarbon shells. TEM characterizations evidenced that the inner shells were made of amorphous carbon with a thickness around 60 nm. The relatively loose and rough outer shells anchored with numerous short carbon nanotubes were found to present a thickness of 40 nm (Figure 1.4b). Beyond using CNTs as the substrate, MOF precursors can also be deposited on a graphene-based support to develop a MOFs/graphene/MOF sandwich structure. Wang and co-workers[46] demonstrated a strategy for preparing a sandwich structure of nanocarbon/graphene oxide (GO)/nanocarbon derived from MOFs/GO/MOF precursors and templates. By introducing GO substrate during the MOFs' synthesis procedure, MOFs went through a homogenous nucleation and confined growth process at both sides of the GO sheets and finally formed MOFs/GO/MOF sandwiches with MOF nanocrystals distributed on GO sheets (Figure 1.5a). The continuous dispersion of MOF nanocrystals on GO sheets effectively prevented the agglomeration of GO and enabled the improved stability of the graphene sheets. Crucially, after high temperature pyrolysis, the MOF-derived N-doped nanocarbon NPs not only served as spacer materials to prevent the agglomeration of graphene, but also were used to functionalize the metal/metal compound NPs. The metal/metal compound NPs derived from MOF precursors were encapsulated in a nanocarbon matrix, which has been considered an attractive structure for energy storage and conversion applications. Wang and co-workers developed a seed-mediated synthesis route to prepare MOF/GO/MOF nano-sandwiches with core-shelled ZIF-8@ZIF-67 particles uniformly dispersed on GO substrate.[47] The GO/core-shelled MOF composites were investigated and found to be excellent precursors for preparing Co-NC/rGO/Co-NC (Figure 1.5b), which were efficient nonprecious metal catalysts for the OER. In addition to the carbon-based substrates, 2D inorganic matrices of layered double hydroxides (LDHs) were investigated as appropriate substrates to fabricate the supported porous carbon architecture.[48] The sandwich-like MOF arrays on both sides of LDH nanoplatelets were first prepared by *in situ* nucleation and growth of ZIF-67 on the surface of LDH. The obtained MOFs arrays were uniformly and tightly locked on the available surface of the LDH substrate,

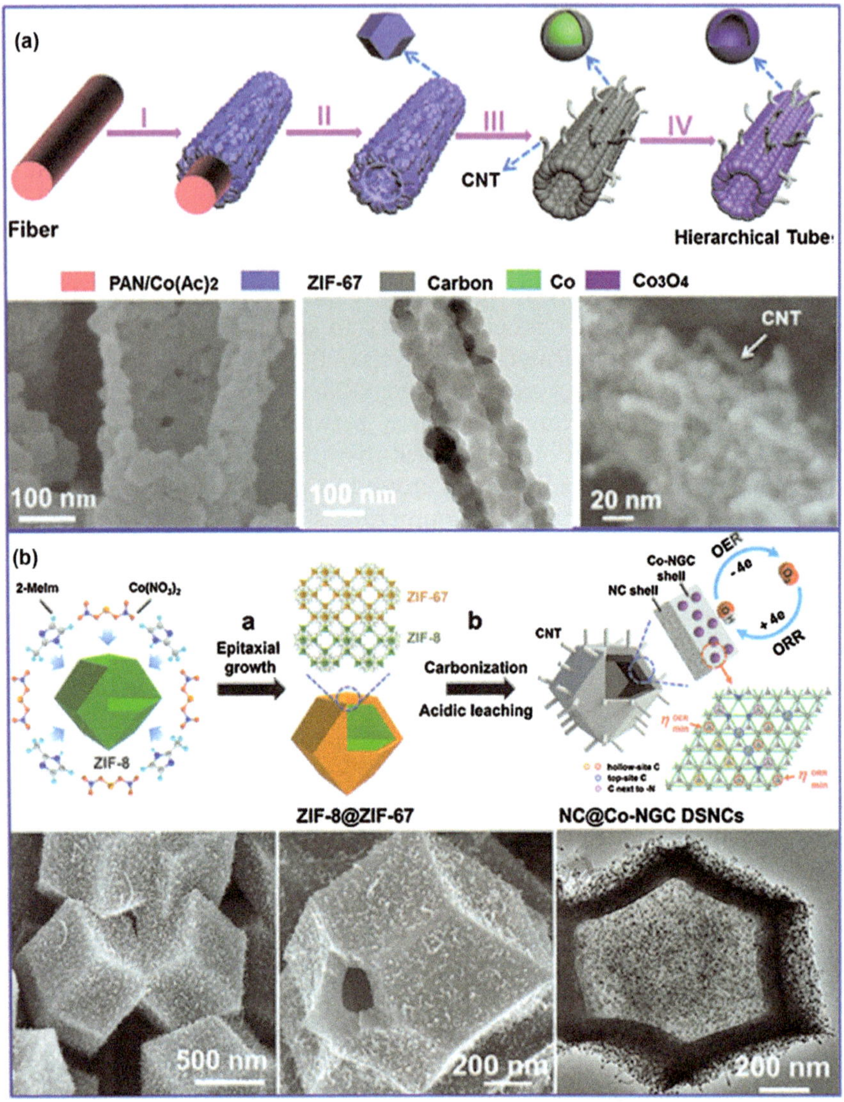

Figure 1.4 (a) Illustration of the formation of CNT/Co_3O_4 microtubes; SEM, TEM and HRTEM images of the synthesized hierarchical CNT/Co_3O_4 microtubes. Reproduced from ref. 39 with permission from John Wiley and Sons, © 2016 WILEY-VCH Verlag GmbH & Co. KGaA, Weinheim. (b) Schematic illustration of the synthesis of NC@Co/NC double-shelled carbon nanocages; SEM and TEM images of a NC@Co/NC nanocage. Reproduced from ref. 45 with permission from John Wiley and Sons, © 2017 WILEY-VCH Verlag GmbH & Co. KGaA, Weinheim.

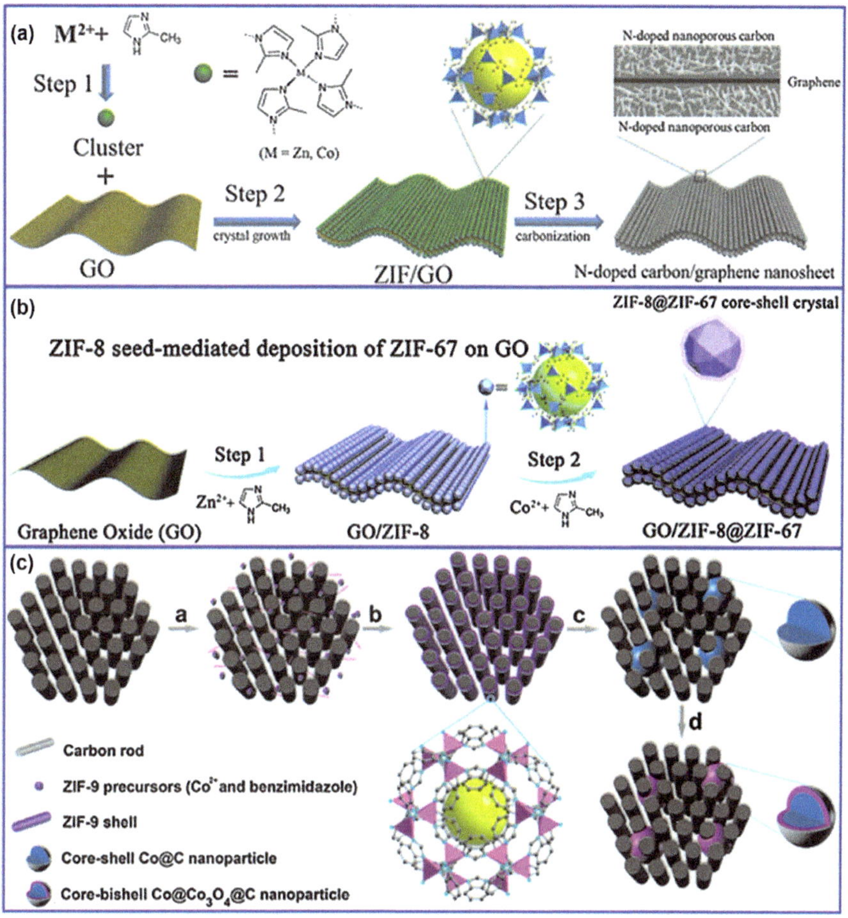

Figure 1.5 (a) Schematic synthesis of N-doped porous carbon/graphene nanosandwiches. Reproduced from ref. 46 with permission from John Wiley and Sons, © 2015 WILEY-VCH Verlag GmbH & Co. KGaA, Weinheim. (b) Schematic of ZIF-8 seed-mediated deposition of ZIF-67 on GO sheet and GO/ZIF-88@ZIF-67-900. Reproduced from ref. 47 with permission from the Royal Society of Chemistry. (c) Schematic illustration of the synthesis of $Co@Co_3O_4@C$-CM. Reproduced from ref. 49 with permission from the Royal Society of Chemistry.

which maintained the structural transformation from a ZIF-67 precursor to nanocarbon arrays with well controlled morphology and composition. Taking advantage of the unique micro/mesoporous nanocarbon matrices, highly efficient active sites (*e.g.*, Nx–C, Co–Nx–C) and high specific surface area, the as-prepared sandwich-like nanocarbon/LDH/nanocarbon architecture showed excellent electrocatalytic activity for the ORR.

Another effective approach to fabricate a dispersed system with a high portion of empty space is to distribution MOF nanocrystals by using 3D conductive porous supports. Xia *et al.*[49] proposed a MOF induced strategy to encapsulate the novel NPs of Co@Co_3O_4@C with a core-bishelled structure into a 3D mesoporous carbon matrix (CM). In their research, ZIF-9, Co-based MOF crystals were first encapsulated into the confined space of the hard template of a CM. The synthesized ZIF-9-CM composite was subjected to heat treatment under Ar gas to design a Co@C core-shell NPs @CM composite. Subsequently, the Co@C was transformed into core@bishell Co@Co_3O_4@C NPs *via* a controlled oxidation procedure (Figure 1.5c), thus, CM-supported Co@Co_3O_4@C NPs with a hierarchical porous structure were achieved. After thermal treatment, the MOF-derived metal/carbon composition was interconnected to the ordered carbon support, yielding a continuous conductivity and strong coupling interaction between the NPs and CM. This research suggests the novel possibility of designing MOF-derived supported nanomaterials by using MOFs as the spacer, while 1D, 2D, and 3D substrate as the building scaffold.

1.2.2.4 MOF-supported Noble Metal NPs

Apart from tailoring the inherent structures, MOFs can also be used as support materials to immobilize highly active precious metal catalysts, by which an advanced catalyst composite with enhanced activity and stability is expected.[50,51] In 2015, Shen and co-workers[52] proposed an MOF-strategy to prepare supported nanocatalysts with a non-noble metal as the core and noble metal as the shell. In Figure 1.6a, Co NPs embedded in N-doped carbon (Co/NC) were synthesized by carbonization of ZIF-67 in an inert environment. Then, by using Co/NC as sacrificial templates, noble metal Pd^{2+} ions under high chemical valence were reduced to metallic Pd^0 through a galvanic displacement reaction, oxidizing Co to Co^{2+}. The reduced Pd^0 was deposited on the surface of Co NPs, forming a thin shell Pd layer outside of the Co core NPs. High-resolution transmission electron microscopy (HRTEM) and X-ray photoelectron spectroscopy (XPS) confirmed that Co@Pd core–shell NPs were highly distributed in the N-doped carbon substrate with particle sizes around 9.0 nm, and Pd loading up to 1.1wt%. The highly exposed and surface-accessible Pd atoms on Co NPs ensured the Co@Pd/NC of an attractive catalyst material. Importantly, the high atom utilization provided the Co@Pd/NC catalyst with superior hydrogenation activity. This MOF-templated strategy gives a typical example of the development of a next-generation catalyst with a relatively low loading of noble metal. In the case of metal NP electrocatalysts, the catalyst's specific activity, normalized by the active metal atoms, increases with the decreasing particle size. In order to develop a highly active electrocatalyst with a low amount of noble metal, one can downsize metal NPs to sub-nanometre or even single atom levels, which is considered to be one of the most efficient ways. However, issues of poor stability and severe aggregation occur for

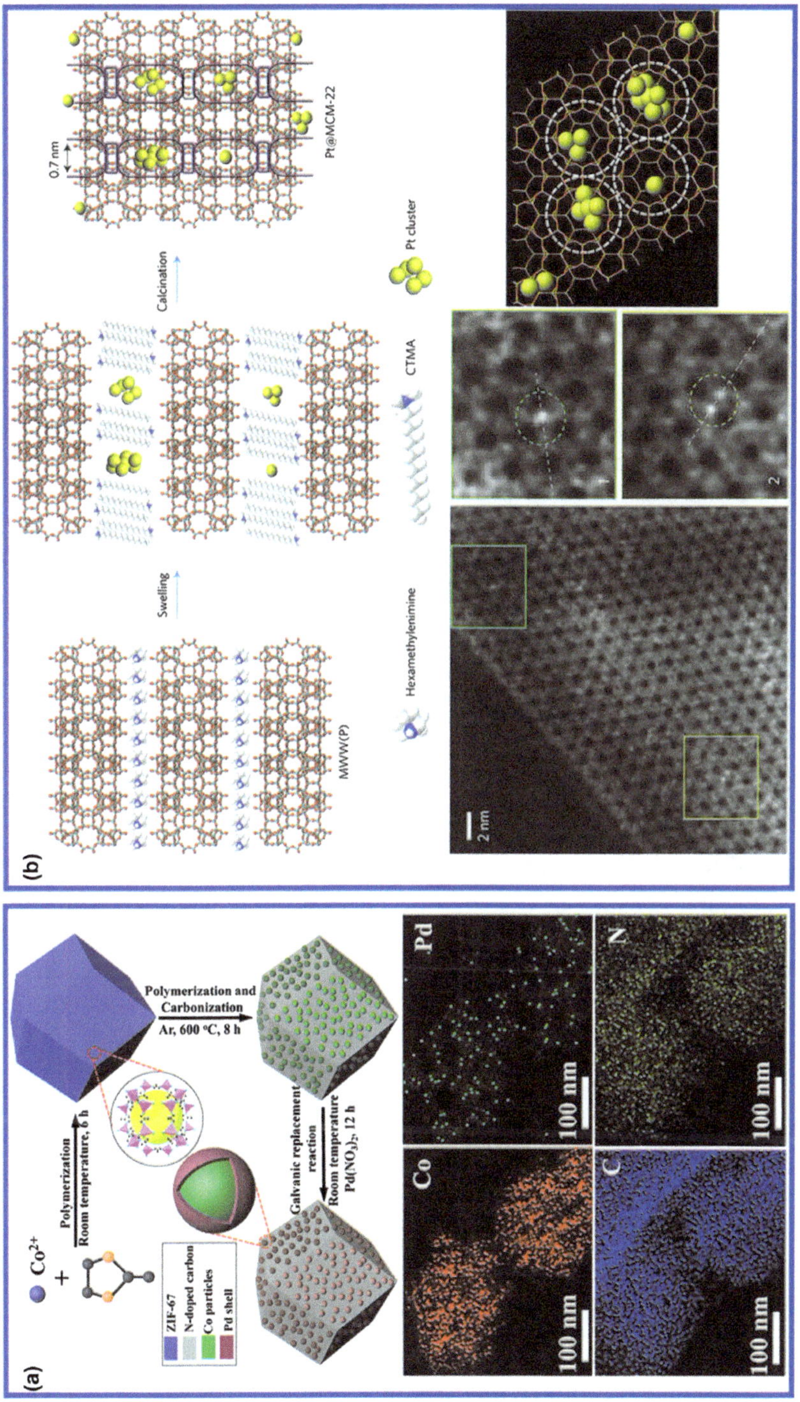

Figure 1.6 (a) Schematic illustration of the process for synthesis of Co@Pd/NC; elemental mapping of Co, Pd, C and N for a Co@Pd/NC particle. Reproduced from ref. 52 with permission from American Chemical Society, Copyright 2015. (b) Illustration of the prepared Pt@MCM-22; TEM images of the atomic structures of Pt@MCM-22. Reproduced from ref. 53 with permission from Springer Nature, Copyright 2017.

smaller metal particles, due to the higher surface energy of active atoms and the tendency of aggregation to lower their surface energy. Therefore, generation of metal sub-nanoclusters or single atoms on a promising support is emergent for developing the next-generation electrocatalysts with both excellent activity and stability. Recently, Liu and co-workers[53] reported a strategy to develop zeolite framework-entrapped Pt single atoms and sub-clusters. Due to the pore size controlled effect, small Pt single atoms and sub-clusters can be entrapped in the pores of a siliceous zeolite framework, by which the high thermal stability of Pt single atoms and sub-clusters can be achieved. The researchers added pre-made Pt sub-clusters into the base materials of a layered 2D zeolite precursor. During the condensation of the 2D precursor into the 3D zeolite framework, the cages and pores with sizes of ~0.7 nm×0.7 nm×1.8 nm entrapped the Pt single atoms and sub-clusters. Pyrolysis and oxidation of the composite precursor in air gave rise to the Pt isolated atoms and sub-clusters incorporated in the cages of the 3D zeolite substrate. HRTEM revealed that both Pt single atoms and sub-clusters of sizes ranging 0.2–0.7 nm (Figure 1.6b) were observed, which were found with quite uniform distribution in the 3D zeolite systems. Extended X-ray absorption fine structure (EXAFS) analysis provided a Pt–Pt coordination number of 4.7, indicating that an average Pt sub-cluster size is less than 1 nm. The Pt sub-clusters confined in the zeolite cavities showed the reactants' selectivity by size effect, extremely high stability and catalytic activity for the hydrogenation of alkenes.

1.3 MOF-derived Materials for Efficient Electrocatalysis

Most state-of-the-art high-performance electrocatalysts are noble metal Pt-based catalysts. Because of the high cost and limited availability of Pt, it is extremely desirable to develop alternative electrocatalysts with high activity and stability under lower Pt loading or even Pt-free. Generally, electrocatalytic performance is mainly affected by the catalytic activity, stability, and conductivity, *etc.* MOFs and their derived materials have attracted great attention for highly efficient electrocatalysis due to their advantages in structural diversity and configurations with nearly infinite designability. Detailed design and application of MOF derivatives as electrocatalysts for the hydrogen evolution reaction (HER), oxygen evolution reaction (OER), oxygen reduction reaction (ORR), and CO_2 reduction are summarized in this section.

1.3.1 MOF-derived Electrocatalysts for the HER

Electrochemical water splitting to generate hydrogen and oxygen is considered to be a promising approach to produce hydrogen. The water-splitting reaction consists of two half-reactions (balanced with acid): the hydrogen evolution reaction at the cathode (HER, $2H^+ + 2e^- \rightarrow H_2$) and the oxygen evolution reaction at the anode (OER, $H_2O \rightarrow 1/2O_2 + 2H^+ + 2e^-$). Electrolysis of water requires excess energy in the form of overpotential to overcome various activation barriers. Although many research efforts have been devoted to the electrolysis of water, state-of-the-art electrocatalysts for the HER and OER are still dependent on Pt-based catalysts. Thus, developing highly active alternatives with low-cost and a non-noble metal is still a significant challenge. A class of transition metal phosphides have attracted significant interest as efficient electrocatalysts for the HER because of the appropriate hydrogen adsorption free energy ($\Delta G\ H\ ^*$), considerable electrochemical activity, and long-term durability. Among them, cobalt phosphide (CoP) is considered to be one of the most promising electrocatalytic materials for the HER as compared to iron, copper, nickel, and tungsten phosphides. For instance, Zou *et al.*[54] reported a strategy to develop CoP-based electrocatalysts through a MOF-controlled method. In their study, a novel bottom-up strategy for the synthesis of boron- and nitrogen-co-doped carbon nanotubes encapsulated CoP NPs (CoP@B/N-CNTs) was reported as the HER electrocatalyst. As shown in Figure 1.7a, two key steps were applied for the preparation of CoP@B/N-CNTs, which were based on pyrolysis and phosphating. ZIF-67-derived Co NPs encapsulated in B/N-CNTs were fabricated first *via* the pyrolysis process. Then phosphorous acid was used as a second precursor for phosphating Co@B/N-CNTs. As a result, a CoP@B/N-CNTs composite was obtained accompanied by the conversion of Co into CoP NPs. The SEM and TEM images showed that CoP NPs of 25 nm in size were uniformly distributed inside B/N-CNTs. As an electrocatalyst for the HER, the CoP@B/N-CNTs exhibited high activity, as indicated by the low overpotential of 87, 215, and 122 mV (*vs.* RHE) at a current density of 10 mA cm^{-2} in acidic, basic and PBS solution, respectively. Moreover, this novel CoP@B/N-CNT catalyst showed outstanding stability for more than 8 h in different electrolytes at varying pH values. The excellent electrocatalytic activity of CoP@B/N-CNTs were attributed to the synergistic structure between the CoP NPs and heteroatoms-doped carbon support. The introduction of B and N dopants in the graphitic nanotubes effectively increased the density of active sites. With encapsulation of CoP NPs into cavities of a B/N-CNTs support, a novel composite structure was obtained, which not only produced active sites with high density, but also prevented the agglomeration of CoP with neighbouring NPs, resulting in composite catalysts with enhanced HER activity and stability. A similar MOF-routed strategy was also proposed for the preparation of CoP NPs encapsulated in N-doped nanocarbon by using ZIF-9 as a metal precursor (Figure 7b).[55] The as-prepared CoP@NC can serve as a superior electrocatalyst for the HER not only in acid solution, but also in alkaline media with excellent electrocatalytic activity and long-term stability. To investigate the direct active sites for the HER, experiments and density functional theory (DFT) calculations were carried

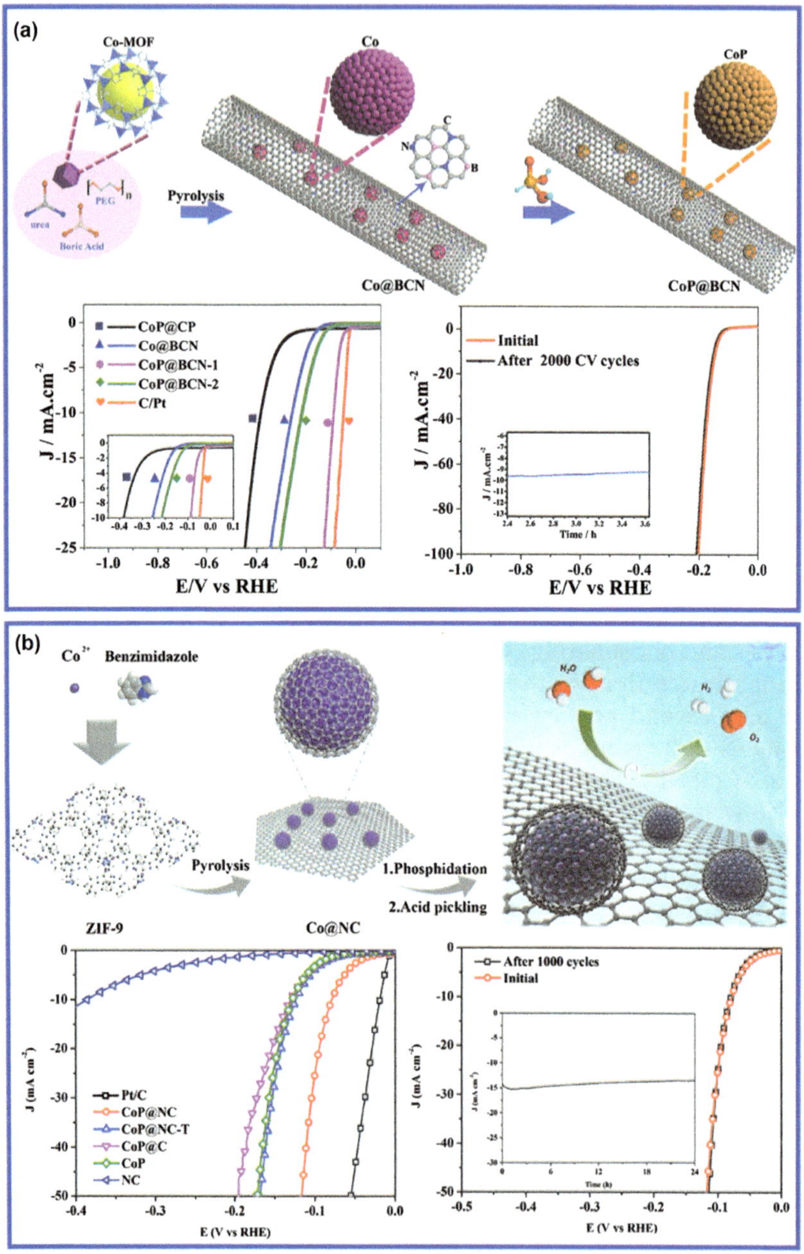

Figure 1.7 (a) Schematic illustration for the fabrication of CoP@BCN nanotubes; TEM image of CoP@BCN nanotubes; The HER performance and stability of CoP@BCN nanotubes in 0.5 m H_2SO_4 solution. Reproduced from ref. 54 with permission from John Wiley and Sons, © 2017 WILEY-VCH Verlag GmbH & Co. KGaA, Weinheim. (b) Schematic illustration of the synthesis of CoP@NC; HER performance and stability in 0.5 m H_2SO_4 solution. Reproduced from ref. 55 with permission from American Chemical Society, Copyright 2017.

out by comparing the performance of CoP with and without N-doped nanocarbon shells. Interestingly, it was revealed that the synergistic effect between N-doped nanocarbon shells and the CoP NPs contributed to the extraordinary electrocatalytic activity for the HER; the carbon atoms adjacent to nitrogen dopants in the carbon shells are the direct active sites for hydrogen evolution.

In addition to Co-based electrocatalysts, other MOF-derived transition metals and compound NPs incorporated with nanoporous carbon were studied as electrocatalysts for the HER. For instance, Wu *et al.*[56] developed molybdenum carbide (MoCx) octahedrons as efficient HER electrocatalysts through a MOF-assisted strategy. Starting from a composite precursor consisting of a Cu-MOFs host and Mo-based polyoxometalates guest, mesoporous MoCx nano-octahedrons were successfully prepared through a pyrolysis and Cu-etching process. Benefiting from a porous and ultrafine nanocrystal structure, the as-proposed MoCx nano-octahedrons exhibited remarkable electrocatalytic activity. Figure 1.8a shows the electrocatalytic activity and durability performance of MoCx nano-octahedrons for the HER in both acid and basic conditions. In acid media, the MoCx electrocatalyst showed a small onset potential of 25 mV, beyond which the hydrogen reduction current increased rapidly. To achieve the high current density of 10 mA cm^{-2}, very low overpotential of 142 mV (*vs.* RHE) was required for the MoCx electrocatalyst, suggesting outstanding catalytic activity. In the basic electrolyte, the MoCx electrocatalyst outperformed the state-of-the-art Pt/C catalyst with a rapidly increasing cathodic current. A small overpotential of 151 mV (*vs.* RHE) was demanded for the MoCx nano-octahedrons to achieve a current density of 10 mA cm^{-2}. To further improve the performance of MOF-derived electrocatalysts, an effective approach is to introduce a conductive template into the MOF-derivatives. Tang and co-workers[57] incorporated a conductive substrate of graphene oxide (GO) into polyoxometalate-based MOFs (POMOFs) to fabricate a POMOFs/GO composite. As polyoxometalate (POM) is rich in phosphorous (P), there is an opportunity to dope P atoms into the resultant product by a post-treatment POMOFs/GO precursor. As a result, the produced hybrid material consisting of Mo-oxide, P-doped porous carbon (PC), and reduced graphene oxide (rGO) substrates (MoO_2@PC-rGO), was achieved through one-step pyrolysis of a POMOFs/GO precursor. The as-prepared MoO_2@PC-rGO composite presented superior electrocatalytic activity for the HER, with a small onset potential of about 0 mV (*vs.* RHE), high exchange current density, and long-term durability in acidic media (Figure 1.8b). The superior activity is due to the synergistic effect arising from the highly distributed MoO_2 particles, P-doped porous carbon, and the conductive rGO substrate. Additionally, the unique structure achieved from POMOFs avoids MoO_2 NP aggregation and thus guarantees the abundant exposed active sites of Mo and oxygen edges, which play a significant role in achieving the excellent HER performance.

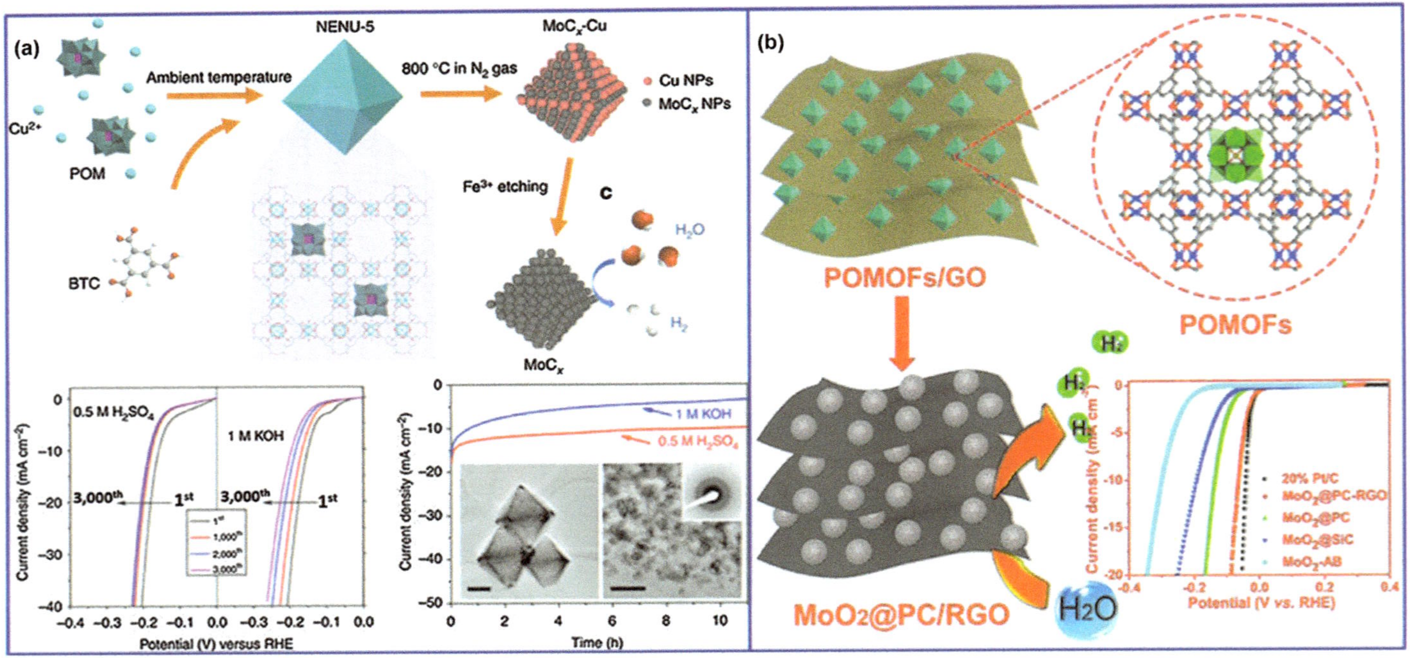

Figure 1.8 (a) Schematic illustration of the synthesis for porous MoCx nano-octahedrons; polarization curves after continuous potential sweeps in 0.5M H_2SO_4 and 1.0M KOH. Reproduced from ref. 56, https://doi.org/10.1038/ncomms7512, under the terms of the CC BY 4.0 licence, https://creativecommons.org/licenses/by/4.0/. (b) Schematic illustration of the preparation process of the MoO_2@PC-rGO nanocomposite; HER polarization curves of MoO_2@PC-RGO and contrast electrocatalysts. Reproduced from ref. 57 with permission from John Wiley and Sons, © 2015 WILEY-VCH Verlag GmbH & Co. KGaA, Weinheim.

1.3.2 MOF-derived Electrocatalysts for the OER

The OER is the counterpart of the HER, in which water is oxidized to produce dioxygen. State-of-the-art catalysts for the OER are generally of noble metal oxides such as ruthenium dioxide (RuO_2) and iridium dioxide (IrO_2). Due to the scarcity and high-cost of noble metal oxides, it is necessary to develop novel alternatives with properties of high activity, abundant source and low cost. Ingenious tailoring of MOF precursors is an effective way to prepare non-precious metal catalysts in order to optimize the OER performance. MOF-derived Co-based materials have been investigated extensively as electrocatalysts for the OER. Wang *et al.*[44] reported that a two-step thermal treatment of Co-based ZIF-67 precursors under Ar and then H_2 atmosphere can be used to fabricate a N-doped carbon nanotube with 3D frameworks (NCNTFs). As shown in Figure 1.9a, the as-prepared NCNTFs well maintained the polyhedral morphology of the parent ZIF-67 precursors. In contrast, a hierarchical shell composed of interconnected crystalline NCNTs was formed after controlled pyrolysis. The H_2 heat treatment played a critical role in dominating the formation of hollow-structured NCNT frameworks. Under H_2 atmosphere, the metallic Co NPs were quickly constructed by reducing the Co-clusters from a ZIF-67 precursor. Subsequently, the reduced Co NPs were used as catalysts to boost the *in situ* growth of N-doped carbon nanotubes by consuming the carbon and nitrogen source from ZIF-67, and hollow-structured NCNT frameworks were finally achieved. With the advantages of optimal graphitic degree, nitrogen-doping active sites, and hollow structures, the resultant NCNTFs exhibited remarkable electrocatalytic activity and stability for both the OER and ORR in alkaline solution. The NCNTF catalyst prepared at 700 °C exhibited the best catalytic activity for the OER with a low oxidation potential of 1.60 V (*vs.* RHE) to generate a current density of 10 mA cm^{-2}, which outperforms the commercial Pt/C catalyst (10 mA cm^{-1}; 1.78 V *vs.* RHE). Moreover, the NCNTF material as an OER electrocatalyst demonstrated remarkable long-time stability. These MOF-derived functional materials have opened up a new avenue for developing highly active and stable electrocatalysts.

Up to now, the vast majority of electrocatalyst materials for the OER are in the form of thin films or particle agglomerates. To optimize the reaction interface with high activity and durability for the OER, the development of novel electrode configurations is urgently needed. Well-aligned nanowire arrays (NWAs) with electrocatalytic active components grown on the current collectors (*e.g.* carbon paper, Ni foam *etc.*) have been identified as a promising direction for designing the advanced electrodes. For example, Ma and colleagues[58] presented a novel strategy to fabricate hybrid Co_3O_4-carbon porous NWAs, which were achieved by carbonization of copper foil support-ZIF-67 NWAs under an inert atmosphere. As illustrated in Figure 1.9b, a Co-based MOF of ZIF-67 with a nanowire configuration was directly grown on Cu foil through a hydrothermal method at 80 °C. Subsequently, the Co-MOF NWAs were converted into a Co_3O_4 incorporated carbon composite

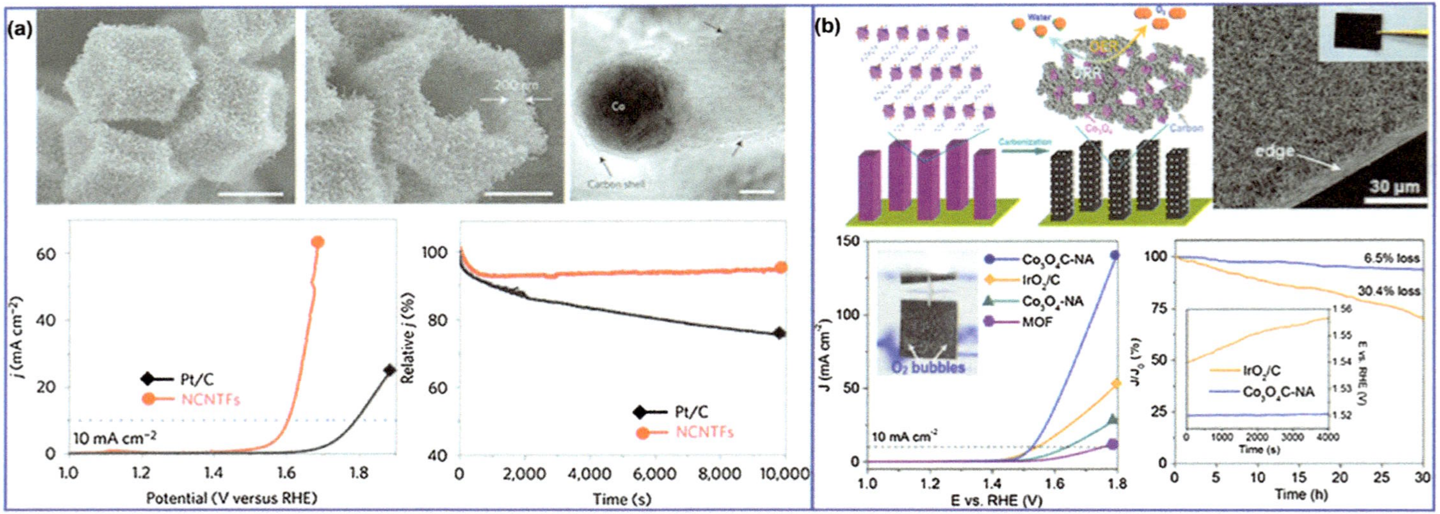

Figure 1.9 (a) SEM, TEM and high resolution-TEM images of the structural characterization of NCNTFs obtained at 700 °C; electrochemical oxygen evolution performance and stability of NCNTFs and Pt/C catalysts. Reproduced from ref. 44 with permission from Springer Nature, Copyright 2016. (b) Schematic illustration of the fabrication of hybrid Co_3O_4-Carbon porous nanowire arrays; SEM image of Co_3O_4-Carbon; polarization curves and stability response of Co_3O_4-Carbon for oxygen evolution. Reproduced from ref. 58 with permission from American Chemical Society, Copyright 2017.

via high temperature pyrolysis under N_2 atmosphere. The as-prepared Co_3O_4/C well maintained the original nanowire morphology and configuration of the parent MOFs. The periodic arrangement of metal ions and organic ligands in a MOF precursor provided a highly uniform distribution of Co_3O_4 NPs and *in situ* formation of porous nanocarbon. Due to the high surface area of Co_3O_4/C porous NWAs and the strong interactions between Co_3O_4/C NWAs and Cu foil, the obtained Cu foil-supported Co_3O_4/C nanowire can be used directly as a working electrode for the OER, without the need for an extra substrate or binder. By analysis of the electrochemical performance of the Co_3O_4/C hybrid, it was found that this novel electrode showed outstanding catalytic activity for the OER in alkaline media, with a relatively low oxidation potential of 1.52 V (*vs.* RHE), able to deliver a current density of 10 mA cm^{-2}, which is lower than that of the Cu foil supported IrO_2/C catalyst, indicating better catalytic activity of the Co_3O_4/C hybrid for the OER. The unique porous nanowire array configuration and *in situ* carbon incorporation contribute to the as-prepared Co_3O_4/C NWAs' electrode with an enlarged active surface area, strong stability and rapid mass transport. All these features lead to the excellent OER performance of the Co_3O_4/C nanowire. The OER performance of the Co_3O_4/C nanowire is superior to that of all the reported nanowire array electrodes and even better than most of the state-of-the-art highly active transition metal and noble metal catalysts. This MOF-derived porous Co_3O_4 carbon nanowire array electrode is promising for direct application in water-splitting devices, fuel cells and other energy conversion systems.

It was found that the change of Gibbs free energy (ΔG) for the reaction pathway of the OER can be decreased significantly under the presence of synergistic bimetallic active centres.[59,60] By introducing a second metal species into the mono-metal-based MOFs, a bi-metallic MOF hybrid precursor can be designed. After controlled post-treatment, the obtained MOF-derived bi-metallic materials are efficient electrocatalysts. The electrocatalytic activity and stability of MOF-derived bi-metallic materials can be obviously enhanced due to the nature of the coordinated bi-metallic centres, which provide a different environment for electrocatalysis compared to the mono-metal centres. Zhao and colleagues[61] reported the synthesis of NiCo-based MOF ultrathin sheets featuring coordinatively unsaturated Ni and Co bi-metallic active sites. The bimetallic NiCo-MOF nanosheets with a thickness of 3 nm were prepared by periodic linking of the organic ligands (benzenedicarboxylic acid) with a Ni^{2+} and Co^{2+} metal source. After loading NiCo-MOFs onto copper foam electrodes, the Cu foam supported NiCo-MOFs as electrocatalysts exhibited remarkable activity for the OER, with a very low overpotential of 189 mV to boost the current density of 10 mA cm^{-2} in alkaline conditions (Figure 1.10). By contrast, the state-of-the-art RuO_2 catalyst required a much higher operating overpotential of 280 mV to achieve a current density of 10 mA cm^{-2} under the same conditions. Moreover, superior stability was demonstrated for NiCo-MOF catalysts with only 2.6% current loss after 200 h of continuous electrolysis. The crystal structure and

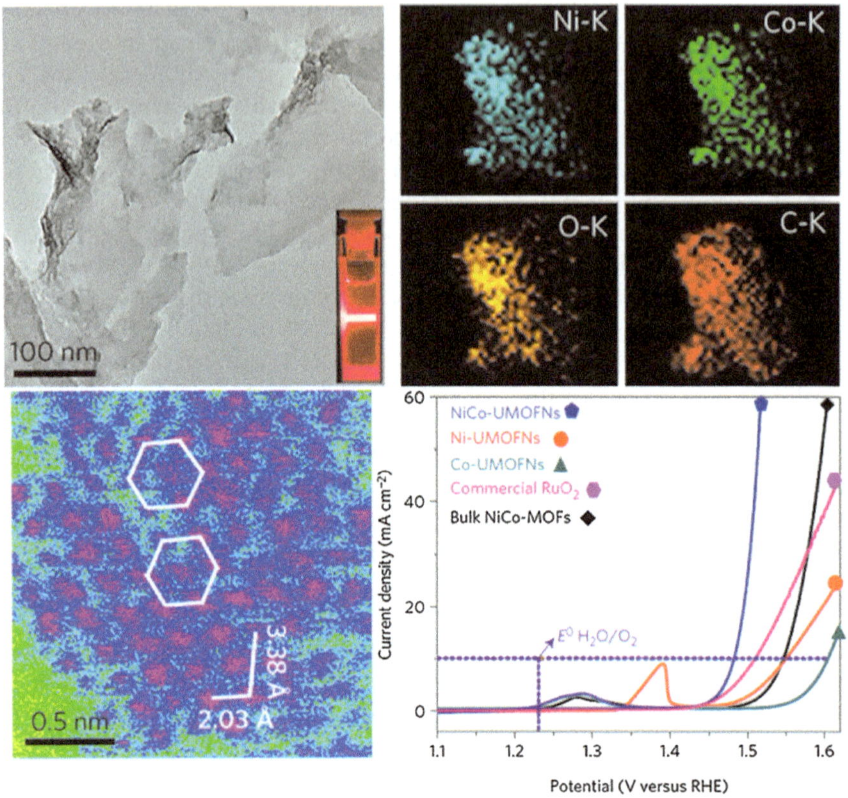

Figure 1.10 TEM and elemental mapping images of NiCo-MOFs; OER electrochemical activity of NiCo-MOF electrocatalysts.
Reproduced from ref. 61 with permission from Springer Nature, Copyright 2016.

catalyst morphology of the NiCo-MOFs were well maintained after the long-time durability test. It was found that the high activity of the NiCo-MOF catalyst may be attributed to the coordinatively unsaturated bimetallic active sites. Interestingly, the coupling effect between Ni and Co, which excites the electron transfer from Ni^{2+} to Co^{2+} sites and increases the density of the unoccupied state for Ni, further enhances the OER activity and stability after the coupling effect. The ability to introduce different metal species into MOF precursors to innovate multi-metal-based functional materials provides great opportunities for adjusting a catalyst's composition with optimized active centres for the OER.

1.3.3 MOF-derived Electrocatalysts for the ORR

The ORR is considered to be technologically important and is one of the most fundamental electrochemical reactions for metal–air batteries[62] and fuel cell devices.[63–65] State-of-the-art Pt-based noble metal catalysts are regarded as the most effective electrocatalysts for promoting the ORR. However, to scale the solution to a manufacturing level is still a great challenge due to the high cost, scarcity, and poor stability of Pt catalysts. Bearing this challenge in mind, extensive efforts have been devoted to developing alternative electrocatalysts for the ORR. In recent years, considerable MOF-derived nanomaterials such as monometal-MOF, bimetal-MOF, and supported-MOF composites have been widely investigated as alternative electrocatalysts for the ORR. Recent research work revealed that MOF-derived nanomaterials exhibit excellent catalytic activity and durability during electrocatalysis for the ORR.

1.3.3.1 Monometal-MOF-derived Electrocatalysts for the ORR

Among the large variety of MOFs, transition metal-based MOFs, such as Co-, Fe-, or Zn-based MOFs, are the most investigated and widely used precursors for the fabrication of non-precious metal catalysts. The MOF-derived porous materials have been demonstrated to be effective electrocatalysts for the ORR. Taking Co-MOFs as an example, Liu and co-workers[66] used Co-MOFs as precursors for preparing non-precious metal electrocatalysts. Upon a pyrolysis strategy under an inert atmosphere at 750 °C, the Co atoms and imidazole ligands from Co-MOFs were converted into Co NPs and N-doped porous carbons, respectively, thus resulting in a hybrid material composed of highly distributed Co NPs in a NC matrix (Co–NC). Working as an electrocatalyst, the Co–NC displayed decent ORR activity, as noted by the much positive reduction onset potential of 0.83 V (*vs.* RHE) and the rapid increase of the cathodic current. Additionally, the oxygen reduction electron numbers over this Co–N–C catalyst were found to be 3.3–3.6, much closer to a four-electron pathway, suggesting a dominative four-electron transfer process with water as the product during oxygen reduction.

To achieve MOF-derived nanostructures with optimized catalytic activity for the ORR, a hierarchical pore structure and active doping sites were designed for MOF-derived materials. Recently, Zhu and co-workers[67] reported a MOF strategy to construct cobalt sulphide NPs (CoS) encapsulated in honeycomb-like porous carbons, which couple-dope with N and S atoms (NS–C). An aluminium-based MOF (MIL-101-NH_2) was chosen as a host material for the encapsulation of thiourea and cobalt chloride to form a hybrid precursor. Interestingly, by pyrolysis of the hybrid precursor *via* a morphology-controlled process, the unique honeycomb-like N/S-co-doped porous carbon with Co_9S_8 NPs' entrapment inside the pores was prepared (as shown in Figure 1.11a). The resultant Co_9S_8@NS-C as an electrocatalyst material showed a high catalytic activity towards the ORR, indicating a very positive onset potential of 0.05 V and half-wave potential of 0.17 V (*vs.* Ag/AgCl), which was comparable to that of a commercial Pt/C (0.04 and 0.15 V *vs.* Ag/AgCl) catalyst. The high current density compared to Pt/C was due to the advantages of the unique honeycomb-like open architecture.

The immobilization of the guest precursors (thiourea, cobalt chloride) inside the MIL-101-NH_2 pores made a great contribution to the uniform dispersion of the active centres in the porous carbon matrix, which are of benefit for promoting the catalytic performance of Co_9S_8@NS-C for the ORR. Moreover, the Co_9S_8@NS-C catalyst exhibited outstanding stability with only a slight activity loss (<10%) over 20 000 s. The superior stability of the Co_9S_8@NS-C material was attributed to the Co_9S_8@NS-C composite with a highly carbon graphitic degree that effectively prevented the catalyst corrosion and Co_9S_8 NPs aggregation, thus ensuring long-term electrochemical stability. Afterward, a new type of hierarchical graphitic porous carbon with atomically dispersed Fe and N doping sites (Fe/N-GPC) was constructed in their group, by using a Fe-based MOF strategy.[68] The combination of hierarchical pores with atomically dispersed Fe/N doping species are of benefit to generate more accessible active sites and facilitate mass transport, which produces Fe/N-GPC with a high catalytic performance for oxygen reduction. The resultant Fe/N-GPC containing 1.1 wt% Fe and 3.3 wt% N, showed extraordinary ORR activity with the most positively shifted onset potential of −0.01 V, half-wave potential of −0.13 V (*vs.* Ag/AgCl) in alkaline media, which outperformed the activity of a Pt/C catalyst (Figure 1.11b). Besides the outstanding ORR activity, the Fe/N-GPC catalyst exhibited prominent durability as well as ORR selectivity and methanol tolerant ability. Furthermore, the high ORR catalytic activity of Fe/N-GPC in an acid electrolyte (0.5 M H_2SO_4) was also investigated in this work. By exploring the difference in porosity, degree of graphitization, and content of active sites, it was revealed that the abundant and highly dispersed active sites, good conductivity, hierarchical porosity, as well as their synergistic effect contribute to the high ORR activity and durability for the resultant Fe/N-GPC catalyst. The MOF-derived non-noble metal catalysts of Fe/N-GPC showed high catalytic activity for the ORR in both acid and alkaline media, which could be promising low-cost candidates for replacing of Pt-catalysts.

1.3.3.2 Bimetal-MOF-derived Electrocatalysts for the ORR

So far, extensive research efforts have been devoted to construct MOF-designed functional carbon materials. However, rational design for the metal NPs' control of the resultant materials is still lacking. Due to the high carbonization temperature employed for the post-treatment, the MOF precursors tend to transform into porous carbons encapsulated with metal-based NPs, which would result in aggregated metal NPs and reduce the efficient active surface area for electrochemical reactions. Designing a bimetal-MOF precursor consisting of both an active metal such as Fe, Co, and Ni, and another kind of inactive metal such as Zn, Al *etc.* would be highly desirable to achieve atomically dispersed active metal centres.

Bimetallic Co/Zn-based MOF crystals (ZIF-67/ZIF-8) can be synthesized by using Co^{2+} and Zn^{2+}; a mixture of a metal source to react with the organic ligands of 2-methylimidazole (Figure 1.12a). Then the ZIF-67/ZIF-8

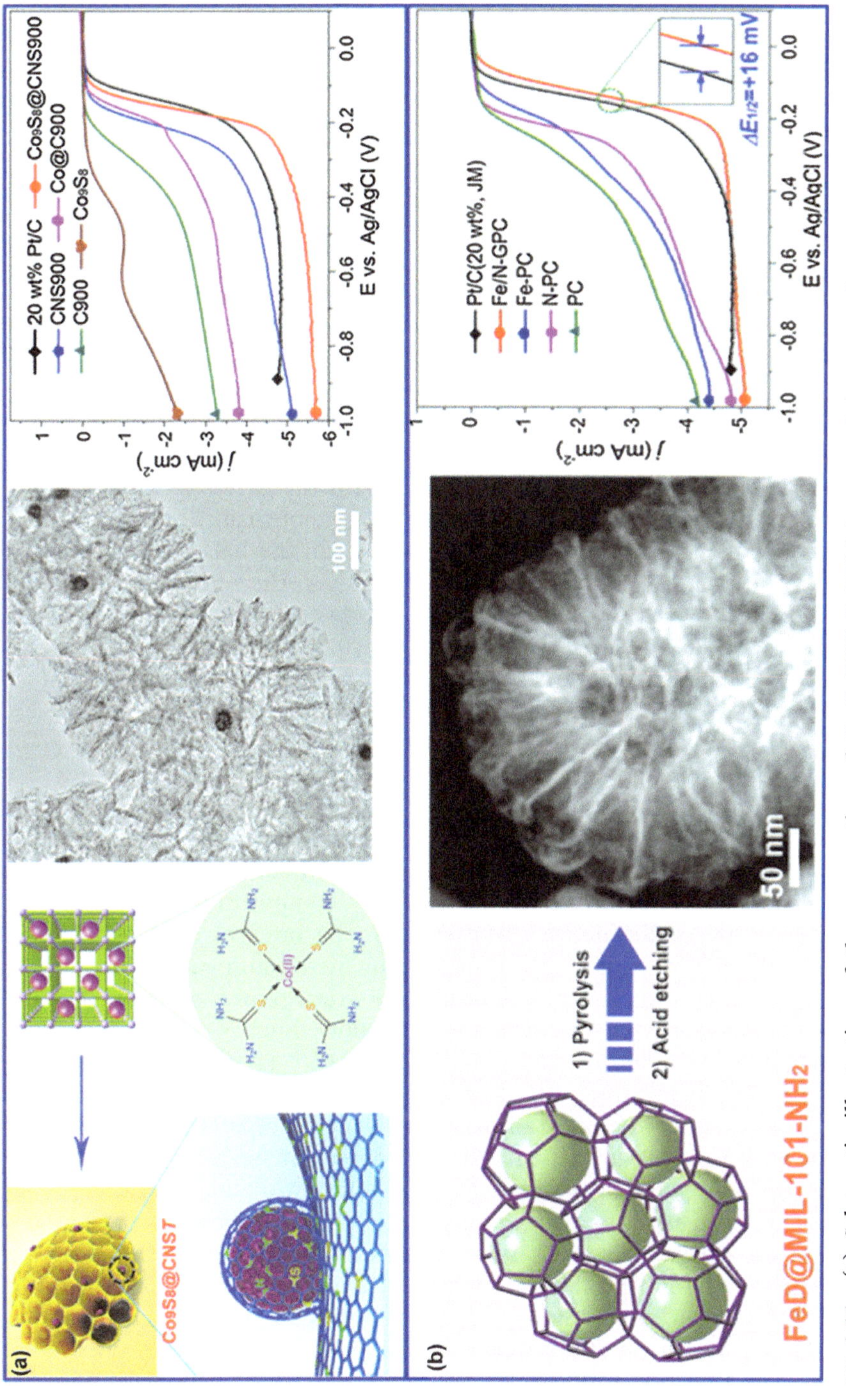

Figure 1.11 (a) Schematic illustration of the preparation of Co_9S_8@NS-C; TEM images of as-prepared Co_9S_8@NS-C; LSV curves of Co_9S_8@NS-C and comparative catalysts at 10 mV s^{-1} and a rotating speed of 1600 rpm. Reproduced from ref. 67 with permission from John Wiley and Sons, © 2017 Wiley-VCH Verlag GmbH & Co. KGaA, Weinheim. (b) Schematic representation for the preparation of the Fe/N-GPC catalyst; TEM images of Fe/N-GPC; LSV curves of Fe/N-GPC; Chronoamperometric responses of Fe/N-GPC and Pt/C in O_2-saturated 0.1 M KOH at −0.2 V. Reproduced from ref. 68 with permission from American Chemical Society, Copyright 2017.

precursors can be converted into functional nanocarbons by a specific high-temperature treatment. During high-temperature pyrolysis, the ZIF-8 component can yield nitrogen-doped carbon with a high specific surface area, while ZIF-67 generates Co/N-co-doped carbon (Co/N/C). This strategy successfully integrates the advantages of both ZIF-8 and ZIF-67, which have been widely used for designing atomically dispersed metal active centres in conductive porous carbon. For example, You *et al.*[69] investigated Co^{2+} and Zn^{2+} in various ratios to synthesize Zn/Co bimetal-MOFs, which were used as precursors for preparing self-supported Co/N/C nano-polyhedrons as electrocatalysts for the ORR. It was revealed that the Zn species not only served as spatial isolations for Co sites to inhibit their sintering issues during pyrolysis, but also helped to achieve the mesoporous and high surface area in the resultant materials through removal of Zn species. By adjusting the molar ratio of Zn *vs.* Co used for the MOF synthesis, the derived Co particle size and overall surface area of Co/N/C was precisely controlled. It was found that an increase of Zn content in the Zn/Co-MOF precursors can result in Co/N/C polyhedrons with increased specific surface area ranging from 270 $m^2 g^{-1}$ to 2148 $m^2 g^{-1}$, but can decrease the graphitization in the corresponding carbons. Due to the balance between porosity, graphitization, and content of active sites, the optimal Co/N/C with the best ORR activity was achieved from the $Zn_{0.8}Co_{0.2}(MeIM)_2$ frameworks (with 80 mol% Zn^{2+}, 20% mol Co^{2+}). The self-supported optimal Co/N/C polyhedrons as superior electrocatalysts showed high activity for the ORR with a positive half-wave potential of 0.871 V (*vs.* RHE), which was 30 mV higher than that of Pt/C (0.841 V *vs.* RHE) in 0.1 M KOH (Figure 1.12a). Benefiting from the uniform dispersion of the active sites and strong interaction between Co and N in the self-supported carbons, this bimetal-MOF-derived Co/N/C also exhibited remarkable electrochemical activity and stability in acidic and neutral media.

By precisely controlling the bimetal-MOFs' composition and post-treatment conditions, the active metal size can be brought down to the single atom level, thus enabling the maximum metal atom utilization in electrocatalysis. Recently, a novel bimetal-MOF strategy was proposed for developing highly dispersed Co single atoms supported in a N-doped porous carbon matrix with Co atoms loading up to 4.0 wt%.[70] The mechanism for synthesis of Co single atoms/NC hybrid (CoSAs/NC) by using MOF precursors is illustrated in Figure 1.12b. Based on a high temperature pyrolysis of bimetallic Zn/Co-MOF precursors, the Co source can be reduced by the decomposition of organic ligands and lead to reduced Co metal. Because of the homogenous distribution of Zn^{2+} in Zn/Co-MOFs—which are present between the Co atoms and serve as space isolation to expand the distance of adjacent Co atoms—atomically dispersed Co SAs/NC can be obtained after Zn is selectively evaporated away. The HRTEM images and EXAFS confirmed that the isolated Co single atoms were atomically dispersed in the N-doped carbon support. The resultant Co SAs/NC as electrocatalysts exhibited outstanding activity for the ORR and superior stability. DFT calculations further

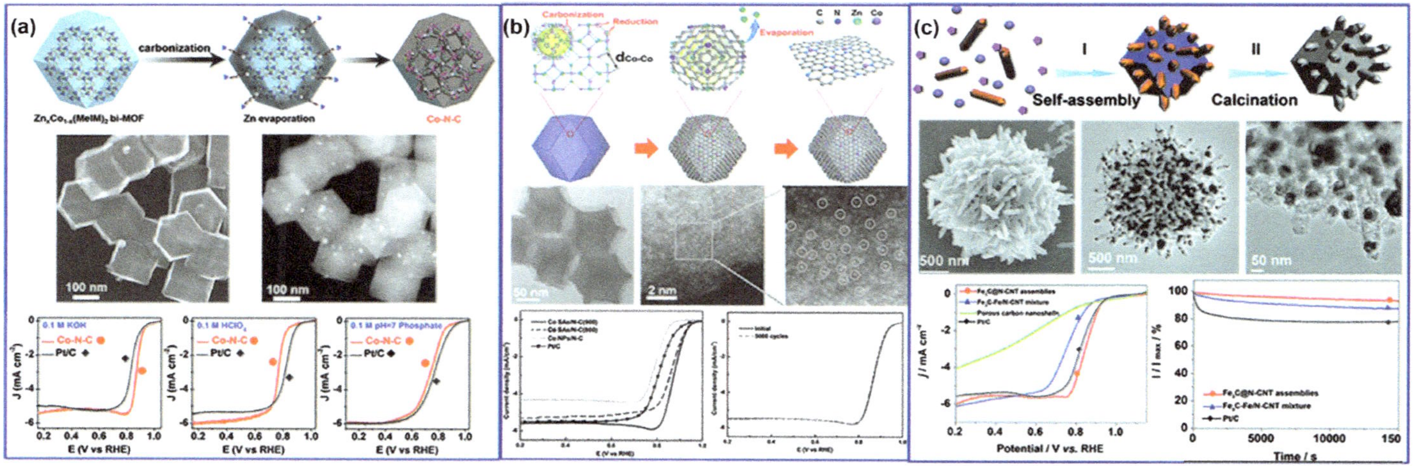

Figure 1.12 (a) Illustration of bi-metal-MOFs self-adjusted synthesis of Co/N/C; SEM and TEM images of Co/N/C derived from bi-metal-MOFs; electrocatalytic ORR results of Co/N/C and Pt/C in different PH media. Reproduced from ref. 69 with permission from American Chemical Society, Copyright 2015. (b) Illustration of the formation of Co SAs/NC; TEM and HAADF-STEM images of Co SAs/NC; RDE polarization curves of Co SAs/NC and comparative catalysts in O_2-saturated 0.1 M KOH; LSV curves of Co SAs before and after 5000 potential cycles in O_2-saturated 0.1 m KOH. Reproduced from ref. 70 with permission from John Wiley and Sons, © 2016 WILEY-VCH Verlag GmbH & Co. KGaA, Weinheim. (c) Schematic illustration of the formation of the Fe_3C@NCNT assembly; SEM, TEM and HRTEM images of morphology characterization of resultant Fe_3C@NCNT; electrochemical characterizations of Fe_3C@N-CNT assemblies as an electrocatalyst for the ORR. Reproduced from ref. 71 with permission from the Royal Society of Chemistry.

revealed that the active sites of Co-N_2 showed strong interaction with the intermediate and promoted the four-electron pathway of the ORR. The highly dispersed Co single atom catalysts with significantly improved catalytic activity and durability, opened up a new avenue for MOF-derived metal single-atom catalysts.

Other than Zn/Co-based bimetal MOF precursors, the dual Zn/Fe-based MOF is another kind of precursor for preparing advanced ORR catalyst materials. Guan and co-workers[71] reported the preparation of iron carbide NPs (Fe_3C)-decorated N-doped carbon nanotube assemblies (NCNTAs) supported by a N-doped porous carbon substrate (Fe_3C@NCNTAs-NC). This synthesis strategy starts from a MOFs-in-MOFs precursor, which is composed of Fe-based MOF nanorods (MIL-88B) embedded in a Zn-based MOF (ZIF-8) polyhedron matrix. The nanostructured composite material of Fe_3C@NCNTAs-NC was then successfully prepared by direct carbonization of this unique MOFs-in-MOFs precursor. The master combining of the MIL-88 nanorods' guests into the ZIF-8 polyhedrons' host ensured the homogeneous distribution of MIL-88 nanorods, which is favoured for the uniform dispersion of Fe_3C@NCNTAs in the porous N-doped carbon matrix (Figure 1.12c). Benefiting from the uniformly dispersed Fe_3C NPs and the porous N-doped carbon matrix, the as-synthesized Fe_3C@NCNTAs-NC composite presented superior catalytic activity for the ORR with a four-electron reduction pathway in alkaline media. This bimetal-MOF strategy offers a new approach in the development of highly active non-noble metal materials in electrocatalysis applications.

1.3.3.3 MOFs/Composite-derived Electrocatalysts for the ORR

To achieve superior stability and avoid the aggregation of MOF nanocrystals, coating MOF precursors on the surface of a functionalized substrate has been thought of as one of the most effective solutions. Embedding MOF nanocrystals onto various conductive supports (*e.g.* carbon nanotube, carbon fibre, carbon cloth, graphene, cellulose aerogel) enables the construction of functional materials with a large surface area and enhanced conductivity. Xia and co-workers[72] developed a graphene aerogel-assisted approach to convert bulk Co-based MOF crystals into monodispersed metal oxide (Co_3O_4) hollow NPs. In a typical synthesis process, N-doped graphene aerogels (NG-As) were first prepared as a support material. Then, Co-MOF crystals were grown *in situ* on N-doped graphene sheets by immersing NG-As in Co ions and an organic ligands solution. Finally, the NG-As-supported Co-MOF composite was dried and thermally activated to obtain Co_3O_4 hollow NP-embedded N-doped graphene aerogels (Co_3O_4/NG-As) (Figure 1.13a). It was interesting to find that the as-obtained Co_3O_4 NPs (with an average particle size of 35 nm) were uniformly dispersed on the graphene network. Additionally, the Co_3O_4 NPs with well-defined hollow structures, which contributed to increasing the numbers of accessible actives sites and defective surfaces, promoted Co_3O_4/NG-A as a potential catalyst with excellent

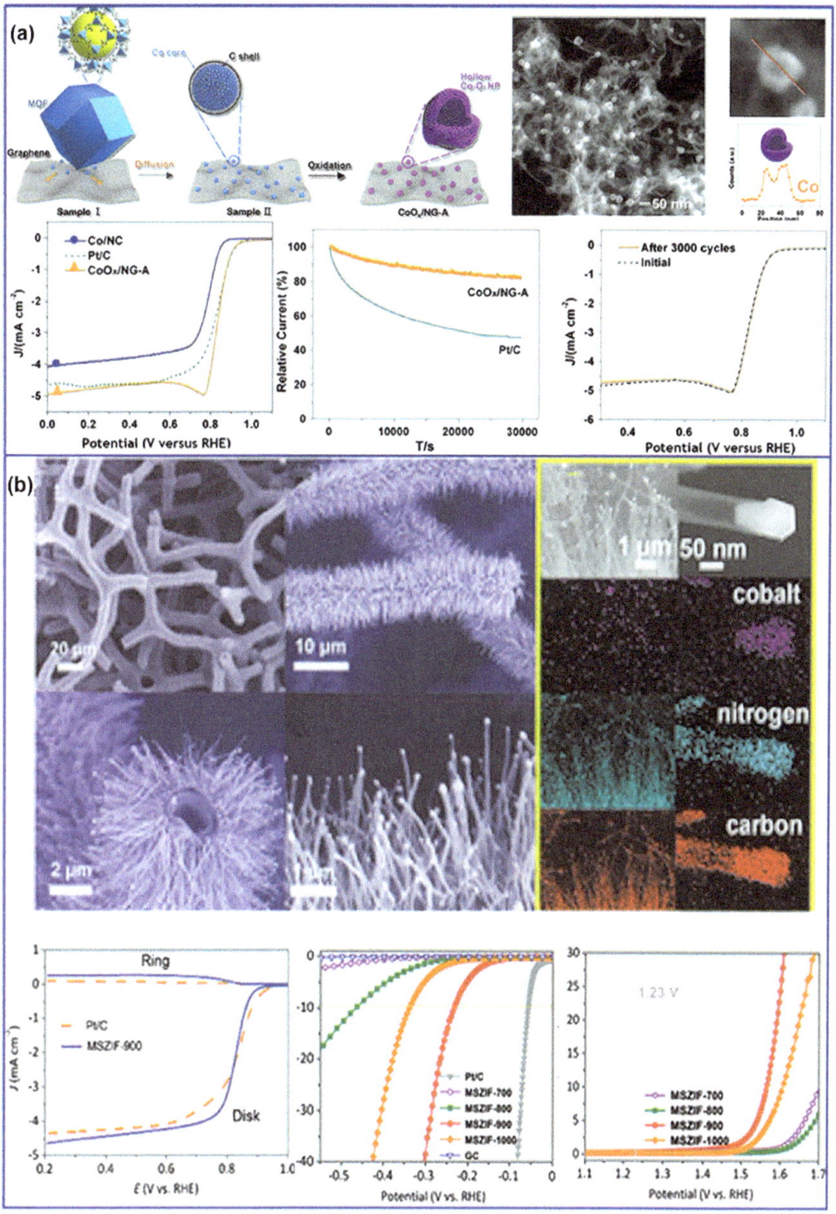

Figure 1.13 (a) Schematic of the formation process of CoOx/NG-A; STEM image of CoOx/NG-A and corresponding EDS profile of Co_3O_4 NP; ORR performance and durability of CoOx/NG-A in 0.1 M KOH. Reproduced from ref. 72 with permission from American Chemical Society, Copyright 2017. (b) FESEM, TEM images and elemental mapping of MSZIF-900; electrochemical performance of the MSZIF-900 catalyst for the ORR, HER, and OER. Reproduced from ref. 73 with permission from John Wiley and Sons, © 2017 Wiley-VCH Verlag GmbH & Co. KGaA, Weinheim.

oxygen absorption and catalytic activity. Serving as an electrocatalyst material for the ORR, Co_3O_4/NG-As indicated a high onset potential of 1.019 V (*vs.* RHE) and an almost comparable half-wave potential to that of a commercial Pt/C catalyst. Besides the excellent catalytic activity for the ORR, Co_3O_4/NG-As also presented superior stability performance during the electrocatalysis (noted as 82%), and higher current retention over 30 000 s of continuous operation at 0.76 V, and almost no negative shift of halfwave potential after 3000 accelerated durability testing (ADT) cycles.

Inspired by the rapid development of MOF-derived materials in electrocatalysis, exploration of an efficient material with catalytic activity for multiple electrochemical reactions such as the ORR, HER and OER, is clearly of great importance. Most recently, Jia *et al.*[73] presented a nickel foam with ZIF-67 thin film method for the scalable synthesis of 3D architectures, which features hierarchical porous structures and highly dispersed active sites and can be applied as advanced trifunctional electrocatalysts for the ORR and water splitting. HRTEM images revealed that the as-prepared melamine sponge-ZIF-900 (MSZIF-900 derived under 900 °C) was characterized with finely distributed Co NPs in the apex of ordered N-doped carbon nanotubes (NCNTs) (Figure 1.13b). In terms of the electrocatalytic performance, the resultant MSZIF-900 showed much more positive onset potential (0.91 V *vs.* RHE) and half-wave potential (0.84 V *vs.* RHE), compared to that of other melamine sponge-based materials. As well as the excellent catalytic performance in alkaline media of 0.1M KOH, MSZIF-900 showed remarkable ORR catalytic activity in the acidic electrolyte. The satisfactory performance of the MSZIF-900 catalyst arises from its unique hierarchical structure. The NCNTs provide good conductivity and Co species are active to capture the reactants, ensuring rapid mass transport for the ORR. Besides the high activity for the ORR, MSZIF-900 exhibited excellent HER and OER activity due to the unique 3D architectures rich with highly disperse active sites. (Figure 1.13b). The MSZIF-900 electrocatalyst required an overpotential of 233 mV and 337 mV to reach a current density of 10 mA cm^{-2} for the HER and OER, respectively, suggesting an outstanding catalytic activity compared with other materials. It was revealed by DFT calculations that the synergistic effect between the Co active sites and N-doped carbon produced a favourable electronic environment, thus enhancing the MSZIF-900 with exceptional trifunctional activity for the HER, OER, and ORR.

1.3.4 MOF-derived Electrocatalysts for CO_2 Reduction

Due to ever-increasing global warming issues and air pollution problems caused by CO_2 emissions to the atmosphere, developing an efficient way of recycling waste CO_2 is in high demand. Numerous efforts have been devoted to developing multifunctional materials for applications of CO_2 capture, storage, and conversion. Electrocatalytic and photocatalytic reduction of CO_2 to carbon monoxide or other carbon products is an important challenge for clean energy research.[74,75] Unlike oxygen reduction, CO_2 electroreduction is a much more complicated process due to the diverse reaction mechanism and multiple products. The complex reaction process and slow kinetics for these processes demand the introduction of highly active and selective catalysts to decrease the overpotential required to promote them. The most significant challenges depend on the fabrication of homogeneous electrocatalyst materials with the features of (i) good electrocatalytic activity for CO_2 reduction with a low overpotential, (ii) high selectivity and Faradaic efficiency toward CO_2 reduction, (iii) superior stability during a long-time reaction, and (iv) facile to synthesis from earth-abundant elements with low-cost. Fortunately, crystalline porous MOF materials could potentially combine the advantages of both homogeneous and molecular selectivity catalysts. Therefore, they have been explored as highly selective catalysts for CO_2 electroreduction. In recent studies, Joseph *et al.*[76] employed Fe-porphyrin-based MOF thin films as electrocatalysts for selective and efficient electrocatalytic reduction of CO_2. The Fe-porphyrin-MOF thin film was deposited on a conductive electrode to achieve the high density of available active sites. Cyclic voltammetry (CV) measurements under both N_2 and CO_2 showed that the Fe-porphyrin-MOF is active for electrocatalytic reduction of CO_2. Electrolysis at a lower overpotential of 650 mV was required in a non-aqueous electrolyte solution to generate the reduction products of 15.3 $\mu mol\,cm^{-2}$ and 14.9 $\mu mol\,cm^{-2}$ for CO and H_2, respectively. The Faradaic efficiency of the CO_2 reduction process is nearly 100%. Mechanism investigations in the catalytic activity of Fe-porphyrin-MOFs indicated that the heterogeneous CO_2 reduction processes are significantly limited by the rate of charge transportation rather than the molecular-scale kinetics for the CO_2 conversion. Therefore, it is highly desirable to promote rapid electron/charge transport for ultimate utilization of Fe-porphyrin-MOF electrocatalysts. The thin film of MOFs allows for the necessary electronic contact of the catalytic materials with the electrode. However, the poor charge carrier mobility limits the catalytic performance. It is therefore necessary to further enhance the mass transport in this system. Considering the MOF system, layered imine-based MOFs are attractive due to their superior charge carrier mobility. COF-366-Co,[77,78] a covalent-organic framework comprising of cobalt porphyrins building blocks, was investigated as an electrocatalyst for CO_2 reduction. COF-366-Co was demonstrated to reduce CO_2 in water at an overpotential of 550 mV, producing 36 $mL\,mg^{-1}$ of CO over the course of 24 h with a Faradaic efficiency of 90%. Interestingly, by means of increasing the pore size of COF-366-Co from 1.8 to 2.3 nm, this enabled a pronounced improvement in the electrocatalysis performance for CO_2 reduction. Expansion of MOFs' pore structure thus contributes to electrocatalysts with high activity and selectivity for CO_2 reduction.

1.4 Summary and Perspectives

In this chapter, we summarized the novel nanostructures and compositions of MOF-derived materials, as well as their applications in electrocatalysis.

With the remarkable advantages of a high surface area, permanent porosity, adjustable pore size, and ordered crystalline properties, MOFs can be used as the most promising precursors to design electrocatalytic materials with various structures and compositions. Among MOF derivates, the materials composed of carbon, transition metal/metal compounds (metal oxide, carbide, sulphide and phosphide), and their composites, with multiple structures such as porous, core–shell, yolk–shell and multi-shelled hollow architectures, can be achieved successfully. These materials exhibited remarkable electrocatalytic activities toward the HER, OER, ORR and CO_2 reduction.

Although MOF-derived functional materials are capable of efficient electrocatalysis for the HER, OER, ORR and CO_2 reduction, the performance of MOF derivatives in most of these applications is not yet satisfactory. Most of the MOF-derived electrocatalysts are based on transition metals and carbon, which provide the remarkable electrocatalytic activity in alkaline media. However, these MOF-derived electrocatalysts still suffer from severe stability problems in acidic conditions, which inhibit their commercial application in polymer electrolyte membrane fuel cells. As most of the commercialized fuel cells are polymer electrolyte membrane fuel cells, their operational condition requires a low PH environment for proton transportation. Therefore, pursuing a MOF-derived electrocatalyst that can work in acidic media with both high activity and durability is highly desirable. To our knowledge, the active sites for driving electrochemical reactions (*e.g.* the HER, OER, ORR, CO_2 reduction) in various PH conditions are quite different due to the changed reaction mechanisms. Thus, an understanding of how to selectively design and synthesize MOF precursors, as well as the subsequent post-treatment impact on yielding effective active sites, is still necessary for clean energy research. One of the future directions in the development of MOF-derived electrocatalysts is the design of superior stable electrocatalysts with advanced electrocatalytic activity under all PH conditions. Moreover, understanding the relationship between the nanostructure and performance is another important problem that needs to be addressed. Once the relationship between the structures and the performance is clear, it will be possible to design superior materials with highly dense, uniformly dispersed and stable active sites. Meanwhile, exploration of unique nanostructured MOF precursors by using multiple transition metals, optimized precursor ratios, suitable doping atoms, in combination with advanced support materials are all anticipated to be possible strategies for innovating next-generation electrocatalysts with the advantages of facile scale-up synthesis, low cost, excellent activity and durability for multi-reactions.

Acknowledgements

This research was supported by the Natural Sciences and Engineering Research Council of Canada (NSERC), Canada Research Chair (CRC) Program,

Canada Foundation for Innovation (CFI), Ontario Research Fund (ORF), Automotive Partnership of Canada, and the University of Western Ontario. Z. Song was supported by the Chinese Scholarship Council.

References

1. H. M. Barkholtz and D.-J. Liu, *Mater. Horizons*, 2017, **4**, 20–37.
2. X. Cao, C. Tan, M. Sindoro and H. Zhang, *Chem. Soc. Rev.*, 2017, **46**, 2660–2677.
3. W. Chaikittisilp, K. Ariga and Y. Yamauchi, *J. Mater. Chem. A*, 2013, **1**, 14–19.
4. J. Chen, K. Shen and Y. Li, *ChemSusChem*, 2017, **10**, 3165–3187.
5. S. Fu, C. Zhu, J. Song, D. Du and Y. Lin, *Adv. Energy Mater.*, 2017, **7**, 1700363.
6. B. Y. Guan, X. Y. Yu, H. B. Wu and X. W. Lou, *Adv. Mater.*, 2017, **29**, 1703614.
7. M. Bosch, S. Yuan, W. Rutledge and H.-C. Zhou, *Acc. Chem. Res.*, 2017, **50**, 857–865.
8. Y.-B. Huang, J. Liang, X.-S. Wang and R. Cao, *Chem. Soc. Rev.*, 2017, **46**, 126–157.
9. T. Kitao, Y. Zhang, S. Kitagawa, B. Wang and T. Uemura, *Chem. Soc. Rev.*, 2017, **46**, 3108–3133.
10. Y. V. Kaneti, J. Tang, R. R. Salunkhe, X. Jiang, A. Yu, K. C.-W. Wu and Y. Yamauchi, *Adv. Mater.*, 2017, **29**, 1604898.
11. A. Mahmood, W. Guo, H. Tabassum and R. Zou, *Adv. Energy Mater.*, 2016, **6**, 1600423.
12. H. Zhang, J. Nai, L. Yu and X. W. D. Lou, *Joule*, 2017, **1**, 77–107.
13. H.-C. Zhou, J. R. Long and O. M. Yaghi, *Chem. Rev.*, 2012, **112**, 673–674.
14. N. Stock and S. Biswas, *Chem. Rev.*, 2011, **112**, 933–969.
15. O. K. Farha and J. T. Hupp, *Acc. Chem. Res.*, 2010, **43**, 1166–1175.
16. K. J. Lee, J. H. Lee, S. Jeoung and H. R. Moon, *Acc. Chem. Res.*, 2017, **50**, 2684–2692.
17. W. Chaikittisilp, M. Hu, H. Wang, H.-S. Huang, T. Fujita, K. C. W. Wu, L.-C. Chen, Y. Yamauchi and K. Ariga, *Chem. Commun.*, 2012, **48**, 7259–7261.
18. J. Tang, R. R. Salunkhe, J. Liu, N. L. Torad, M. Imura, S. Furukawa and Y. Yamauchi, *J. Am. Chem. Soc.*, 2015, **137**, 1572–1580.
19. L. Yu, H. Hu, H. B. Wu and X. W. Lou, *Adv. Mater.*, 2017, **29**, 1604563.
20. B. Y. Guan, A. Kushima, L. Yu, S. Li, J. Li and X. W. Lou, *Adv. Mater.*, 2017, **29**, 1605902.
21. W. Guo, W. Sun and Y. Wang, *ACS Nano*, 2015, **9**, 11462–11471.
22. B. Y. Guan, L. Yu and X. W. Lou, *Adv. Sci.*, 2017, **4**, 1700247.
23. C.-H. Kuo, Y. Tang, L.-Y. Chou, B. T. Sneed, C. N. Brodsky, Z. Zhao and C.-K. Tsung, *J. Am. Chem. Soc.*, 2012, **134**, 14345–14348.
24. H. Hu, B. Guan, B. Xia and X. W. Lou, *J. Am. Chem. Soc.*, 2015, **137**, 5590–5595.

25. S. Pandiaraj, H. B. Aiyappa, R. Banerjee and S. Kurungot, *Chem. Commun.*, 2014, **50**, 3363–3366.
26. S. J. Yang, T. Kim, J. H. Im, Y. S. Kim, K. Lee, H. Jung and C. R. Park, *Chem. Mater.*, 2012, **24**, 464–470.
27. L. Zhang, X. Wang, R. Wang and M. Hong, *Chem. Mater.*, 2015, **27**, 7610–7618.
28. B. Liu, H. Shioyama, T. Akita and Q. Xu, *J. Am. Chem. Soc.*, 2008, **130**, 5390–5391.
29. L. Zhang, Z. Su, F. Jiang, L. Yang, J. Qian, Y. Zhou, W. Li and M. Hong, *Nanoscale*, 2014, **6**, 6590–6602.
30. Z. Song, W. Liu, N. Cheng, M. Norouzi Banis, X. Li, Q. Sun, B. Xiao, Y. Liu, A. Lushington, R. Li, L. Liu and X. Sun, *Mater. Horizons*, 2017, **4**, 900–907.
31. X.-Y. Yu, L. Yu, H. B. Wu and X. W. Lou, *Angew. Chem.*, 2015, **127**, 5421–5425.
32. L. Zhang, H. B. Wu and X. W. Lou, *J. Am. Chem. Soc.*, 2013, **135**, 10664–10672.
33. Y. Hou, J. Li, Z. Wen, S. Cui, C. Yuan and J. Chen, *Nano Energy*, 2015, **12**, 1–8.
34. N. L. Torad, M. Hu, S. Ishihara, H. Sukegawa, A. A. Belik, M. Imura, K. Ariga, Y. Sakka and Y. Yamauchi, *Small*, 2014, **10**, 2096–2107.
35. L. Li, T. Tian, J. Jiang and L. Ai, *J. Power Sources*, 2015, **294**, 103–111.
36. B. You, N. Jiang, M. Sheng, S. Gul, J. Yano and Y. Sun, *Chem. Mater.*, 2015, **27**, 7636–7642.
37. H. Hu, L. Han, M. Yu, Z. Wang and X. W. D. Lou, *Energy Environ. Sci.*, 2016, **9**, 107–111.
38. S. H. Ahn, M. J. Klein and A. Manthiram, *Adv. Energy Mater.*, 2017, 7, 1601979.
39. Y. M. Chen, L. Yu and X. W. Lou, *Angew. Chem., Int. Ed.*, 2016, **55**, 5990–5993.
40. G. Huang, F. Zhang, L. Zhang, X. Du, J. Wang and L. Wang, *J. Mater. Chem. A*, 2014, **2**, 8048–8053.
41. Y. Hou, T. Huang, Z. Wen, S. Mao, S. Cui and J. Chen, *Adv. Energy Mater.*, 2014, **4**, 1400337.
42. Y. Hou, Z. Wen, S. Cui, S. Ci, S. Mao and J. Chen, *Adv. Funct. Mater.*, 2015, **25**, 872–882.
43. L. Ge, Y. Yang, L. Wang, W. Zhou, R. De Marco, Z. Chen, J. Zou and Z. Zhu, *Carbon*, 2015, **82**, 417–424.
44. B. Y. Xia, Y. Yan, N. Li, H. B. Wu, X. W. D. Lou and X. Wang, *Nat. Energy*, 2016, **1**, 15006.
45. S. Liu, Z. Wang, S. Zhou, F. Yu, M. Yu, C.-Y. Chiang, W. Zhou, J. Zhao and J. Qiu, *Adv. Mater.*, 2017, **29**, 1700874.
46. J. Wei, Y. Hu, Y. Liang, B. Kong, J. Zhang, J. Song, Q. Bao, G. P. Simon, S. P. Jiang and H. Wang, *Adv. Funct. Mater.*, 2015, **25**, 5768–5777.
47. J. Wei, Y. Hu, Y. Liang, B. Kong, Z. Zheng, J. Zhang, S. P. Jiang, Y. Zhao and H. Wang, *J. Mater. Chem. A*, 2017, **5**, 10182–10189.

48. Z. Li, M. Shao, L. Zhou, R. Zhang, C. Zhang, M. Wei, D. G. Evans and X. Duan, *Adv. Mater.*, 2016, **28**, 2337–2344.
49. W. Xia, R. Zou, L. An, D. Xia and S. Guo, *Energy Environ. Sci.*, 2015, **8**, 568–576.
50. H. Liu, L. Chang, C. Bai, L. Chen, R. Luque and Y. Li, *Angew. Chem., Int. Ed.*, 2016, **55**, 5019–5023.
51. M. Moliner, J. E. Gabay, C. E. Kliewer, R. T. Carr, J. Guzman, G. L. Casty, P. Serna and A. Corma, *J. Am. Chem. Soc.*, 2016, **138**, 15743–15750.
52. K. Shen, L. Chen, J. Long, W. Zhong and Y. Li, *ACS Catal.*, 2015, **5**, 5264–5271.
53. L. Liu, U. Diaz, R. Arenal, G. Agostini, P. Concepcion and A. Corma, *Nat. Mater.*, 2017, **16**, 132–138.
54. H. Tabassum, W. Guo, W. Meng, A. Mahmood, R. Zhao, Q. Wang and R. Zou, *Adv. Energy Mater.*, 2017, 7, 1601671.
55. F. Yang, Y. Chen, G. Cheng, S. Chen and W. Luo, *ACS Catal.*, 2017, 7, 3824–3831.
56. H. B. Wu, B. Y. Xia, L. Yu, X.-Y. Yu and X. W. Lou, *Nat. Commun.*, 2015, **6**, 6512.
57. Y. J. Tang, M. R. Gao, C. H. Liu, S. L. Li, H. L. Jiang, Y. Q. Lan, M. Han and S. H. Yu, *Angew. Chem., Int. Ed.*, 2015, **54**, 12928–12932.
58. T. Y. Ma, S. Dai, M. Jaroniec and S. Z. Qiao, *J. Am. Chem. Soc.*, 2014, **136**, 13925–13931.
59. C. Guan, X. Liu, W. Ren, X. Li, C. Cheng and J. Wang, *Adv. Energy Mater.*, 2017, 7, 1602391.
60. J. Duan, S. Chen and C. Zhao, *Nat. Commun.*, 2017, **8**, 15341.
61. S. Zhao, Y. Wang, J. Dong, C.-T. He, H. Yin, P. An, K. Zhao, X. Zhang, C. Gao, L. Zhang, J. Lv, J. Wang, J. Zhang, A. M. Khattak, N. A. Khan, Z. Wei, J. Zhang, S. Liu, H. Zhao and Z. Tang, *Nat. Energy*, 2016, **1**, 16184.
62. C. Zhao, C. Yu, S. Liu, J. Yang, X. Fan, H. Huang and J. Qiu, *Adv. Funct. Mater.*, 2015, **25**, 6913–6920.
63. Z. Song, B. Wang, N. Cheng, L. Yang, D. Banham, R. Li, S. Ye and X. Sun, *J. Mater. Chem. A*, 2017, **5**, 9760–9767.
64. Y. Zhao, Z. Song, X. Li, Q. Sun, N. Cheng, S. Lawes and X. Sun, *Energy Storage Mater.*, 2016, **2**, 35–62.
65. Z. Song, N. Cheng, A. Lushington and X. Sun, *Catalysts*, 2016, **6**, 116.
66. S. Ma, G. A. Goenaga, A. V. Call and D. J. Liu, *Chem. – A Eur. J.*, 2011, **17**, 2063–2067.
67. Q.-L. Zhu, W. Xia, T. Akita, R. Zou and Q. Xu, *Adv. Mater.*, 2016, **28**, 6391–6398.
68. Q.-L. Zhu, W. Xia, L.-R. Zheng, R. Zou, Z. Liu and Q. Xu, *ACS Energy Lett.*, 2017, **2**, 504–511.
69. B. You, N. Jiang, M. Sheng, W. S. Drisdell, J. Yano and Y. Sun, *ACS Catal.*, 2015, **5**, 7068–7076.
70. P. Yin, T. Yao, Y. Wu, L. Zheng, Y. Lin, W. Liu, H. Ju, J. Zhu, X. Hong and Z. Deng, *Angew. Chem., Int. Ed.*, 2016, **55**, 10800–10805.
71. B. Y. Guan, L. Yu and X. W. Lou, *Energy Environ. Sci.*, 2016, **9**, 3092–3096.

72. W. Xia, C. Qu, Z. Liang, B. Zhao, S. Dai, B. Qiu, Y. Jiao, Q. Zhang, X. Huang and W. Guo, *Nano Lett.*, 2017, **17**, 2788–2795.
73. G. Jia, W. Zhang, G. Fan, Z. Li, D. Fu, W. Hao, C. Yuan and Z. Zou, *Angew. Chem., Int. Ed.*, 2017, **56**, 13781–13785.
74. C. S. Diercks, Y. Liu, K. E. Cordova and O. M. Yaghi, *Nat. Mater.*, 2018, **17**, 301–307.
75. A. Fateeva, P. A. Chater, C. P. Ireland, A. A. Tahir, Y. Z. Khimyak, P. V. Wiper, J. R. Darwent and M. J. Rosseinsky, *Angew. Chem.*, 2012, **124**, 7558–7562.
76. I. Hod, M. D. Sampson, P. Deria, C. P. Kubiak, O. K. Farha and J. T. Hupp, *ACS Catal.*, 2015, **5**, 6302–6309.
77. S. Wan, F. Gándara, A. Asano, H. Furukawa, A. Saeki, S. K. Dey, L. Liao, M. W. Ambrogio, Y. Y. Botros, X. Duan, S. Seki, J. F. Stoddart and O. M. Yaghi, *Chem. Mater.*, 2011, **23**, 4094–4097.
78. S. Lin, C. S. Diercks, Y.-B. Zhang, N. Kornienko, E. M. Nichols, Y. Zhao, A. R. Paris, D. Kim, P. Yang, O. M. Yaghi and C. J. Chang, *Science*, 2015, **349**, 1208–1213.

CHAPTER 2

Two-dimensional Layered Materials for High-performance Lithium-ion Batteries

XIANGBO MENG*[a] AND JIANLIN LI*[b]

[a] Department of Mechanical Engineering, University of Arkansas, Fayetteville, Arkansas 72701, USA; [b] Energy and Transportation Science Division, Oak Ridge National Laboratory, Oak Ridge, Tennessee 37831, USA
*Email: xbmeng@uark.edu; lij4@ornl.gov

2.1 Introduction

Since the delivery of graphene by mechanical exfoliation in 2004,[1] its exotic properties have been reported, such as a high surface area, high Young's modulus, and excellent thermal and electrical conductivity.[2] Thus, graphene has been regarded as the first two-dimensional (2D) wonder material and has attracted an extraordinary amount of interest from both academia and industry. In this context, the applications of graphene have been extensively explored to address various challenges, ranging from microelectronics[3–5] to filtration,[6–8] heavy metal adsorption,[9] biomedical,[4,10–12] gas sensing,[13,14] catalysis,[15–17] smart materials,[18,19] and clean energies.[20–25] In addition, the exceptional properties of graphene make it an ideal scaffold to fabricate composite materials and this has further dramatically widened its applications. In the meantime, the discovery of graphene has also prompted efforts to pursue alternative 2D materials. In return, a variety of 2D layered materials beyond graphene has been discovered and can be divided into five major groups: Xenes (*e.g.*, silicone, germanene, and phosphorene),[26–28]

Smart Materials No. 34
Layered Materials for Energy Storage and Conversion
Edited by Dongsheng Geng, Yuan Cheng and Gang Zhang

Published by the Royal Society of Chemistry, www.rsc.org

MXenes (*e.g.*, Ti_3C_2 and Ta_4C_3),[29] transition metal chalcogenides (TMCs, *e.g.*, MoS_2, MoO_3, $MoSe_2$, WS_2, and WSe_2),[25,27,30–32] nitrides (*i.e.*, hBN, GaN, and Ca_2N),[27] and organic materials.[33,34] All of these research outcomes have greatly enriched our knowledge and advanced the search for new solutions to real challenges.

One of the biggest challenges facing society is the ever-aggravating energy crisis, mainly due to depleting fossil fuels as well as the environmental issues caused by their combustion.[35] To this end, renewable clean energies (*e.g.*, solar radiation, wind, biomass, and hydrogen) are undergoing intensive investigation as alternatives. In order to widely implement these new energies, electrical energy storage (EES) devices are essentially needed, such as batteries, supercapacitors, and fuel cells. Among them, rechargeable battery technologies are very promising, and to date, lithium-ion batteries (LIBs) enable the highest energy density. Using graphite as the anode and a metal oxide (*e.g.*, $LiCoO_2$, $LiMn_2O_4$, or $LiFePO_4$) as the cathode, state-of-the-art LIBs could realize an energy density of up to 250 Wh kg^{-1} and are dominating consumer electronics (*e.g.*, cell phones and laptops). LIBs have been applied to the market of electric vehicles (EVs). However, deep penetration into the EV market requires a battery with an energy density >300 Wh kg^{-1} [36] or ideally 500 Wh kg^{-1}.[37] To this end, new battery materials with high energy densities as well as tailored electrode engineering and architectures are vital. Given its many excellent characteristics, graphene is versatile in next-generation LIBs. Since the discovery of graphene, there has been an increasing number of publications using graphene in LIBs. At the same time, the applications of graphene in other battery technologies have also been intensively studied. Besides serving as LIB anodes, graphene has been more extensively synthesized into advanced composite materials such as anodes and cathodes with improved LIB performance. With continuous studies on other 2D materials, however, there have been more and more reports on the different electrochemical behaviors in LIBs.

In this chapter, we focus on summarizing the applications of graphene and other 2D layered materials in LIBs. Following the introductory part, the role of graphene in LIBs and a summary on recent studies of other 2D materials in LIBs are presented in the second and third sections, respectively. Finally, we summarize this chapter and provide an outlook on future studies.

2.2 Graphene and Its Composites as Electrodes in LIBs

Since first isolated in 2004 by Novoselov and co-workers[1] using scotch tape, many different fabrication methods have been developed. In this regard, Bonaccorso *et al.* have made an excellent review.[38] Among the different methods, mechanical exfoliation, chemical vapor deposition (CVD) processes, and some other bottom-up techniques are not suitable for mass production.[39,40] Liquid-phase exfoliation and reduction of graphene oxide (GO) are two methods that are widely employed for the bulk production of graphene.[41] During liquid-phase exfoliation, pristine or expanded graphite particles, obtained by thermal expansion of graphite intercalation compounds (usually known as "expandable graphite"), are first dispersed in a solvent to reduce the strength of the van der Waals attraction between the graphene layers.[40] Then, an external driving force (*e.g.*, ultrasonication, electric field or shearing) is applied to perform the exfoliation of graphite into high-quality graphene sheets. In another method of GO reduction, pristine graphite was first strongly oxidized and then turned into a highly defective form of graphene with a disrupted sp^2-bonding network with a subsequent stirring or ultrasonication in liquid media. In order to restore its π network (*i.e.*, the characteristic of conductive graphene), GO must be reduced. This reduction can be performed *via* chemical, thermal and electrochemical processes, leading to reduced graphene oxide (rGO). These two methods enable the mass production of graphene with low-to-medium quality, due to the presence of both intrinsic defects (edges and deformations) and extrinsic defects (O- and H-containing groups).[40] Due to their scalability and cost-effectiveness, these two methods are also the primary processes for graphene used in the devices of EES devices.

2.2.1 Anodes

2.2.1.1 Graphene Anodes

Graphite, the first commercialized anode material in LIBs, features reliable cyclability, good conductivity, and little volume change. However, graphite has a medium capacity of 372 mAh g^{-1} and this hinders LIBs from higher energy density. Differing from the intercalation-based bulk-type graphite, a single layer graphene may have a different electrochemical mechanism and this prompted the first investigation on graphene-based anodes. Yoo *et al.*[42] prepared graphene nanosheets (GNS) *via* a solution-based chemical reduction process and tested the GNS as a LIB anode. In the same study, Yoo *et al.*[42] also fabricated two other anodes using a mixture of 1 : 9 CNT + GNS and 1 : 9 C_{60} + GNS. This study identified the relationship between the *d*-spacing and the graphene layer stacking number of the GNS families (Figure 2.1a). TEM analyses revealed that the pure GNS had a smaller average *d*-spacing of ~0.365 nm compared with those of mixtures of GNS with CNT and C_{60}. It was demonstrated that the *d*-spacing of GNS increased with the decreased number of graphene layers. Specifically, the *d*-spacing of the GNS with 17 stacking graphene layers is ~0.34 nm, while the one with only four layers increases up to 0.385 nm. In contrast, the average *d*-spacing of the GNS + CNT and GNS + C_{60} is ~0.40 nm, regardless of the stacking layer numbers. Yoo *et al.*[42] believed that the presence of the π-electron system macromolecules of CNT and C_{60} must have significantly enhanced the *d*-spacing of the GNS + CNT and GNS + C_{60} structures.

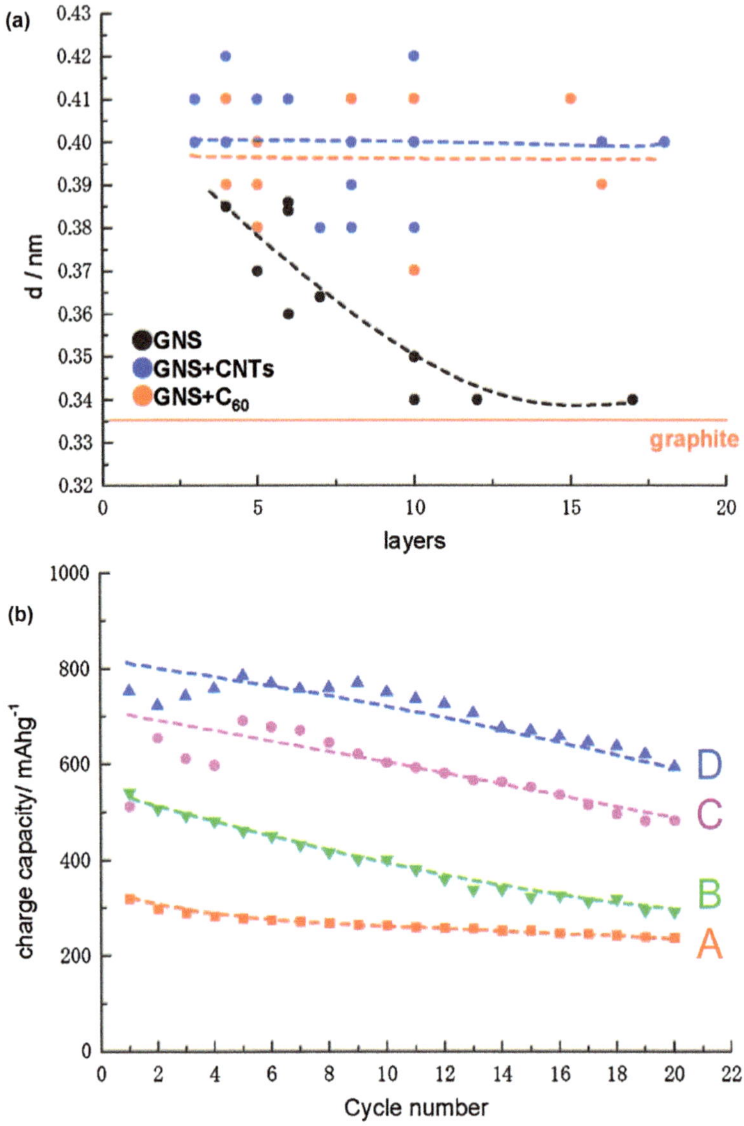

Figure 2.1 (a) Relationship between the number of graphene stacking sheets in the GNS families and the *d*-spacing perpendicular to the basal plane. (b) Charge/discharge cycling performance of (A) graphite, (B) GNS, (C) GNS + CNT, and (D) GNS + C_{60} at a current density of 50 mA g^{-1} in the range of 0.01–3.5 V *vs.* Li/Li^+.[42]
Reproduced from ref. 42 with permission from American Chemical Society, Copyright 2008.

This study further revealed the charge/discharge performance for graphite, GNS, GNS + CNT, and GNS + C_{60} (Figure 2.1b). Apparently, GNS, GNS + CNT, and GNS + C_{60} exhibited higher specific capacities than the graphite anode, *i.e.*, 290, 480, and 600 mAh g^{-1} after 20 cycles, respectively. In comparison, the graphite anode sustained a capacity of 240 mAh g^{-1} after 20 cycles. Yoo *et al.*[42] postulated two possible reasons for the enhanced capacities of GNS, GNS + CNT, and GNS + C_{60}. First, the electronic structures of GNS, GNS + CNT, and GNS + C_{60} must be different from that of graphite. Second, the expansion in the *d*-spacing of the graphene layers may cause additional sites to accommodate Li^+ ions.

Following the study by Yoo *et al.*,[42] Guo *et al.*[43] used a similar chemical process and synthesized GNS with only 2–3 layers of graphene sheets, having an interplanar distance of ~0.37 nm larger than that of graphite (0.34 nm). The as-received GNS powders contain –H and –OH groups. Therefore, Guo *et al.*[43] treated the GNS powders at 500 °C in argon. Raman analyses indicated that the GNS was partially disordered, for there was a strong D band as well as an evident G band. The former is due to the reduction in size of the in-plane sp^2 domains due to the extensive oxidation, while the latter is indicative of the in-phase vibration of the graphite lattice.[44] Compared to the well crystalline graphite, the D/G intensity ratio of the GNS significantly increased. Electrochemical testing revealed that the cyclability of the GNS electrode demonstrated a good cyclic performance and reversibility, sustaining a specific capacity of 460 mAh g^{-1} after 100 cycles at a current density of 1 °C (*i.e.*, 372 mA g^{-1}) in the potential range of 0.01–3.0 V *vs.* Li/Li^+. The enhanced performance of the GNS electrode was believed to be possible due to two reasons: (i) the two layers of lithium for each graphene sheet (744 mAh g^{-1} theoretically) and (ii) nano-cavities contributing to higher lithium storage capacity.

The intriguing electrochemical performance of the GNS materials prompted many more efforts to understand the underlying lithium storage mechanism. In this regard, Pan *et al.*[45] conducted a systematic study using disordered GNS. In their work, a series of GNS samples were prepared from GO *via* three different routes including hydrazine reduction, pyrolytic deoxidation and electron beam irradiation. Pan *et al.*[45] disclosed that the 300 °C pyrolytic and electron-beam-irradiated GO sheets showed exceptionally high reversible capacities and cyclability, compared to the natural graphite, the pristine GO, and the hydrazine-reduced GO. The 300 °C pyrolytic GO sheets enabled a reversible capacity in the range of 1013–834 mAh g^{-1} and a capacity retention of 82% in 15 cycles. Except for the first cycle, the Coulombic efficiencies (CEs) of the other cycles reached 90–95%. The 600 °C pyrolytic GO also showed good cyclability with CEs ~95% after the first cycle. For the electron-beam-irradiated GO, the reversible capacity is in the range of 1054–784 mAh g^{-1} with a capacity retention of 74% during 15 cycles. To better understand the electrochemical performance, Pan *et al.*[45] analyzed four structural parameters, *i.e.*, surface functional groups, specific surface area, interlayer spacing, and Raman intensity ratio of D/G. Based on their analyses of these parameters, Pan *et al.*[45] asserted that the greatly enhanced reversible capacities of the highly disordered GNSs (*via* the low-temperature 300 °C pyrolysis and electron beam irradiation) might arise from their numerous defects (confirmed by Raman and HRTEM). They believed that a larger specific surface area might contribute to larger irreversible capacity due to more SEI formation. In addition, their analyses did not show any clear implication on the relations between the (002) interlayer spacing and the reversible capacity. Pan *et al.*[45] believed that some other potential factors might also have influenced reversible capacity. Among the four factors examined, Pan *et al.*[45] believed that the defects at the interface between the GNS and the electrolyte might have accommodated irreversible Li in SEI films, whereas the defects at the edge sites and internal (basal-plane) defects (vacancies *etc.*) of the nanodomains embedded in GNSs might have been involved in the reversible Li storage. The defects embedded are not exposed to the electrolyte. In addition, some Li^+ ions might be stored reversibly between (002) planes, but the defect-based reversible storage might predominate. It was believed that the large voltage hysteresis was related to active defects in the disordered GNSs. The reaction of Li with the active defects in the discharge processes occurred at low voltages, but the break of the relatively strong bonds of Li with the defects in the charge processes required higher voltages. Furthermore, the notable capacity fading observed in the highly disordered samples might be ascribed to the disorder-induced structure instability. In the study, the D/G ratio was identified as a key parameter to evaluate reversible capacity in carbon anodes. The D/G ratio describes the disorder degree in carbonaceous materials. The D band is ascribed to disordered carbon, edge defects, and other defects (sp^3 bonded carbon, dangling bonds, vacancies, and topological defects), while the G band signifies ordered sp^2 carbon. The larger D/G ratio of the electron-beam-reduced GO, in comparison to that of the hydrazine-reduced GO, suggested more disorder, more defects, and smaller sp^2 domains (~3 nm). In addition, graphene has two types of edges such as chair edges and zigzag edges. Consequently, the concentration of edge sites at the periphery of such small nanodomains may be quite high in comparison to the basal plane sites. The 300 °C pyrolytic sample has substantial disorder, confirmed by the disappearance of the ordered G band and the observation of highly distorted (002) planes by HRTEM. Besides lots of edge sites at the periphery of the sp^2 nanodomains, other defects, including vacancies and topological defects, might exist in the highly distorted basal planes in quantity. Thus, edges and other defects are believed to be among the main contributors to higher reversible capacities.

In a following study, Tong *et al.*[46] developed a technical route to synthesize GNSs with controllable layers, in which the oxidation degree of GOs could be tuned. The higher the oxidation degree of the GOs, the fewer the layers of the resultant GNSs. Using GNSs with different layers (1, 3, and 5), Tong *et al.*[46] revealed that the reversible capacity of GNSs varied regularly with their layer numbers and the GNSs with fewer layers have larger reversible capacity. Tong *et al.*[46] believed that two main reasons were responsible for the higher capacity of the GNSs with fewer layers. First, the specific surface area becomes larger as the layer number decreases. Second, the GNSs with fewer layers can provide more lithium insertion active sites, such as edge-type sites, to store lithium ions. Tong *et al.*[46] confirmed that the GNSs with fewer layers have more nanocavities due to higher oxidation. A similar work has also been reported by Wang *et al.*[47]

The contribution of graphene defects to reversible capacity was further corroborated by doped graphene. In one work, Reddy *et al.*[48] grew graphene and nitrogen-doped graphene (NG) layers on copper foils using CVD processes, respectively. The NG layers were incorporated with numerous N-related defects, such as pyrodine-like N atoms. Reddy *et al.*[48] verified that the NG layers performed much better in sustaining higher reversible capacity over the graphene layers. More studies have been conducted on doped graphene and they clearly demonstrated the benefits of doped elements. In one work, Wu *et al.*[49] developed two doped graphene, *i.e.* NG and boron-doped graphene (BG) having layers fewer than 3 (Figure 2.2). At a low charge-discharge rate of 50 $mA\,g^{-1}$, the doped graphene electrodes could sustain a very high discharge capacity of 896 $mAh\,g^{-1}$ for NG and 1327 $mAh\,g^{-1}$ for BG after 30 cycles. Furthermore, the NG and BG exhibited excellent rate capability, enabling a capacity of 199 $mAh\,g^{-1}$ for NG and 235 $mAh\,g^{-1}$ for BG at a current density of 25 $A\,g^{-1}$ (*i.e.*, 126 °C). These results are superior to those of the pristine graphene, *i.e.*, a capacity of ~100 $mAh\,g^{-1}$ at a current density of 25 $A\,g^{-1}$. Wu *et al.*[49] also verified that the rate capability of the doped graphene could be further improved by increasing the N- or B-dopant concentrations. Wu *et al.*[49] verified that N- and B-doped graphene have multiple advantages over pristine graphene. These combined benefits favored the doped graphene to achieve an exceptional electrochemical performance. First, the N- and B-doped graphene have better electrical conductivity and electrochemical activity at high rates. These were confirmed by the measurements of electrochemical impedance spectroscopy (EIS) and electronic conductivity. Second, the N- and B-doped graphene have increased disordered surface morphology (corrugations and scrolling) (Figure 2.2). These features and related defects are favorable to lithium storage. Third, the N- and B-doped graphene have higher hydrophobicity and better wettability toward organic electrolytes, promoting ion diffusion in the interface between the electrode and electrolyte. Furthermore, the N- and B-doped graphene also showed better thermal stability and kept the characteristics of pristine graphene such as a high surface area, porous structure, mechanical flexibility, and chemical stability. Simulations also provided some insightful views on doped graphene. Ma *et al.*[50] conducted first-principles calculations and revealed that pyridinic graphene was the most suitable for Li storage, while the graphitic structure was not suitable. In other words, a higher concentration of the pyridinic structure would contribute a higher capacity of Li storage. In this regard, S-doped[51–53] and even both S- and N-co-doped graphene[54] were investigated. All of these types of doped graphene have demonstrated improved performance in Li storage.

In addition to the aforementioned studies, the exceptional performance of graphene and doped graphene has also been reported somewhere else.[55–58] Graphene has been fabricated into different structures such as microspheres,[57]

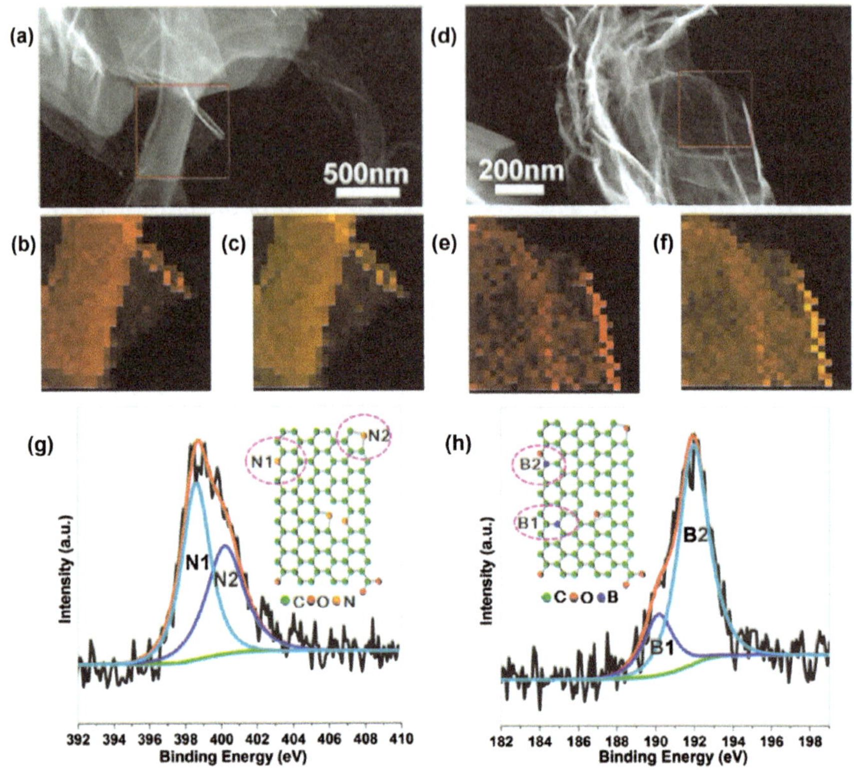

Figure 2.2 (a) STEM image of the N-doped graphene sheets and (b) C- and (c) N-elemental mapping of the square region in (a). (d) STEM image of the B-doped graphene sheets and (e) C- and (f) B-elemental mapping of the square region in (d). (g) N1s XPS spectrum of the N-doped graphene. Inset: schematic structure of the binding conditions of N in a graphene lattice showing the pyridinic N (N1) and pyrrolic N (N2), indicated by magenta dotted rings. (h) B1s XPS spectrum of the B-doped graphene. Inset: schematic structure of the binding conditions of B in a graphene lattice showing BC_3 (B1) and BC_2O (B2), indicated by magenta dotted rings.[49]
Reproduced from ref. 49 with permission from American Chemical Society, Copyright 2011.

vertically-aligned graphene sheets,[56] and graphene-constructed carbon fibers.[58] Furthermore, graphene has been used to combine with other carbon materials. For example, carbon nanotubes have been vertically grown on graphene.[59] All these efforts commonly demonstrated that graphene and its derivatives are promising anode materials for LIBs.

2.2.1.2 Graphene-based Composite Anodes

Graphene features some exceptional mechanical, electrical, and other physical properties. These features distinguish graphene from other carbon materials and make it an outstanding scaffold for establishing new composite materials as anodes in LIBs. Among the anodes known to date, conversion-based (*e.g.*, Fe_2O_3 and CuO) and alloy-based (*e.g.*, Si, Ge, Sn, and SnO_2) materials are two classes enabling high capacities. However, these materials commonly suffer from high volume changes. Many of these materials also have low electrical conductivity. These inherent drawbacks make them very difficult to be utilized in practice. To this end, graphene offers a new platform to address these challenges. Consequently, there have been a huge number of studies reported using graphene to constitute new graphene-based composites as LIB anodes. In this chapter, we exemplify the graphene-based composite anodes by discussing strategies to construct graphene-based Si nanocomposites; for Si as an anode material is very promising but poses the most challenging issues for practical applications. Si has an extremely high capacity of 3579 mAh g^{-1} at room temperature,[60] ~10 times higher than a commercial graphite anode (372 mAh g^{-1}). In addition, Si is cost-effective and abundant on the Earth. Thus, it is one of the most attractive anode materials. However, Si has been being hindered from practical uses due to its huge volume change of 300% and low electrical conductivity of 10^{-3} S cm^{-1}.[60,61] Furthermore, these downsides are prone to cause many more harmful effects on Si anodes, *e.g.*, continuous formation of SEI, electrochemical pulverization, low CE, and even cell failure.[61] Thus, it is not viable to apply pure Si as an anode in LIBs.

The simplest method for Si–graphene composites is *via* a mechanical mixing or solution-filtration process,[62–64] in which Si particles are mixed with graphene in a certain ratio. The resultant Si–graphene composites did exhibit noticeable improvements over the pure graphene and Si performance.[62] However, these composites also demonstrated several shortcomings: (i) high surface area leading to large irreversible capacity loss; (ii) low CE; and (iii) questionable stability. The reason lies in the lack of a strong interaction between graphene and Si particles in these composites. In order to improve the interaction between the graphene and Si particles, Yang *et al.*[65] developed a process to covalently bind Si nanoparticles on graphene (Figure 2.3a). First, graphene (G) was functionalized with $PhNH_2$. Then, the received G-$PhNH_2$ was further fabricated into a Si–Ph–G nanocomposite. The Si–Ph–G nanocomposite was confirmed to have an exceptional electrochemical performance, *i.e.*, an initial charge capacity of 1079 mAh g^{-1} and a sustainable charge capacity of 828 mAh g^{-1} after 50 cycles at a current density of 300 mA g^{-1} (Figure 2.3b). Furthermore, the resultant Si–Ph–G nanocomposite exhibited excellent rate capabilities (Figure 2.3c). In a following work by Hassan *et al.*,[66] Si nanoparticles (SiNP, ~60%), S-doped graphene (SG), GO, and cyclized polyacrylonitrile (PAN) were mixed in dimethylformamide (DMF) to form a homogenous mixture under ultrasonic radiation (Figure 2.3d). Then the received slurry was cast on a Cu current collector and dried in a convection oven (Figure 2.3e). The received electrodes were cut and pressed. Then, the electrodes were subject to a sluggish heat treatment (SHT) by slow heating in inert gas to 450 °C for

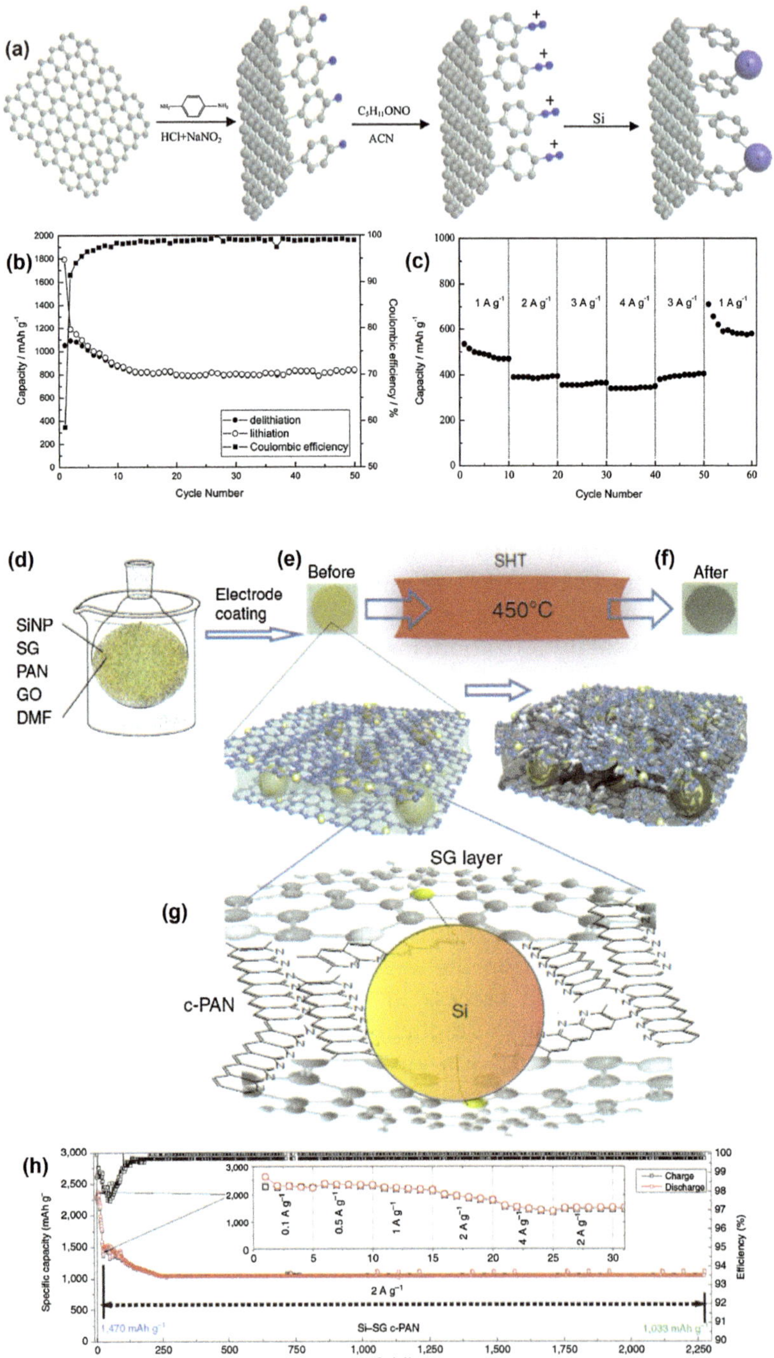
(a)
HCl+NaNO₂
C₅H₁₁ONO
ACN
Si
(b)
Capacity / mAh g⁻¹
Coulombic efficiency / %
delithiation
lithiation
Coulombic efficiency
Cycle Number
(c)
1 A g⁻¹
2 A g⁻¹
3 A g⁻¹
4 A g⁻¹
3 A g⁻¹
1 A g⁻¹
Capacity / mAh g⁻¹
Cycle Number
(d)
SiNP
SG
PAN
GO
DMF
Electrode coating
(e)
Before
SHT
450°C
(f)
After
SG layer
(g)
c-PAN
Si
(h)
Specific capacity (mAh g⁻¹)
Efficiency (%)
Charge
Discharge
0.1 A g⁻¹
0.5 A g⁻¹
1 A g⁻¹
2 A g⁻¹
4 A g⁻¹
2 A g⁻¹
2 A g⁻¹
1,470 mAh g⁻¹
Si–SG c-PAN
1,033 mAh g⁻¹
Cycle No.

10 min before cooling down. After the SHT, the resultant Si–SG electrodes (Figure 2.3f) changed from yellow to black, indicating the PAN was partially carbonized. Hassan *et al.*[66] speculated that the SiNPs were preferentially adsorbed on the sulfur and defect sites in graphene as well as being coated with PAN (Figure 2.3g). This speculation has been supported by DFT simulations, *i.e.*, some Si atoms formed covalent interactions with sulfur atoms in SG and two adjacent carbon atoms. These Si atoms did not participate in alloy formation with lithium, but provided anchoring sites for the majority of Si atoms within the Si nanoparticle that were readily available for alloying-dealloying. It has also been verified that Si bound more strongly to SG than on graphene (*i.e.*, G), which has been ascribed to the strong covalent interaction between the Si atoms with the sulfur atom. Hassan *et al.*[66] confirmed that the Si–SG electrodes performed well as an anode (Figure 2.3h), *i.e.*, retaining a high capacity of 1033 $mAh\,g^{-1}$ after 2275 cycles at 2 $A\,g^{-1}$. In addition, the Si–SG electrodes also exhibited an excellent rate capability (see the inset of Figure 2.3h). In another work, Zhang *et al.*[67] reported an *in situ* facile and low-cost bottom-up strategy to synthesize a novel N-doped graphene frame supported silicon@graphitic carbon composite (NGSi@G) *via* a one-step solid-state approach. N-doped graphene (NG) was grown *in situ* around the SiNPs by pyrolyzing and re-carbonizing metal phthalocyanine (or their derivatives) precursors. A cross-linked graphene frame network was gradually formed. Simultaneously, SiNPs were coated by a thin graphitic carbon layer during the same pyrolysis and re-carbonization processes. The strong interconnections among the graphene, graphitic carbon, and silicon led to a spontaneous self-assembly process, resulting in the formation of μm scale composite granules with available nanoporosities and irregular channels. Zhang *et al.*[67] further confirmed that the resultant NGSi@G composite exhibited a much better performance over pristine SiNPs and graphene. In a subsequent work, Mi *et al.*[68] developed a facile one-step carbon-thermal method to coat SiNPs with N-doped graphene-like nanosheets derived from a liquid acrylonitrile homopolymer (LPAN) precursor. The LPAN possessed unique fluid properties and enabled a uniform coating of the SiNPs. In addition, LPAN contains nitrile groups and has excellent graphitization properties. This makes it possible for LPAN to transform into N-doped graphene at a temperature of 750 °C. The resultant

Figure 2.3 (a) Illustration for the covalent anchoring of Si nanoparticles to graphene nanosheets *via* achromatic linkers, (b) cycling performance at the current density of 300 $mA\,g^{-1}$, and (c) rate capability of Si–Ph–G electrodes.[65] Reproduced from ref. 65 with permission from the Royal Society of Chemistry. (d) Components mixing under ultrasonic irradiation, (e) an optical image of the as-fabricated electrode made of SiNP, SG, and PAN, (f) the electrode after SHT, and (g) schematic of the atomic scale structure of the electrode. (h) Rate capability of the SG–Si anode followed by cycle stability at 2 $A\,g^{-1}$.[66] Reproduced from ref. 66, https://doi.org/10.1038/ncomms9597, under the terms of the CC BY 4.0 licence, https://creativecommons.org/licenses/by/4.0/.

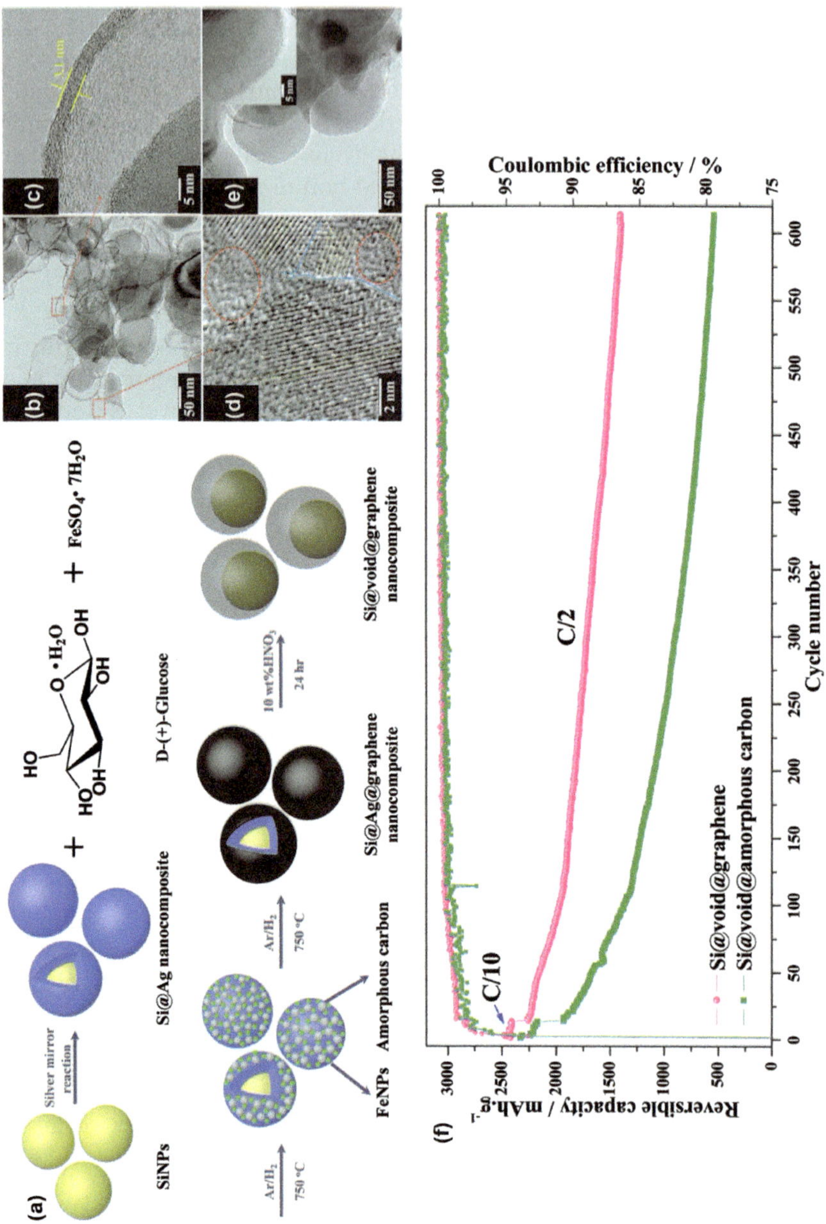
(a)
SiNPs
Silver mirror reaction
Si@Ag nanocomposite
+
D-(+)-Glucose
+
FeSO4• 7H2O
Ar/H2
750 °C
FeNPs
Amorphous carbon
Ar/H2
750 °C
Si@Ag@graphene nanocomposite
10 wt%HNO3
24 hr
Si@void@graphene nanocomposite
(b)
50 nm
(c)
5 nm
(d)
2 nm
(e)
50 nm
5 nm
(f)
Reversible capacity / mAh.g-1
Coulombic efficiency / %
Cycle number
C/10
C/2
Si@void@graphene
Si@void@amorphous carbon
0
500
1000
1500
2000
2500
3000
75
80
85
90
95
100
0
50
100
150
200
250
300
350
400
450
500
550
600

NG-coated Si exhibited dramatically improved cycling stability and much higher sustainable capacity over SiNPs and graphene.

In order to construct a permanently stable structural and electrical environment for Si, a promising strategy is to embed and isolate Si nanostructures into sealed hollow conducting spheres with sufficient voids inside. In this regard, Zhang *et al.*[69] developed a bottom-up route to controllably synthesize a polycrystalline graphene hollow sphere isolated SiNP nanocomposite (Si@void@graphene) through an *in situ* pyrolysis and metal-catalyzed graphitization reaction (Figure 2.4a). In order to control the void space between the graphene shell and the Si core, a silver (Ag) sacrificial conformal layer with the desired thickness was formed on the SiNP surface *via* a well-known silver mirror reaction. To further grow the graphene shell, Zhang *et al.*[69] used glucose and a trace amount of $FeSO_4$ with strictly controlled content and ratio as the carbon source and catalyst precursor, respectively. Under the catalysis of the *in situ* produced Fe nano-islands, a compact and uniform polycrystalline graphene shell with a thickness of *ca.* 3.1 nm (*i.e.*, 9 layers) was grown on the Ag surface through a metal-catalyzed transformation of amorphous carbon, *i.e.*, Si@Ag@graphene nanocomposites. The amorphous carbon was formed from the pyrolysis of carbon sources. Zhang *et al.*[69] further etched the Ag layer away from the Si@Ag@graphene using a 10 wt% nitric acid solution. Then, the ultimate product Si@void@graphene nanocomposites were achieved. In addition, the Si@void@amorphous carbon nanocomposites were also synthesized under the same processing conditions in the absence of a $FeSO_4{\cdot}7H_2O$ catalyst precursor. Zhang *et al.*[69] confirmed the formation of the polycrystalline graphene layers (Figure 2.4b–d) and amorphous carbon layers (Figure 2.4e) using TEM. Given the high degree of crystallinity, excellent electronic conductivity, and superior mechanical strength of the graphene layers, the developed Si@void@graphene nanocomposites showed a much better performance than that of Si@void@amorphous carbon nanocomposites (Figure 2.4f). In addition, Zhang *et al.*[69] pointed out that the graphene overlayer could also effectively prevent the electrolyte from reaching the

Figure 2.4 (a) Schematic diagram of the synthesis process of Si@void@graphene nanocomposites. The synthetic procedure mainly consists of a liquid phase reaction (*i.e.*, Ag coating) and a solid-state reaction (polycrystalline graphene shell formation). TEM analyses: (b) TEM image of the Si@void@graphene nanocomposites, (c) HRTEM image of the edge of the graphene hollow sphere, (d) HRTEM image of a microdomain of the graphene shell surface (the in-plane lattice constant of the synthetic graphene has been determined to be 2.5 ± 0.05 Å, perfectly matching the theoretical value of graphene (2.46 Å)), and (e) TEM image of the Si@void@amorphous carbon nanocomposites, the inset of (e) shows a HRTEM image of the edge of the amorphous carbon spherical surface. (f) Reversible charge capacity *vs.* cycle number for the Si@void@amorphous carbon and Si@void@graphene electrodes.[69]
Reproduced from ref. 69 with permission from the Royal Society of Chemistry.

SiNPs inside and therefore enabled the inhibition of excessive growth of the SEI layer on the Si surface under repeated volume change.

In addition to the different nanostructured graphene-based Si composites discussed above, there are still many more graphene-related Si anodes reported in the literature. In this section, we focus on graphene-based Si anodes to exemplify that graphene is a very useful material to tackle the most challenging issues in LIB anodes, and how graphene could be integrated with Si or other anode materials for better electrochemical performance of LIBs. Besides Si, graphene has also been widely used for a long list of other anode materials, such as Sn, Ge, SnO_2, SiO_2, and Fe_2O_3. There are, to date, many reviews available in the literature.[40,70–75] All of these are ascribed to the excellent properties of graphene. It should be pointed out that the Li storage capability of graphene-based composite anodes is partially contributed by the integrated graphene.

2.2.2 Graphene-based Composite Cathodes

State-of-the-art LIBs are based on intercalation chemistry. On the cathode side, intercalation-based cathodes are mainly three types, *i.e.*, layered $LiCO_2$ (LCO),[76] spinel $LiMn_2O_4$ (LMO),[77] and olivine $LiFePO_4$ (LFP).[78] Each type has benefits and drawbacks. Among them, LFP has a high theoretical capacity of 170 mAh g^{-1}, first reported in 1997.[78] LFP is also low-cost and environmentally friendly. Compared to LCO (10^{-4} S cm^{-1}) and LMO (10^{-6} S cm^{-1}), LFP has a much lower electronic conductivity of 10^{-9} S cm^{-1},[79] adversely hindering it from achieving high performance. To this end, researchers have been inspired to develop different strategies to synthesize various LFP-graphene composites with improved properties for high performance in LIBs. In an early study,[80] a graphene-LFP composite (Figure 2.5a) was fabricated *via* a co-precipitation method, in which 100-nm LFP particles were attached on rGO of ~1.5 wt%. Electrochemical evaluation revealed that the LFP-graphene composite could realize a capacity of 160 mAh g^{-1} *vs.* 113 mAh g^{-1} for LFP at 0.2 °C, which has been ascribed to the smaller size of LFP in the composite and improved conductivity related to the addition of rGO. In a later study,[81] 5 wt% graphene was integrated into LFP *via* a spray drying process, leading to microscale spherical LFP-graphene particles (Figure 2.5b) consisting of graphene-wrapped LFP nanoparticles (Figure 2.5c). The graphene formed a network connecting all the nanoscale LFP particles together, leading to a capacity of 70 mAh g^{-1} at 60 °C. In another work, Wu *et al.*[82] mixed amorphous LFP nanoparticles, graphene, and sucrose in ethanol with the assistance of ultrasonication. Stirring at 60 °C to achieve dryness, the resulting products were then ground to ensure that the LFP was dispersed on graphene. In this step, sucrose molecules transferred the non-uniformly scattered bonding sites on the graphene surface into uniformly functional groups with the assistance of hydrogen bonds. Then, hydroxyl groups were able to bind LFP nanoparticles and anchor them onto the surface of the graphene with the assistance of grind.

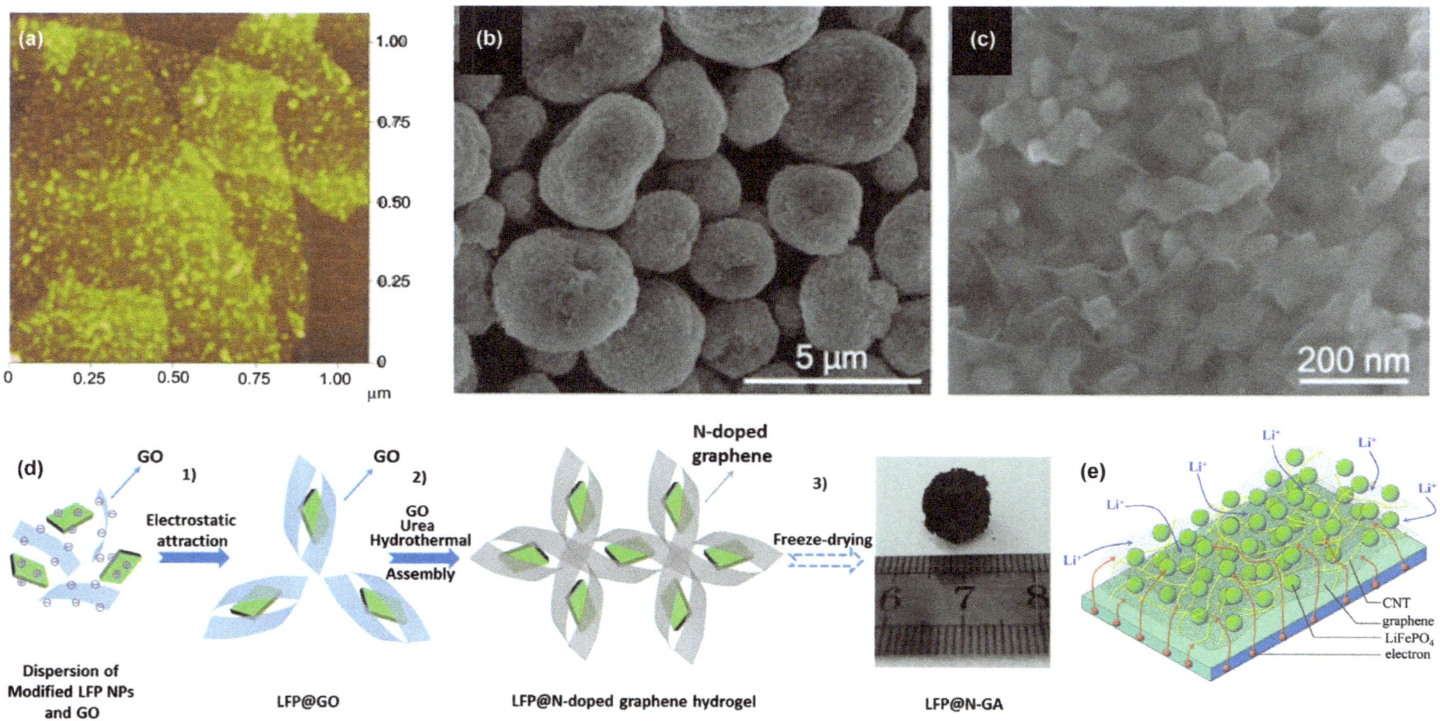

Figure 2.5 Various LFP-graphene composites by different methods. (a) Morphological characteristics of a co-precipitated LFP-graphene composite by atomic force microscopy.[81] Reproduced from ref. 80 with permission from Elsevier, Copyright 2010. (b,c) SEM images of a LFP-graphene composite by a spray drying process.[81] Reproduced from ref. 81 with permission from the Royal Society of Chemistry. (d) The preparation process of LFP@N-GA.[84] Reproduced from ref. 84 with permission from the Royal Society of Chemistry. (e) The network structure of a LFP–CNT-G composite.[85] Reproduced from ref. 85 with permission from Elsevier, Copyright 2015.

The dried products were then annealed at 700 °C under vacuum to improve the crystallinity of LFP in the graphene-sucrose-LFP composites. In the resultant graphene-sucrose-LFP, sucrose greatly improved the stability of the structure for long-term cyclability, while graphene increased the electrical conductivity and ion diffusion. Consequently, the graphene-sucrose-LFP showed highly sustainable capacity and exceptional rate capability, *e.g.*, 115 mAh g^{-1} at 10 °C.

In a following study, Ha and Lee[83] reported a fabrication process for synthesizing a carbon-coated LFP-graphene hybrid. This fabrication process featured three factors: (i) *in situ* polymerization of polypyrrole enabling the LFP nanoparticle size to be constrained in the range of 30–40 nm, (ii) enhanced dispersion of the conducting 2D networks endowed by colloidal stability of GO, and (iii) intimate contact between the active materials and rGO. The fabricated hybrid cathodes showed superior rate capability and cyclability with rates from 0.05 to 60 °C, specifically, a discharge capacity of 168 mAh g^{-1} at 0.05 °C and 72 mAh g^{-1} at 60 °C. The carbon occupied ~10 wt% in the hybrid. Through this study, Ha and Lee[83] demonstrated that the synergistic combination of nanoengineering with efficient conducting templates enabled the facile transport of both Li^+ ions and electrons for high power applications. More inspiringly, Wang *et al.*[84] developed a process enabling one to wrap (010) facet-oriented LFP NPs with N-doped graphene aerogel (N-GA) (Figure 2.5d). The resultant LFP@N-GA composite exhibited several benefits in the following aspects. First, the formed 3D porous network of N-GA facilitated rapid Li^+ transport. Second, the LFP NPs were 150–200 nm in length, 70–100 nm in width, and 30–40 nm in thickness, enabling a short Li^+ diffusion length in the LFP NPs with an enhanced solid-state ion diffusion rate. Third, a large electrode/electrolyte interfacial area was formed to boost the number of active sites for electrochemical reactions. Furthermore, the resultant LFP@N-GA composite had high electronic conductivity to increase the electron transfer rate in the electrode. Wang *et al.*[84] demonstrated that the LFP@N-GA composite electrode realized a high rate capability (78 mAh g^{-1} at 100 °C) and a stable cyclability (89% capacity retention over 1000 cycles at 10 °C).

Among the efforts in developing LFP-graphene composites, there was also a study[85] reported to integrate both graphene and CNTs with LFP NPs, resulting in a composite of LFP–CNT-graphene (Figure 2.5e). The LFP–CNT-graphene composite was synthesized by a solid-state reaction, in which wet ball milling was performed with stoichiometric amounts of Li_2CO_3, $FeC_2O_4{\cdot}2H_2O$, $NH_4H_2PO_4$, 3 wt% CNTs, and 1 wt% graphene in ethanol (>99.9%) for 12 h. The precursor mixture after ball milling was dried at 90 °C for 8 h in a vacuum drying oven. The dried powders were subject to a two-step heat treatment process, *i.e.*, 350 °C for 4 h under Ar atmosphere and 650 °C for 10 h. The first step was to decompose the carbonate, oxalate, and ammonium mixture of the starting materials, while the second step was to form the LFP–CNT-graphene composite. The LFP–CNT-graphene electrode enabled a reversible capacity of 168.9 mAh g^{-1} at 0.2 °C and 115.8 mAh g^{-1} at

20 °C, much higher than LFP–CNT and LFP-graphene composites fabricated *via* the same method. The researchers[85] believed that graphene nanosheets and CNTs constructed a 3D conducting network facilitating a faster electron transfer and lower resistance during the Li^+ reversible reaction, and that the synergistic effect between the graphene nanosheets and CNTs improved the rate capability and cycling stability of the LFP–CNT-graphene composite. In summary, the integration of graphene into these LFP composites could evidently improve the electrodes' conductivity for Li^+ and electron transport. Associated with the integration of graphene, the LFP nanoparticles in the composites have also been controlled. This is beneficial to short the pathways of Li^+ ions and electrons, and increase the surface areas for achieving high rate capabilities. In addition, the excellent mechanical property of graphene is also favorable for sustaining electrode integrity for long-term stability. Thus, the benefits of graphene in LFP composites are multiple, as discussed above.

In comparison to LCO, LMO is favorable for cost and nontoxicity.[86] However, LMO has encountered fast capacity fading, which is mainly ascribed to two problems:[86,87] (i) dissolution of Mn^{2+} into the electrolyte generated by the disproportional reaction $2Mn^{3+} \rightarrow Mn^{4+} + Mn^{2+}$, and (ii) generation of new phases during cycling and the related micro-strains. To address these issues, there have been intensive investigations with different strategies such as surface coatings[88] and element doping.[89] Using a CVD method, Jaber-Ansari *et al.*[90] grew a single-layer graphene (<15% double-layer graphene islands) on Cu foils. Then, the researchers transferred the single-layer graphene onto a sputtered LMO film. The LMO thin films were evaluated electrochemically in Li-ion coin cells in both half-cell and full-cell geometries. Jaber-Ansari *et al.*[90] found that the graphene-coated LMO cathodes showed superior capacity retention in both half-cell testing and full cells, in comparison with uncoated LMO cathodes. Specifically, the uncoated LMO cathodes in a half-cell had ~25% capacity fade after ~300 cycles and failed completely after ~350 cycles. In comparison, the graphene-coated LMO cathodes had a capacity retention of ~90% after more than 750 cycles. Jaber-Ansari *et al.*[90] revealed that the uncoated LMO cathode failure was due to significant manganese depletion. In addition, the single-layer graphene remarkably reduced the formation SEI and it remained mostly intact after 750 cycles. Based on their experimental data, Jaber-Ansari *et al.*[90] proposed two mechanisms for the effects of single-layer graphene on LMO films. First, the interaction between the graphene and the LMO film suppressed the reaction: $2Mn^{3+} \rightarrow Mn^{4+} + Mn^{2+}$. Second, both the graphene layer and the thin SEI layer served as a barrier for manganese ion migration and dissolution into the electrolyte, while concurrently allowing rapid lithiation/delithiation. Using DFT simulations, Jaber-Ansari *et al.*[90] first confirmed that the graphene acted as a physical barrier to Mn diffusion, while allowing the diffusion of Li. In their following work, Chen *et al.*[91] further explored the effects of graphene on LMO cathodes, in which graphene nanoflakes (GNF) were used to wrap nanoscale LMO particles (Figure 2.6a), and the received nano-LMO-graphene

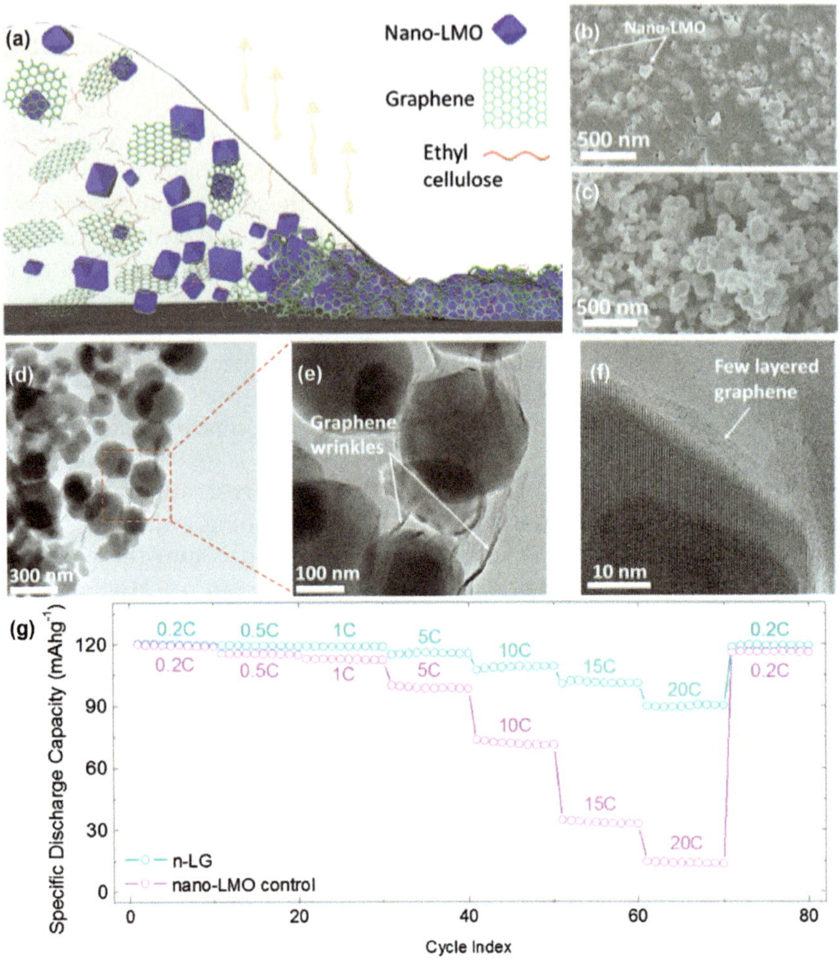

Figure 2.6 Schematic fabrication and characterization of a n-LG cathode: (a) schematic illustration of ethyl cellulose-stabilized nano-LMO and GNF dispersion (left) that forms high packing density n-LG (right), SEM images of (b) n-LG and (c) nano-LMO control, and (d–f) TEM images of n-LG at different magnifications. (g) Rate capability measurements of n-LG and nano-LMO control with various current rates from 0.2 to 20 °C.[91] Reproduced with permission from ref. 91 with permission from American Chemical Society, Copyright 2017.

composite (n-LG) cathode delivered dramatically improved electrochemical performance in LIBs, even at low (−20 °C) operating temperature. In the work, a unique cathode slurry fabrication was presented. The slurry comprised nano-LMO and GNF stabilized by ethyl cellulose that offered improved packing density. Subsequently, the slurry was cast onto aluminum foil and dried in a vacuum oven at 110 °C for 12 h to evaporate the solvent of *N*-methyl-2-pyrrolidone (NMP). Then, ambient annealing at 285 °C for 3h decomposed the ethyl cellulose, resulting in a binder-free n-LG cathode containing 90 wt% of active material loading and 10 wt% of conductive graphene additive. The resultant n-LG film had little aggregation (Figure 2.6b) while a nano-LMO control cathode prepared with nano-LMO, carbon black, and polyvinylidene fluoride (PVDF) binder in an 8 : 1 : 1 weight ratio, respectively, showed substantial aggregation (Figure 2.6c). TEM observations confirmed that the GNFs formed uniform coverage on n-LMO particles (Figure 2.6d–f). The rate capability of the n-LG and nano-LMO was investigated at room temperature (25 °C) and the results showed that, in the range of 0.2–20 °C, the n-LG retained 100%, 95%, 90%, 85%, 75% capacity at 1, 5, 10, 15, and 20 °C, respectively (Figure 2.6g). In contrast, the nano-LMO control cathode exhibited a significantly inferior rate capability, likely due to its much lower electrical conductivity. In particular, the excellent rate capability of n-LG was further verified at 0 and −20 °C. All of this strongly demonstrated that the GNFs played a remarkable role in boosting the n-LG cathode performance.

In addition to the aforementioned cases, graphene has also been used to form composites with many more cathode materials, such as $LiNi_{0.5}Mn_{1.5}O_4$,[92] $LiNi_{0.6}Co_{0.2}Mn_{0.2}O_2$,[93] Li_2FeSiO_4 (LFS),[94,95] LiV_3O_8,[96,97] and others.[98–100] In all of the cases, graphene has evidently improved the graphene-based cathodes' performance with a higher rate capability and longer-term stability.

2.3 2D Layered Anode Materials Beyond Graphene

2.3.1 Xenes

Xenes beyond graphene have been theoretically predicted as being promising LIB anodes, however, there are no experimental reports. One of the main reasons lies in the difficulty of their mass production. Borophene was deposited on silver substrate in 2015.[101] Based on DFT calculations, Jiang *et al.*[102] predicted that the full lithiated phase of borophene would be $Li_{0.75}B$, accounting for, theoretically, a specific capacity of 1860 mAh g^{-1}, four times larger than that of commercial graphite. In particular, lithium diffusion on borophene can be extremely fast.[102] In comparison, borophane (hydrogenated borophene) can enable a lithium-storage capacity of 504 mAh g^{-1}.[103] DFT calculations have also been performed for phosphorene and silicene, and predicted a fully lithiated configuration of $PLi_{0.5}$[104] and $Li_{0.5}Si_{0.5}$,[105,106] respectively. As a result, the corresponding specific capacity is 432.79 mAh g^{-1} [104] for phosphorene and 954 mAh g^{-1} [105,106] for silicene.

2.3.2 MXenes

MXenes represent an important class of 2D layered materials and they are among the latest research outcomes, including transition metal carbides,[107–111] carbonitrides,[108,109] and nitrides.[29] The first MXene, Ti_3C_2, was reported by Gogotsi and Barsoum in 2011.[107] Since then, to date, there have been about twenty MXenes synthesized and dozens have been theoretically predicted.[29] MXenes are made from layered precursors, such as MAX phases. The term MAX phases reflects the chemical composition: $M_{n+1}AX_n$, where $n = 1$, 2, or 3 (M_2AX, M_3AX_2, or M_4AX_3, *etc.*), "M" is an early transition metal, "A" is an A group (mostly groups 13 and 14) element, and "X" is C and/or N.[108] The dominant synthesis is selective etching of certain layers from layered precursors (Figure 2.7a).[108] MXenes can also be intercalated with a variety of polar organic molecules, such as isopropylamine.[112] In their applications in LIBs, M_2X materials (*e.g.*, Ti_2C, Nb_2C, V_2C, and Sc_2C) have lower formula weights than those of M_3X_2 and M_4X_3. Consequently, M_2X electrodes are expected to have higher gravimetric capacities. However, the formula weight is not the only factor affecting the capacity. The surface termination is one key factor related to the performance of MXenes. Because the bonds between M an X are too strong to be broken readily, it was assumed that lithium ions penetrate only between the MXene sheets.[29] The electrochemical performance of some MXenes have been investigated (Figure 2.7b). V_2C shows the highest lithium storage of all the MXenes tested under similar conditions to date (280 mAh g^{-1} at 1 °C and 125 mAh g^{-1} at 10 °C).[29,110]

MXenes have a hydrophilic surface, metallic conductivity, and rich surface chemistries. However, the poor ion transport of MXenes significantly inhibits their full exploitation. To solve this issue, Ma *et al.*[113] constructed a 3D porous structure of MXene-rGO hybrids (Figure 2.7c). Specifically, the Ti_3C_2-rGO hybrids could be tuned with different ratios. The hybrids could effectively facilitate rapid diffusion and transport of electrolyte ions in the electrodes while still maintaining high electrical conductivity. As a result, the hybrids exhibited a high specific capacity of 335.5 mAh g^{-1} at 0.05 A g^{-1} and good rate capability (Figure 2.7d). Additionally, the hybrids enabled excellent cycling stability without capacity decay after 1000 cycles under a high rate of 1 A g^{-1}. Ma *et al.*[113] confirmed that the novel 3D porous microstructure played an important role in the fast reaction kinetics and high capacity.

2.3.3 Transition Metal Chalcogenides

TMCs are an important family of 2D layered materials. A typical atomic ratio in layered TMCs is one transition metal (M) to two chalcogen (X) atoms to create MX_2 (*e.g.*, MoS_2, TiS_2, TaS_2, WS_2, $MoSe_2$, and WSe_2), except for several cases with 2 : 3 quintuple layers (M_2X_3) and 1 : 1 metal chalcogenides (MX).[114] 2D TMCs have great potentials for a wide variety of applications such as energy storage, gas sensing, electronics, photonics, catalysis, and many others.[31,114,115]

Among TMCs, MoS_2 is a very popular material. As LIB anodes, MoS_2 suffers from low electrical and ionic conductivity. To this end, various MoS_2 composites[116–121] have been developed and demonstrated improved performance, such as microspheres,[116] nanotubes,[117,118] and other nanostructured comosites.[119–121] For instance, Chen *et al.*[116] fabricated unique hierarchical

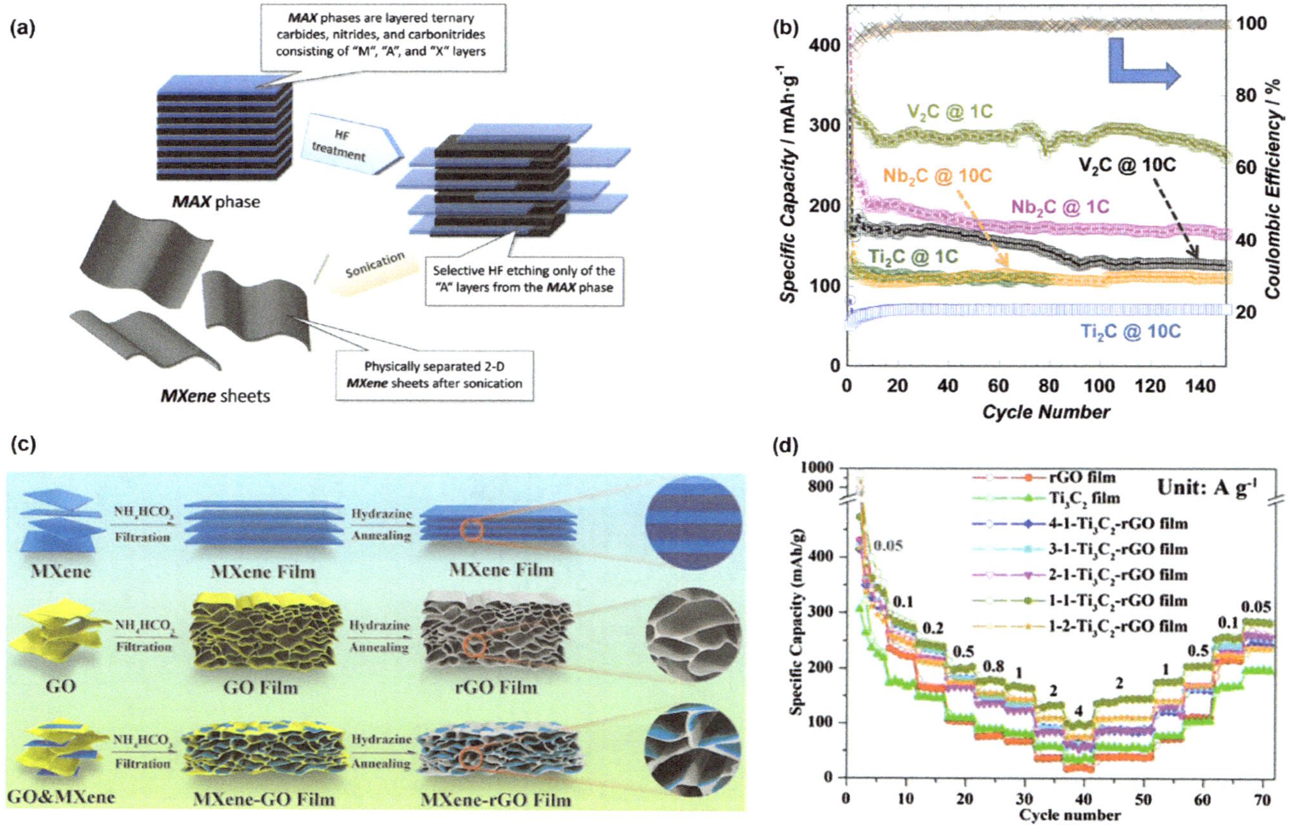

Figure 2.7 Synthesis and lithium storage of MXenes. (a) The exfoliation process of MAX phases and formation of MXenes.[108] Reproduced from ref. 108 with permission from American Chemical Society, Copyright 2012. (b) Electrochemical performance of MXenes.[110] Reproduced from ref. 110 with permission from American Chemical Society, Copyright 2013. (c) The fabrication of the MXene, rGO, and MXene-rGO hybrids, and (d) their rate capability.[113] Reproduced from ref. 113 with permission from American Chemical Society, Copyright 2018.

MoS_2-carbon (MoS_2-C) microspheres *via* a facile continuous and scalable ultrasonic nebulization route (Figure 2.8a). The hierarchical MoS_2-C microspheres consisted of few-layer MoS_2 nanosheets bridged by carbon. Thus, the exfoliated MoS_2 layers were separated in order to prevent them from aggregation and restacking. Consequently, the as-prepared MoS_2-C microspheres demonstrated an excellent electrochemical performance. Specifically, the hierarchical MoS_2-C microspheres exhibited a discharge capacity of 917, 886, 855, 803, and 731 $mAh\,g^{-1}$ at current densities of 200, 400, 800, 1600, and 3200 $mA\,g^{-1}$, respectively. In particular, a resetting to 200 $mA\,g^{-1}$ has led to a

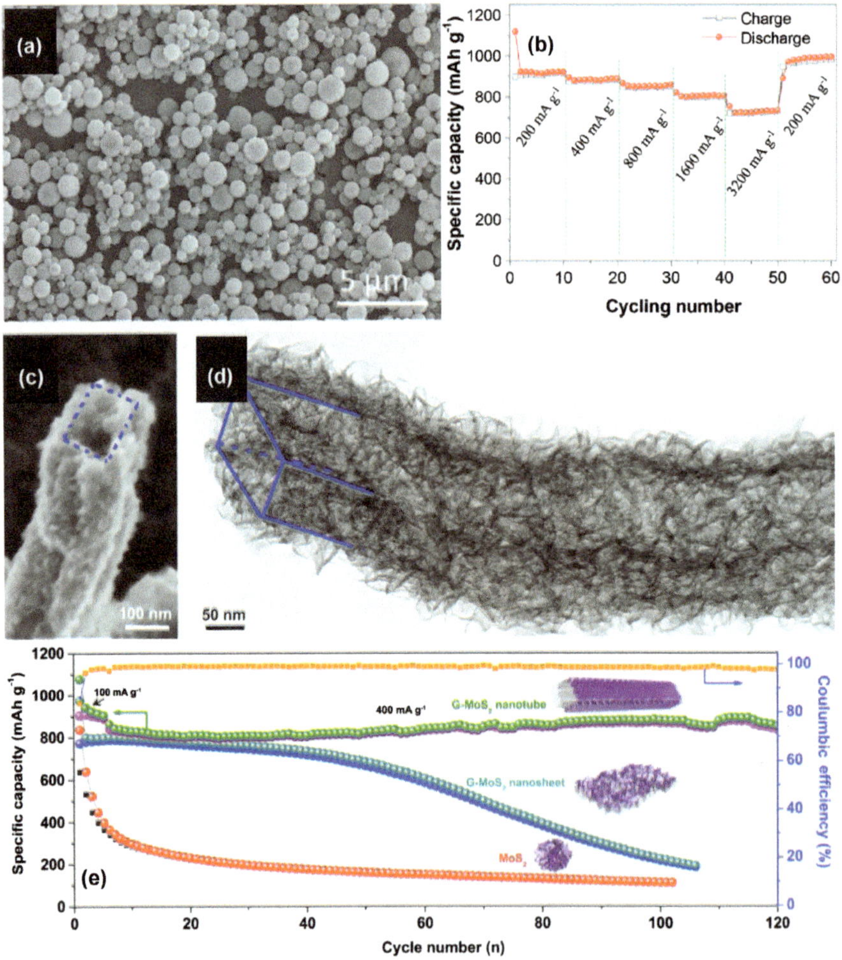

Figure 2.8 MoS_2-carbon microspheres as LIB anodes:[116] (a) SEM image and (b) rate capability. Reproduced from ref. 116 with permission from the Royal Society of Chemistry. The graphene@MoS_2 nanotubes as LIB anodes: (c) cross-sectional SEM image, (d) typical TEM image, and (e) cyclability at 400 $mA\,g^{-1}$.[117] Reproduced from ref. 117 with permission from Elsevier, Copyright 2017.

capacity of ~990 mAh g^{-1}, indicating the robust nature of hierarchical MoS_2-C microspheres (Figure 2.8b). In addition, MoS_2-C microspheres have been further demonstrated for long-term cyclability at a high current density of 1000 mAh g^{-1}, enabling a sustainable capacity of 780 mAh g^{-1} after 160 cycles. In another work, Li *et al.*[117] developed a unique hierarchical hollow structure, in which MoS_2 nanosheets were grown on assembled graphene nanotubes (*i.e.*, graphene@MoS_2 nanotubes, Figure 2.8c,d) *via* a template-sacrificed approach. The graphene@MoS_2 nanotubes as LIB anodes have enabled a capacity of 830 mAh g^{-1} and a retention of 96.5% at 400 mA g^{-1} up to 120 cycles (Figure 2.8e). They have also exhibited excellent rate capability, realizing 502 mAh g^{-1} at 2000 mA g^{-1}. The graphene@MoS_2 nanotubes outperformed both the pure MoS_2 nanosheets and graphene@-MoS_2 nanosheets. Besides carbon materials (*e.g.*, graphene and CNTs), other layered materials have also been used to construct composites with MoS_2, such as SnS[120] and MXenes.[121]

MoS_2 has a unique structure with an interlayer distance of ~0.62 nm (much larger the 0.34 nm spacing of graphite), facilitating a fast diffusion of Li^+ ions. This promises to make MoS_2 an attractive anode material for LIBs. However, the exact mechanism of lithiation and delithiation of MoS_2 had not been clarified, until, using *in situ* TEM, Li *et al.*[117] explored the lithiation mechanism of MoS_2. The TEM results indicated that Li^+ ions initially intercalated into the layered MoS_2 to form Li_xMoS_2 intercalates at the beginning of the lithiation. As the deep lithiation proceeded, a layer-by-layer conversion reaction occurred with the products of Mo and Li_2S. Li *et al.*[117] therefore proposed an atomistic lithiation mechanism (Figure 2.9a). Combining a suite of techniques (electrochemical investigation, morphology characterization, *in situ* and *ex situ* X-ray absorption spectroscopy, and DFT) in a recent work, Zhang *et al.*[122] clarified the mechanism of lithiation and delithiation of MoS_2 as follows (Figure 2.9b).

$$MoS_2 + xLi^+ + xe^- \leftrightarrow Li_xMoS_2 \tag{2.1}$$

$$Li_xMoS_2 + (4-x)Li^+ + (4-x)e^- \rightarrow Li_2S + Mo \tag{2.2}$$

$$Li_2S \leftrightarrow S + 2Li^+ + 2e^- \tag{2.3}$$

where x is in the range of $0 \leq x \leq 1$. Zhang *et al.*[122] explained that the intercalation reaction (eqn 2.1) is reversible. However, the conversion reaction (eqn 2.2) is irreversible. The discharge product Li_2S is oxidized to sulfur (eqn 2.3) in the first charge. This work clarified the debate about the reversibility of the conversion reaction of MoS_2. This may be also inspiring for understanding the electrochemical reactions of other layered TMCs.

In addition to MoS_2 as the most studied TMC material, WS_2 has a higher intrinsic electrical conductivity and is potentially promising as an anode material. In this regard, there have been an increasing number of studies on WS_2 LIB anodes. Using a facile one-pot hydrothermal process (Figure 2.10a), for instance, Chen *et al.*[123] developed a graphene-based hierarchical architecture of WS_2-NTs/GS. This composite enabled fast electron/ion transfer and therefore led to an excellent electrochemical performance, *e.g.*, maintaining a capacity of 318.6 mA g^{-1} over 500 cycles at a current density of 1 A g^{-1} (Figure 2.10b). In particular, this hybrid anode did not require any additional binder, conductive additives, or a separate metal current-collector. In another effort, Zeng *et al.*[124] synthesized a hierarchical nano-composite, in which ultrathin WS_2 nanosheets were uniformly attached on the surface of hollow N-doped carbon spheres (HNCS) (Figure 2.10c). This composite exhibited a high specific capacity of 801.4 mAh g^{-1} at 0.1 A g^{-1} and excellent rate capability, accounting for 545.6 mAh g^{-1} at 2 A g^{-1}. Interestingly, Wang *et al.*[125] reported a hybrid nano-architectured aerogel composed of WS_2 nanosheets and CNT-rGO with an ordered microchannel

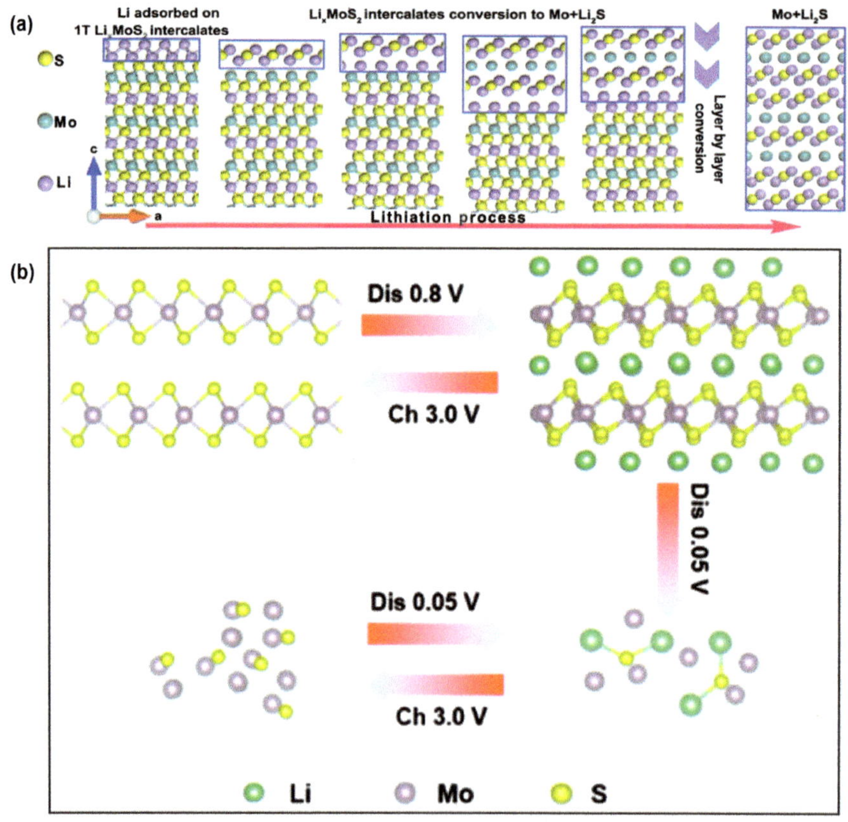

Figure 2.9 Mechanism of lithiation and delithiation of MoS_2. (a) The proposed schematic atomistic models of illustrated the layer-by-layer lithiation mechanism of a MoS_2 electrode.[117] Reproduced from ref. 117 with permission from Elsevier, Copyright 2017. (b) The proposed electrochemical reaction mechanism of the MoS_2 electrode.[122] Reproduced from ref. 122 with permission from, American Chemical Society, Copyright 2018.

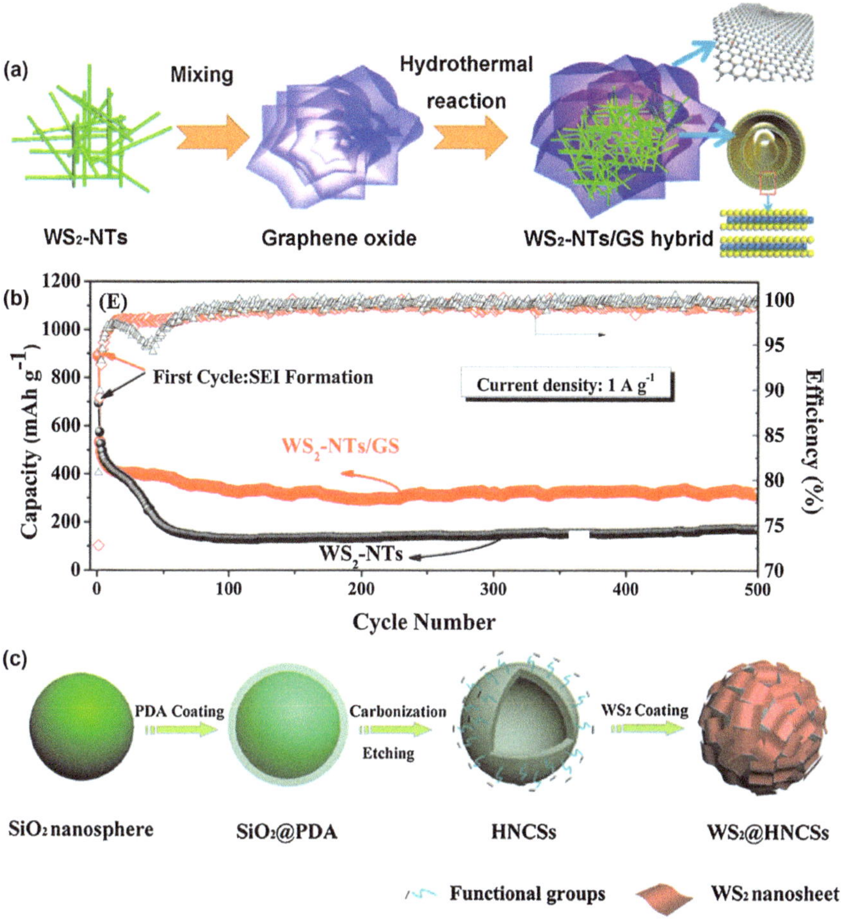

Figure 2.10 WS_2 composite LIB anodes. (a) Schematic illustration of the fabrication and (b) electrochemical performance of a 3D hierarchically structured WS_2-NTs/GS hybrid.[123] Reproduced from ref. 123 with permission from American Chemical Society, Copyright 2014. (c) Synthesis procedure for a WS_2@HNCSs nanocomposite.[124] Reproduced from ref. 124 with permission from American Chemical Society, Copyright 2016.

3D scaffold structure *via* a simple solvothermal method, followed by freeze-drying and a post annealing process. The WS_2/CNT-rGO aerogel was confirmed as being much better than both the commercial WS_2 and CNT-rGO, enabling a specific capacity of 749 $mAh\,g^{-1}$ at 100 $mAh\,g^{-1}$. Wang *et al.*[125] believed that the excellent electrochemical performance of this WS_2/CNT-rGO aerogel is closely related to the synergistic effect between the WS_2 nanosheets and CNT-rGO scaffold network.

Besides MoS_2 and WS_2, there are also other TMCs reported as being suitable electrode materials in LIBs, such as VS_2,[126,127] GeP_3,[128] $NbSe_2$,[129] and $MoSe_2$.[130] All of these layered materials offer a large class of LIB electrodes.

2.4 Conclusions

In this chapter we have summarized the applications of different 2D layer-structured materials in LIBs. Among the 2D layered materials, graphene itself can serve as LIB anodes and has played versatile roles for constructing advanced composite anodes and cathodes. It has been identified that, in graphene-based composite electrodes, graphene enables higher conductivity, accommodates volume change, enhances mechanical integrity, and inhibits SEI formation. All of these factors are related to the exceptional properties of graphene. In addition to graphene, we have also introduced studies on Xenes, MXenes, and TMCs as anodes. The studies on Xenes are still at an early stage and there is a lack of experimental results. MXenes are very new and they were first reported in 2011. Based on the studies to date, MXenes generally have a limited capacity. In comparison, TMCs may offer many high-capacity anode materials, such as MoS_2 and WS_2. With the increasing interest in TMCs, their electrochemical reaction mechanism has also been undergoing investigation. However, TMC composites enable improved properties and high performance in LIBs. Along with our persistent search for new electrode materials for next-generation LIBs, we believe that these 2D layer materials are among the most attractive candidates.

Acknowledgements

X.M. acknowledges partial support from the Center for Advanced Surface Engineering, under the National Science Foundation Grant No. OIA-1457888 and the Arkansas EPSCoR Program, ASSET III. X.M. also appreciates the financial research support from University of Arkansas, Fayetteville, AR, USA. Part of the research at Oak Ridge National Laboratory, managed by UT Battelle, LLC, for the U.S. Department of Energy under contract DE-AC05-00OR22725 was sponsored by the Office of Energy Efficiency and Renewable Energy Vehicle Technologies Office (VTO) Applied Battery Research (ABR) subprogram (Program Managers: Peter Faguy).

References

1. K. S. Novoselov, A. K. Geim, S. V. Morozov, D. Jiang, Y. Zhang, S. V. Dubonos, I. V. Grigorieva and A. A. Firsov, *Science*, 2004, **306**, 666–669.
2. K. S. Novoselov, V. I. Fal′ko, L. Colombo, P. R. Gellert, M. G. Schwab and K. Kim, *Nature*, 2012, **490**, 192.
3. S. J. Kim, K. Choi, B. Lee, Y. Kim and B. H. Hong, *Annu. Rev. Mater. Res.*, 2015, **45**, 63–84.
4. A. Dubey, S. Dave, M. Lakhani and A. Sharm, *International Conference on Electrical, Electronics, and Optimization Techniques (ICEEOT)*, 2016, 2435–2439.

5. J. Basu, J. K. Basu and T. K. Bhattacharyya, *Int. J. Smart Nano Mater.*, 2010, **1**, 201–223.
6. A. Anand, B. Unnikrishnan, J.-Y. Mao, H.-J. Lin and C.-C. Huang, *Desalination*, 2018, **429**, 119–133.
7. M.-M. Cheng, L.-J. Huang, Y.-X. Wang, J.-G. Tang, Y. Wang, Y.-C. Zhao, G.-F. Liu, Y. Zhang, M. J. Kipper, L. A. Belfiore and W. S. Ranil, *RSC Adv.*, 2017, 7, 47886–47897.
8. S. Homaeigohar and M. Elbahri, *NPG Asia Mater.*, 2017, **9**, e427.
9. J. Xu, Z. Cao, Y. Zhang, Z. Yuan, Z. Lou, X. Xu and X. Wang, *Chemosphere*, 2018, **195**, 351–364.
10. R. Kumar, R. Singh, D. Hui, L. Feo and F. Fraternali, *Composites, Part B*, 2018, **134**, 193–206.
11. M. Silva, N. M. Alves and M. C. Paiva, *Polym. Adv. Technol.*, 2018, **29**, 687–700.
12. R. Geetha Bai, N. Ninan, K. Muthoosamy and S. Manickam, *Prog. Mater. Sci.*, 2018, **91**, 24–69.
13. A. Nag, A. Mitra and S. C. Mukhopadhyay, *Sens. Actuators, A*, 2018, **270**, 177–194.
14. W. Yuan and G. Shi, *J. Mater. Chem. A*, 2013, **1**, 10078–10091.
15. A. Nikokavoura and C. Trapalis, *Appl. Surf. Sci.*, 2018, **430**, 18–52.
16. X.-K. Kong, C.-L. Chen and Q.-W. Chen, *Chem. Soc. Rev.*, 2014, **43**, 2841–2857.
17. B. F. Machado and P. Serp, *Catal. Sci. Technol.*, 2012, **2**, 54–75.
18. X. Yu, H. Cheng, M. Zhang, Y. Zhao, L. Qu and G. Shi, *Nat. Rev. Mater.*, 2017, **2**, 17046.
19. B. Aïssa, N. K. Memon, A. Ali and M. K. Khraisheh, *Front. Mater.*, 2015, 2.
20. E. L. Lim, C. C. Yap, M. H. H. Jumali, M. A. M. Teridi and C. H. Teh, *Nano-Micro Lett.*, 2017, **10**, 27.
21. P. Huo, P. Zhao, Y. Wang, B. Liu, G. Yin and M. Dong, *Energies*, 2018, **11**, 167.
22. K. Chen, Q. Wang, Z. Niu and J. Chen, *J. Energy Chem.*, 2018, **27**, 12–24.
23. F. Bonaccorso, L. Colombo, G. Yu, M. Stoller, V. Tozzini, A. C. Ferrari, R. S. Ruoff and V. Pellegrini, *Science*, 2015, 347.
24. Z.-S. Wu, G. Zhou, L.-C. Yin, W. Ren, F. Li and H.-M. Cheng, *Nano Energy*, 2012, **1**, 107–131.
25. M. Chhowalla, H. S. Shin, G. Eda, L.-J. Li, K. P. Loh and H. Zhang, *Nat. Chem.*, 2013, **5**, 263.
26. S. Balendhran, S. Walia, H. Nili, S. Sriram and M. Bhaskaran, *Small*, 2015, **11**, 640–652.
27. M. Xu, T. Liang, M. Shi and H. Chen, *Chem. Rev.*, 2013, **113**, 3766–3798.
28. G. R. Bhimanapati, Z. Lin, V. Meunier, Y. Jung, J. Cha, S. Das, D. Xiao, Y. Son, M. S. Strano, V. R. Cooper, L. Liang, S. G. Louie, E. Ringe, W. Zhou, S. S. Kim, R. R. Naik, B. G. Sumpter, H. Terrones, F. Xia, Y. Wang, J. Zhu, D. Akinwande, N. Alem, J. A. Schuller, R. E. Schaak, M. Terrones and J. A. Robinson, *ACS Nano*, 2015, **9**, 11509–11539.

29. B. Anasori, M. R. Lukatskaya and Y. Gogotsi, *Nat. Rev. Mater.*, 2017, **2**, 16098.
30. M. Chhowalla, Z. Liu and H. Zhang, *Chem. Soc. Rev.*, 2015, **44**, 2584–2586.
31. C. Tan and H. Zhang, *Chem. Soc. Rev.*, 2015, **44**, 2713–2731.
32. S. Z. Butler, S. M. Hollen, L. Cao, Y. Cui, J. A. Gupta, H. R. Gutiérrez, T. F. Heinz, S. S. Hong, J. Huang, A. F. Ismach, E. Johnston-Halperin, M. Kuno, V. V. Plashnitsa, R. D. Robinson, R. S. Ruoff, S. Salahuddin, J. Shan, L. Shi, M. G. Spencer, M. Terrones, W. Windl and J. E. Goldberger, *ACS Nano*, 2013, **7**, 2898–2926.
33. D. N. Bunck and W. R. Dichtel, *J. Am. Chem. Soc.*, 2013, **135**, 14952–14955.
34. Z. Zheng, L. Opilik, F. Schiffmann, W. Liu, G. Bergamini, P. Ceroni, L.-T. Lee, A. Schütz, J. Sakamoto, R. Zenobi, J. VandeVondele and A. D. Schlüter, *J. Am. Chem. Soc.*, 2014, **136**, 6103–6110.
35. M. Höök and X. Tang, *Energy Policy*, 2013, **52**, 797–809.
36. M. M. Thackeray, C. Wolverton and E. D. Isaacs, *Energy Environ. Sci.*, 2012, **5**, 7854–7863.
37. A. Manthiram, *ACS Cent. Sci.*, 2017, **3**, 1063–1069.
38. F. Bonaccorso, A. Lombardo, T. Hasan, Z. Sun, L. Colombo and A. C. Ferrari, *Mater. Today*, 2012, **15**, 564–589.
39. Y. Zhang, L. Zhang and C. Zhou, *Acc. Chem. Res.*, 2013, **46**, 2329–2339.
40. R. Raccichini, A. Varzi, S. Passerini and B. Scrosati, *Nat. Mater.*, 2014, **14**, 271.
41. Y. L. Zhong, Z. Tian, G. P. Simon and D. Li, *Mater. Today*, 2015, **18**, 73–78.
42. E. Yoo, J. Kim, E. Hosono, H.-S. Zhou, T. Kudo and I. Honma, *Nano Lett.*, 2008, **8**, 2277–2282.
43. G. Wang, X. Shen, J. Yao and J. Park, *Carbon*, 2009, **47**, 2049–2053.
44. X. Meng, D. Geng, J. Liu, M. N. Banis, Y. Zhang, R. Li and X. Sun, *J. Phys. Chem. C*, 2010, **114**, 18330–18337.
45. D. Pan, S. Wang, B. Zhao, M. Wu, H. Zhang, Y. Wang and Z. Jiao, *Chem. Mater.*, 2009, **21**, 3136–3142.
46. X. Tong, H. Wang, G. Wang, L. Wan, Z. Ren, J. Bai and J. Bai, *J. Solid State Chem.*, 2011, **184**, 982–989.
47. L. Wan, Z. Ren, H. Wang, G. Wang, X. Tong, S. Gao and J. Bai, *Diamond Relat. Mater.*, 2011, **20**, 756–761.
48. A. L. M. Reddy, A. Srivastava, S. R. Gowda, H. Gullapalli, M. Dubey and P. M. Ajayan, *ACS Nano*, 2010, **4**, 6337–6342.
49. Z.-S. Wu, W. Ren, L. Xu, F. Li and H.-M. Cheng, *ACS Nano*, 2011, **5**, 5463–5471.
50. C. Ma, X. Shao and D. Cao, *J. Mater. Chem.*, 2012, **22**, 8911–8915.
51. X. Wang, G. Li, F. M. Hassan, J. Li, X. Fan, R. Batmaz, X. Xiao and Z. Chen, *Nano Energy*, 2015, **15**, 746–754.
52. Y. Yan, Y.-X. Yin, S. Xin, Y.-G. Guo and L.-J. Wan, *Chem. Commun.*, 2012, **48**, 10663–10665.

53. Y. S. Yun, V.-D. Le, H. Kim, S.-J. Chang, S. J. Baek, S. Park, B. H. Kim, Y.-H. Kim, K. Kang and H.-J. Jin, *J. Power Sources*, 2014, **262**, 79–85.
54. D. Sun, J. Yang and X. Yan, *Chem. Commun.*, 2015, **51**, 2134–2137.
55. X. Li, D. Geng, Y. Zhang, X. Meng, R. Li and X. Sun, *Electrochem. Commun.*, 2011, **13**, 822–825.
56. X. Xiao, P. Liu, J. S. Wang, M. W. Verbrugge and M. P. Balogh, *Electrochem. Commun.*, 2011, **13**, 209–212.
57. T. Wei, F. Wang, J. Yan, J. Cheng, Z. Fan and H. Song, *J. Electroanal. Chem.*, 2011, **653**, 45–49.
58. Z.-J. Fan, J. Yan, T. Wei, G.-Q. Ning, L.-J. Zhi, J.-C. Liu, D.-X. Cao, G.-L. Wang and F. Wei, *ACS Nano*, 2011, **5**, 2787–2794.
59. S. Li, Y. Luo, W. Lv, W. Yu, S. Wu, P. Hou, Q. Yang, Q. Meng, C. Liu and H. M. Cheng, *Adv. Energy Mater.*, 2011, **1**, 486–490.
60. Y. Jin, B. Zhu, Z. Lu, N. Liu and J. Zhu, *Adv Energy Mater*, 2017, 7, 1700715.
61. S. Chae, M. Ko, K. Kim, K. Ahn and J. Cho, *Joule*, 2017, **1**, 47–60.
62. S.-L. Chou, J.-Z. Wang, M. Choucair, H.-K. Liu, J. A. Stride and S.-X. Dou, *Electrochem. Commun.*, 2010, **12**, 303–306.
63. J.-Z. Wang, C. Zhong, S.-L. Chou and H.-K. Liu, *Electrochem. Commun.*, 2010, **12**, 1467–1470.
64. J. K. Lee, K. B. Smith, C. M. Hayner and H. H. Kung, *Chem. Commun.*, 2010, **46**, 2025–2027.
65. S. Yang, G. Li, Q. Zhu and Q. Pan, *J. Mater. Chem.*, 2012, **22**, 3420–3425.
66. F. M. Hassan, R. Batmaz, J. Li, X. Wang, X. Xiao, A. Yu and Z. Chen, *Nat. Commun.*, 2015, **6**, 8597.
67. L. Zhang, W. Hao, H. Wang, L. Zhang, X. Feng, Y. Zhang, W. Chen, H. Pang and H. Zheng, *J. Mater. Chem. A*, 2013, **1**, 7601–7611.
68. H. Mi, Y. Li, P. Zhu, X. Chai, L. Sun, H. Zhuo, Q. Zhang, C. He and J. Liu, *J. Mater. Chem. A*, 2014, **2**, 11254–11260.
69. J. Zhang, L. Zhang, P. Xue, L. Zhang, X. Zhang, W. Hao, J. Tian, M. Shen and H. Zheng, *J. Mater. Chem. A*, 2015, **3**, 7810–7821.
70. Y. Zhao, X. Li, B. Yan, D. Li, S. Lawes and X. Sun, *J. Power Sources*, 2015, **274**, 869–884.
71. X. Huang, X. Qi, F. Boey and H. Zhang, *Chem. Soc. Rev.*, 2012, **41**, 666–686.
72. D. Chen, L. Tang and J. Li, *Chem. Soc. Rev.*, 2010, **39**, 3157–3180.
73. W. Sun and Y. Wang, *Nanoscale*, 2014, **6**, 11528–11552.
74. J. Zhu, D. Yang, Z. Yin, Q. Yan and H. Zhang, *Small*, 2014, **10**, 3480–3498.
75. C. Xu, B. Xu, Y. Gu, Z. Xiong, J. Sun and X. S. Zhao, *Energy Environ. Sci.*, 2013, **6**, 1388–1414.
76. K. Mizushima, P. C. Jones, P. J. Wiseman and J. B. Goodenough, *Mater. Res. Bull.*, 1980, **15**, 783–789.
77. M. M. Thackeray, W. I. F. David, P. G. Bruce and J. B. Goodenough, *Mater. Res. Bull.*, 1983, **18**, 461–472.
78. A. K. Padhi, K. S. Nanjundaswamy and J. B. Goodenough, *J. Electrochem. Soc.*, 1997, **144**, 1188–1194.

79. G. Kucinskis, G. Bajars and J. Kleperis, *J. Power Sources*, 2013, **240**, 66–79.
80. Y. Ding, Y. Jiang, F. Xu, J. Yin, H. Ren, Q. Zhuo, Z. Long and P. Zhang, *Electrochem. Commun.*, 2010, **12**, 10–13.
81. X. Zhou, F. Wang, Y. Zhu and Z. Liu, *J. Mater. Chem.*, 2011, **21**, 3353–3358.
82. Y. Wu, Z. Wen, H. Feng and J. Li, *Chem. – A Eur. J.*, 2013, **19**, 5631–5636.
83. S. H. Ha and Y. J. Lee, *Chem. – A Eur. J.*, 2015, **21**, 2132–2138.
84. B. Wang, W. Al Abdulla, D. Wang and X. S. Zhao, *Energy Environ. Sci.*, 2015, **8**, 869–875.
85. X. Lei, H. Zhang, Y. Chen, W. Wang, Y. Ye, C. Zheng, P. Deng and Z. Shi, *J. Alloy Compd.*, 2015, **626**, 280–286.
86. H. Xia, Z. Luo and J. Xie, *Progr. Nat. Sci.: Mater. Int.*, 2012, **22**, 572–584.
87. B. Xu, D. Qian, Z. Wang and Y. S. Meng, *Mater. Sci. Eng.: R: Rep.*, 2012, **73**, 51–65.
88. T.-F. Yi, Y.-R. Zhu, X.-D. Zhu, J. Shu, C.-B. Yue and A.-N. Zhou, *Ionics*, 2009, **15**, 779–784.
89. J. W. Fergus, *J. Power Sources*, 2010, **195**, 939–954.
90. L. Jaber-Ansari, K. P. Puntambekar, S. Kim, M. Aykol, L. Luo, J. Wu, B. D. Myers, H. Iddir, J. T. Russell, S. J. Saldaña, R. Kumar, M. M. Thackeray, L. A. Curtiss, V. P. Dravid, C. Wolverton and M. C. Hersam, *Adv. Energy Mater.*, 2015, **5**, 1500646.
91. K.-S. Chen, R. Xu, N. S. Luu, E. B. Secor, K. Hamamoto, Q. Li, S. Kim, V. K. Sangwan, I. Balla, L. M. Guiney, J.-W. T. Seo, X. Yu, W. Liu, J. Wu, C. Wolverton, V. P. Dravid, S. A. Barnett, J. Lu, K. Amine and M. C. Hersam, *Nano Lett.*, 2017, **17**, 2539–2546.
92. X. Tang, S. S. Jan, Y. Qian, H. Xia, J. Ni, S. V. Savilov and S. M. Aldoshin, *Sci. Rep-UK*, 2015, **5**, 11958.
93. J.-H. Shim, Y.-M. Kim, M. Park, J. Kim and S. Lee, *ACS Appl. Mater. Intfaces*, 2017, **9**, 18720–18729.
94. L.-L. Zhang, S. Duan, X.-L. Yang, G. Peng, G. Liang, Y.-H. Huang, Y. Jiang, S.-B. Ni and M. Li, *ACS Appl. Mater. Interfaces*, 2013, **5**, 12304–12309.
95. H. Zhu, X. Wu, L. Zan and Y. Zhang, *ACS Appl. Mater. Interfaces*, 2014, **6**, 11724–11733.
96. R. Mo, Y. Du, D. Rooney, G. Ding and K. Sun, *Sci. Rep-UK*, 2016, **6**, 19843.
97. Z.-K. Wang, J. Shu, Q.-C. Zhu, B.-Y. Cao, H. Chen, X.-Y. Wu, B. M. Bartlett, K.-X. Wang and J.-S. Chen, *J. Power Sources*, 2016, **307**, 426–434.
98. Y. Yang, L. Li, H. Fei, Z. Peng, G. Ruan and J. M. Tour, *ACS Appl. Mater. Interfaces*, 2014, **6**, 9590–9594.
99. X. Zhu, Z. Yan, W. Wu, W. Zeng, Y. Du, Y. Zhong, H. Zhai, H. Ji and Y. Zhu, *Sci. Rep-UK*, 2014, **4**, 5768.
100. Z. Guo, D. Zhang, H. Qiu, T. Zhang, Q. Fu, L. Zhang, X. Yan, X. Meng, G. Chen and Y. Wei, *ACS Appl. Mater. Interfaces*, 2015, 7, 13972–13979.

101. A. J. Mannix, X.-F. Zhou, B. Kiraly, J. D. Wood, D. Alducin, B. D. Myers, X. Liu, B. L. Fisher, U. Santiago, J. R. Guest, M. J. Yacaman, A. Ponce, A. R. Oganov, M. C. Hersam and N. P. Guisinger, *Science*, 2015, **350**, 1513–1516.
102. H. R. Jiang, Z. Lu, M. C. Wu, F. Ciucci and T. S. Zhao, *Nano Energy*, 2016, **23**, 97–104.
103. N. K. Jena, R. B. Araujo, V. Shukla and R. Ahuja, *ACS Appl. Mater. Interfaces*, 2017, **9**, 16148–16158.
104. S. Zhao, W. Kang and J. Xue, *J. Mater. Chem. A*, 2014, **2**, 19046–19052.
105. G. A. Tritsaris, E. Kaxiras, S. Meng and E. Wang, *Nano Lett.*, 2013, **13**, 2258–2263.
106. J. Zhuang, X. Xu, G. Peleckis, W. Hao, S. X. Dou and Y. Du, *Adv. Mater.*, 2017, **29**, 1606716.
107. M. Naguib, M. Kurtoglu, V. Presser, J. Lu, J. Niu, M. Heon, L. Hultman, Y. Gogotsi and M. W. Barsoum, *Adv. Mater.*, 2011, **23**, 4248–4253.
108. M. Naguib, O. Mashtalir, J. Carle, V. Presser, J. Lu, L. Hultman, Y. Gogotsi and M. W. Barsoum, *ACS Nano*, 2012, **6**, 1322–1331.
109. O. Mashtalir, M. Naguib, V. N. Mochalin, Y. Dall'Agnese, M. Heon, M. W. Barsoum and Y. Gogotsi, *Nat. Commun.*, 2013, **4**, 1716.
110. M. Naguib, J. Halim, J. Lu, K. M. Cook, L. Hultman, Y. Gogotsi and M. W. Barsoum, *J. Am. Chem. Soc.*, 2013, **135**, 15966–15969.
111. J. Come, M. Naguib, P. Rozier, M. W. Barsoum, Y. Gogotsi, P.-L. Taberna, M. Morcrette and P. Simon, *J. Electrochem. Soc.*, 2012, **159**, A1368–A1373.
112. O. Mashtalir, M. R. Lukatskaya, M. Q. Zhao, M. W. Barsoum and Y. Gogotsi, *Adv. Mater.*, 2015, **27**, 3501–3506.
113. Z. Ma, X. Zhou, W. Deng, D. Lei and Z. Liu, *ACS Appl. Mater. Interfaces*, 2018, **10**, 3634–3643.
114. W. Choi, N. Choudhary, G. H. Han, J. Park, D. Akinwande and Y. H. Lee, *Mater. Today*, 2017, **20**, 116–130.
115. S. Jayabal, G. Saranya, J. Wu, Y. Liu, D. Geng and X. Meng, *J. Mater. Chem. A*, 2017, **5**, 24540–24563.
116. G. Chen, S. Wang, R. Yi, L. Tan, H. Li, M. Zhou, L. Yan, Y. Jiang, S. Tan, D. Wang, S. Deng, X. Meng and H. Luo, *J. Mater. Chem. A*, 2016, **4**, 9653–9660.
117. Y. Li, Z. Liu, X. Cheng, X. Liu, B. Zhang, D. Sun, R. Wang and Y. Zhang, *Energy Storage Mater.*, 2017, **9**, 188–194.
118. Z. Zhang, H. Zhao, Y. Teng, X. Chang, Q. Xia, Z. Li, J. Fang, Z. Du and K. Świerczek, *Adv. Energy Mater.*, 2018, **8**, 1700174.
119. R. Wang, S. Wang, D. Jin, Y. Zhang, Y. Cai, J. Ma and L. Zhang, *Energy Storage Mater*, 2017, **9**, 195–205.
120. Q. Pan, F. Zheng, Y. Wu, X. Ou, C. Yang, X. Xiong and M. Liu, *J. Mater. Chem. A*, 2018, **6**, 592–598.
121. C. Chen, X. Xie, B. Anasori, A. Sarycheva, T. Makaryan, M. Zhao, P. Urbankowski, L. Miao, J. Jiang and Y. Gogotsi, *Angew. Chem., Int. Ed.*, 2018, **57**, 1846–1850.

122. L. Zhang, D. Sun, J. Kang, J. Feng, H. A. Bechtel, L.-W. Wang, E. J. Cairns and J. Guo, *Nano Lett.*, 2018, **18**, 1466–1475.
123. R. Chen, T. Zhao, W. Wu, F. Wu, L. Li, J. Qian, R. Xu, H. Wu, H. M. Albishri, A. S. Al-Bogami, D. A. El-Hady, J. Lu and K. Amine, *Nano Lett.*, 2014, **14**, 5899–5904.
124. X. Zeng, Z. Ding, C. Ma, L. Wu, J. Liu, L. Chen, D. G. Ivey and W. Wei, *ACS Appl. Mater. Interfaces*, 2016, **8**, 18841–18848.
125. Y. Wang, D. Kong, W. Shi, B. Liu, G. J. Sim, Q. Ge and H. Y. Yang, *Adv. Energy Mater.*, 2016, **6**, 1601057.
126. Y. Jing, Z. Zhou, C. R. Cabrera and Z. Chen, *J. Phys. Chem. C*, 2013, **117**, 25409–25413.
127. W. Fang, H. Zhao, Y. Xie, J. Fang, J. Xu and Z. Chen, *ACS Appl. Mater. Interfaces*, 2015, **7**, 13044–13052.
128. C. Zhang, Y. Jiao, T. He, F. Ma, L. Kou, T. Liao, S. Bottle and A. Du, *Phys. Chem. Chem. Phys.*, 2017, **19**, 25886–25890.
129. E. Hitz, J. Wan, A. Patel, Y. Xu, L. Meshi, J. Dai, Y. Chen, A. Lu, A. V. Davydov and L. Hu, *ACS Appl. Mater. Interfaces*, 2016, **8**, 11390–11395.
130. Z. Luo, J. Zhou, L. Wang, G. Fang, A. Pan and S. Liang, *J. Mater. Chem. A*, 2016, **4**, 15302–15308.

CHAPTER 3

Intercalation-based Layered Materials for Rechargeable Sodium-ion Batteries

BIJOY KUMAR DAS AND R. GOPALAN*

Centre for Automotive Energy Materials, International Advanced Research Centre for Powder Metallurgy and New Materials (ARCI), IIT Madras Research Park, Chennai- 600113, India
*Email: gopy@arci.res.in

3.1 Introduction to Sodium-ion Batteries (SIBs)

Recently, electrical energy storage (EES) devices such as batteries have played a great role in storing the energy produced from renewable sources to meet the energy demand during peak hours.[1] In the past, various rechargeable batteries were developed and are now available on the market. Among them, lithium-ion batteries (LIBs) are becoming more popular to power electronic devices, electric vehicles (EVs) and hybrid electric vehicles (HEVs) due to their high specific energy and power density.[2,3] With the large production in EVs/HEVs, the demand for batteries is expected to grow exponentially in the coming years. However, the limited lithium resources will have a great impact on the cost of a battery, which will rise to a high value in the future. Hence, low-cost alternative storage devices with a similar performance to that of LIBs are required to cope with the increasing demand for batteries.[4]

Sodium-ion batteries (SIBs) are regarded as a potential competitor to LIBs, owing to their low cost and the wide availability of sodium (~2.5% of the

Smart Materials No. 34
Layered Materials for Energy Storage and Conversion
Edited by Dongsheng Geng, Yuan Cheng and Gang Zhang

Published by the Royal Society of Chemistry, www.rsc.org

Earth's content, especially available in seawater).[5] The similar standard reduction potential of sodium ($E = -2.71$ V) to that of lithium and its non-alloy forming nature with Al adds to its credentials. This would be favorable for the development of SIBs using aluminium instead of copper as a current collector, which would be cost effective. Development of low-cost SIBs with high specific capacity, specific energy (~500 Wh kg^{-1}) and rate capability are the main challenges for modern-day researchers. SIBs are safer to operate compared to LIBs, as they can be stored and transported in a short-circuit state, which is less risky.[6] There are fewer diffusion barriers for the Na^+ ion in layered oxide than the Li^+ ion as seen from computational studies. This enables us to develop suitable sodium-host materials by altering the chemical composition.[7]

The intercalation/de-intercalation mechanism of Na^+ ion in SIBs is similar to that of LIBs, where the Na^+ ions travel between a positive (cathode) and negative (anode) electrode during charge and discharge processes. Sodium salt containing an electrolyte acts as a medium for the migration of Na^+ ions. When the battery is being charged, Na^+ ions are extracted from the cathode and inserted into the anode *via* an electrolyte, whereas the reverse occurs during discharge. The insertion/de-insertion mechanism is known as the rocking chair mechanism.[7] The working principle of a SIB is shown in Figure 3.1.

R&D activities of SIBs began in the early 1980s in parallel with that of LIBs. However, its poor electrochemical performance and commercialization of LIBs put an interim stop on the research activities of SIBs. SIBs suffer from a lower specific capacity and specific energy because of the larger size and higher atomic weight of a sodium ion in comparison to a lithium ion.

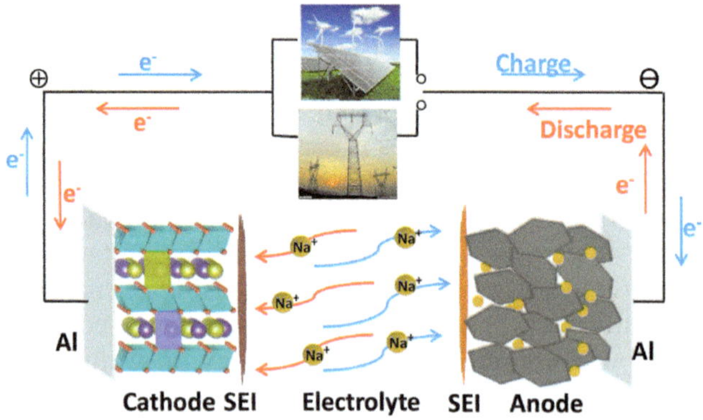

Figure 3.1 Working principle of a rechargeable sodium ion battery (SIB). Reproduced from ref. 7 with permission from the Royal Society of Chemistry.

Moreover, the large volume change associated with sodium ion intercalation/de-intercalation affects the structural stability, resulting in a poor electrochemical performance compared to LIBs.[6,7]

3.2 Intercalation Based Layered Electrode Materials for SIBs

Among the different Na^+ ion intercalation host materials explored so far, sodium transition metal oxides, having a composition of Na_xMO_2 ($0.5 \leq x \leq 1$) (where M represents the transition metal) are found to be the most suitable cathode materials for SIBs. Layered Na_xMO_2 consists of edge-sharing MO_6 octahedra forming transition metal layers; whereas Na^+ ions occupy the interstitial site between two metal layers.[8] There are two different classifications of layered Na_xMO_2 based on the Na^+ ion environment: the first one is the prismatic (P) type where Na^+ ions occupy trigonal prismatic coordination and the other is the octahedral (O) type where Na^+ ions occupy octahedral coordination according to the classification made by Delmas *et al.*[9] The P2-type Na_xMO_2 is formed when $0.5 \leq x \leq 0.8$; whereas O3-Na_xMO_2 is formed at a higher concentration of Na^+ ions ($0.85 \leq x \leq 1$) (Figure 3.2).[9] Here, the numbers (2 and 3) represent the number of metal layers present per unit cell. The unit cell of a P2-type structure has two forms of MO_6 layers, namely AB and BA with all the Na^+ ions positioned at prismatic sites, as shown in Figure 3.2d. Two different P sites are available for sodium ions occupancy: Na_f (Na1), which connects with two MO_6 at the face; and Na_e (Na2), which connects with six MO_6 at the edges.[10] P2-type Na_xMO_2 has a space group of $P6_3/mmc$ and falls under the 2H phase. The unit cell of O3-$NaMO_2$ consists of AB, CA and BC MO_6 layers with various O stacking as shown in Figure 3.2a. This structure has a space group of $R3m$ and is also known as the 3R phase.[11]

There are certain drawbacks encountered by layered Na_xMO_2 electrodes, such as storage instability, irreversible phase transition and inadequate battery performance. The layered Na_xMO_2 absorbs moisture when exposed to air, leading to phase instability, electrochemical performance deterioration and difficulty in preparation and storage.[11] When exposed to the environment, H_2O/CO_2 intercalate between MO_2 slabs leading to the formation of insulating $NaOH/Na_2CO_3$ and causing a poor cycle life.[12] Capacity degradation also occurs due to the phase transformation in P2 and O3 during electrochemical cycling. The O3 phase is known to undergo severe phase changes ($O3 \rightarrow O3' \rightarrow P3 \rightarrow P3' \rightarrow P3''$) compared to that of the P2 phase ($P2 \rightarrow O2$). The layered materials containing high voltage redox couples (Fe^{3+}/Fe^{4+} and Ni^{2+}/Ni^{4+}) favor high voltage and specific energy. However, the de-intercalation of Na^+ ions at higher voltage leads to irreversible phase changes and hence causes a poor cycle life. Problems such as irreversible phase transformation and Jahn–Teller distortions have to be tackled to yield good electrode materials for high power applications with a long cycle life.[9–12]

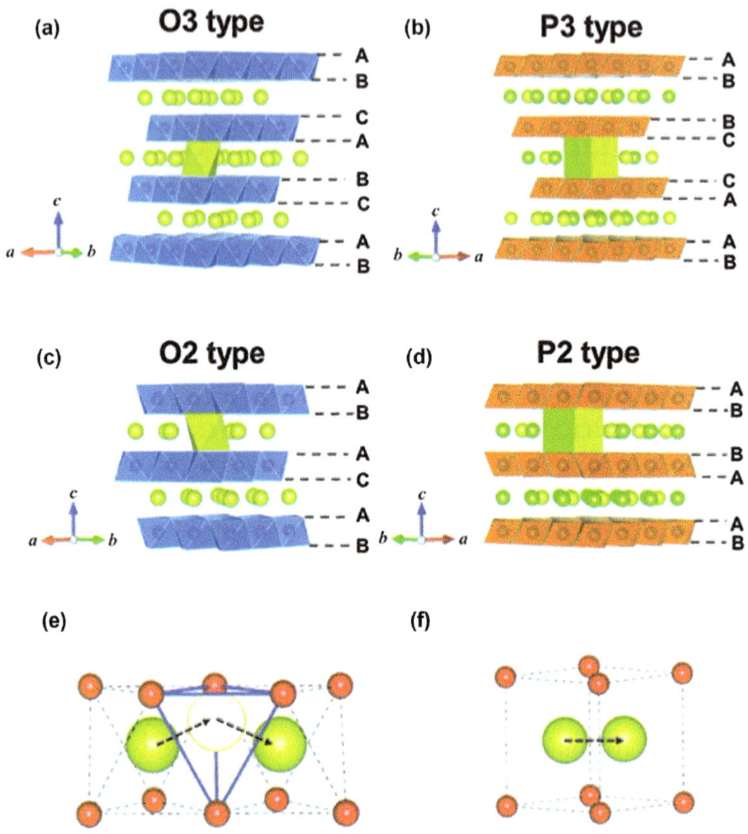

Figure 3.2 Crystal structure of (a) O3; (b) P3; (c) O2 and (d) P2-type layered Na_xMO_2. Deep orange and blue show the transition metal, and yellow and green show the sodium ions in the crystal structure. (e) Octahedral and (f) prismatic arrangement of a sodium ion, showing the sodium ion migration path.
Reproduced from ref. 11 with permission from the Royal Society of Chemistry.

3.2.1 Layered Sodium Cobalt Oxide and Its Derivatives as Cathodes for SIBs

The lithium-containing layered transition metal oxide, $LiCoO_2$ cathode for LIBs, was first reported by Sony in 1991, due to its excellent electrochemical performance.[1,2] In 1980, Delmas *et al.* were the first to report the sodium analogues transition metal oxide with general formulae, Na_xCoO_2 ($x \leq 1$).[13] Na_xCoO_2 has various phases under different synthesis conditions. For example, it can have P3′ at $x = 0.55$–0.60, P2 at $x = 0.64$–0.74, O3′ at $x = 0.77$ and O3 at $x = 1$, respectively, when temperature is varied from 500–800 °C under an oxygen environment of 1 bar.[14] The phases O3, O3′, and P3 are stable at a 400–600 °C temperature range. However, the P2 phase is only stable at 700 °C in the presence of O_2.[14] Na_xCoO_2 with all possible phases consists of sheets with edge-sharing CoO_6 octahedra, where Na^+ ions occupy either the trigonal prismatic (P) or octahedral (O) environment between two layers of CoO_6.[13] While using O3-$NaCoO_2$ in a cell, it undergoes a reversible structural transition of O3↔O3′↔P3′.[13] However, the distortion in the P3′ phase disappears at high operating potential. Low specific capacity and a poor cycle life is observed in the case of O3-$NaCoO_2$ due to the large energy barriers, where the direct hopping from one octahedral site to another requires high activation energy (Figure 3.2e). O3-$NaCoO_2$ showed a high discharge capacity at 30 C rate, when a solid solution was formed with its isostructural $NaFeO_2$, resulting in $Na[Fe_{0.5}Co_{0.5}]O_2$.[15]

P2-Na_xCoO_2 is considered to be a superior cathode material, where the Na^+ ions occupy trigonal prismatic coordination within the layers of CoO_6 octahedra and hence shows a lower Na^+ ion diffusion barrier compared to that of the O3 phase (Figure 3.2e and f).[1,2] In addition, it remains structurally stable during electrochemical cycling over a large range of x. The electrochemical behavior for the P2 phase is almost similar to that of P3 and O3 under the same cyclic conditions. However, a complex phase transition was involved with the P2 phase compared to P3 and O3 when the lower operating voltage was dropped below 2.7 V. The difference in the electrochemical behavior below 2.7 V is due to the difference in Na^+ ion ordering.[14] Better capacity retention was noticed for the P2 phase compared to the P3 and O3 phases. P2-$Na_{0.7}CoO_2$ showed better cycling stability compared to P3-$Na_{0.7}CoO_2$. The former showed 80% capacity retention for 300 cycles against 60% capacity retention for 250 cycles for the latter. The low crystallinity of the P3 phase, which is usually formed at a lower temperature than the P2 phase, is the possible reason for such a poor cycle life.[14,16]

P2-Na_xCoO_2 ($0.5 \leq x \leq 0.8$) is sodium deficient in the slab, leading to an increase in cobalt oxidation to 4+. For such a peculiar composition, the electronic localization was noticed and proved from electron diffraction and NMR studies. The distribution of metal cations in P2-Na_xCoO_2 depends on the energy minimization between (i) the electrostatic repulsion that tends to separate Na^+ ions, (ii) the Na^+–Co^{3+} repulsion through the common face between the NaO_6 and CoO_6 polyhedra, and (iii) the electron–electron interaction in the Co layer.[1,2] These parameters are very sensitive to sodium content and hence many possible cationic distributions have been noticed. Delmas' group investigated the complex phase transformation behavior of P2-$Na_{0.74}CoO_2$ during charge/discharge cycling using an *in situ* XRD measurement.[17] They argued that the weak voltage polarization due to the high conductivity of P2-$Na_{0.74}CoO_2$ enabled one to perform *in situ* XRD in each voltage drop and to understand the associated phase transformation. The presence of two-phase and single-phase reactions were confirmed from the XRD patterns. The split of the peaks is due to a two-phase reaction (voltage plateau region), whereas the fixed peaks are due to single-phase reactions (voltage drops).[17]

Electrochemical improvement in P2-Na_xCoO_2 was performed by various metal cations doping, which suppresses the Na^+/vacancy ordering. Various studies revealed that Mn doping to P2-Na_xCoO_2 has been shown to improve the electrochemical performance.[18–20] No stepwise voltage plateaus were seen in P2-$Na_x[Co_{2/3}Mn_{1/3}]O_2$ for x in the range of 0.5–0.83 due to the elimination of Na^+/vacancy ordering.[19] Yang *et al.* reported the phase stabilization of P2-$Na_{2/3}[Co_{1-x}Mn_x]O_2$ at a cobalt content of 0.5 and showed a reversible capacity of 120 mAh g^{-1} due to both a $Co^{3+/2+}$ and $Mn^{4+/3+}$ redox reaction, when cycled between 1.5–2.1 V.[20] Further improvement was achieved by synthesizing a bi-phasic compound of composition P2/P3-$Na_{2/3}$-$[Co_{0.5}Mn_{0.5}]O_2$, which forms relatively at a low temperature (~700 °C). The bi-phasic compound showed a high specific capacity of 180 and 125 mAh g^{-1} at 0.1 and 5.0 C rate in a voltage range of 1.5–4.3 V compared to pure P2-$Na_{2/3}[Co_{0.5}Mn_{0.5}]O_2$.[21] The improved electrochemical storage noticed for the bi-phasic compound is due to the better structural stabilization during cycling. Recently, doping of Ca^{2+} to Na_xCoO_2 demonstrated improved electrochemical performance by suppressing the phase instability at the cost of decreased capacity. However, the large Ca^{2+} ion, having a similar ionic size to Na^+ (1.02 Å), occupied the Na^+ sites and blocked the Na^+ ion transport. Matsui *et al.* demonstrated that Ca^{2+} doping into Na^+ sites in $Na_{2/3}CoO_2$ to form $Na_{2/3-x}Ca_xCoO_2$ improved the electrochemical performance by suppressing the multiple phase transformation to a great extent. However, the observed capacity was decreased due to the incorporation of Ca^{2+} into the Na^+ site.[22]

3.2.2 Layered Sodium Nickel Oxide and Its Derivatives as Cathodes for SIBs

Among the layered sodium metal oxides, $NaNiO_2$ is widely investigated because of its high specific capacity and redox potential. It has two stable polymorphs based on the synthesis temperature range; O3 with a monoclinic structural distortion is stable at low temperature and a rhombohedral phase is stable at a higher temperature.[1,2] The O3 phase consists of Ni–O layers formed by the edge-sharing of NiO_6 octahedra containing elongated Jahn–Teller Ni^{3+} ions (low spin configuration) and the Na^+ ions occupying the interstitial sites between the NiO_2 slabs, exhibiting a distorted octahedral coordination by oxygen atoms.[1,2,23] At high temperature (~450 °C), the symmetry reduction caused by the Jahn–Teller effect is overcome by thermal motion leading to phase transformation from monoclinic distorted O3 to orthorhombic.[24] The synthesis of O3′-$NaNiO_2$ is easy, unlike $LiNiO_2$, due to the larger ionic radius of Na^+ ions. However, it requires an oxidative environment to convert Ni^{2+} (used as a starting material) to Ni^{3+}. The first report on the electrochemical performance of O3′-$NaNiO_2$ was published by Braconnier *et al.*, where they showed reversibly extraction/insertion of 0.2 moles of Na^+ ions corresponding to 50 mAh g^{-1}, when cycled between

1.8–3.4 V *via* a multiple phase transformation of O3′↔P3′↔P3″↔O3″.[14] The phase transformation corresponds to the redox reaction of $Ni^{3+/4+}$. A significant increase in specific capacities was noticed corresponding to the extraction of 0.85 moles and the insertion of 0.62 moles of Na^+ ions during the first cycle, when the potential window was changed to 2.0–4.5 V. Such a dramatic increase in capacity is due to the formation of new phases confirmed from the XRD patterns obtained at highly charged states (3.75 and 4.5 V).[23] Vassilaras *et al.* reported a very high specific capacity of ~145 mAh g^{-1} under a similar cyclic condition.[23] However, a better capacity retention of 94% was achieved when only cycling to a cut-off voltage of 3.75 V.

Ni^{3+} in O3′-$NaNiO_2$ being a Jahn–Teller ion causes crystal structure distortion and leads to a poor cycle life of $NaNiO_2$. In order to reduce the distortion, partial substitution of metal ions (M^{n+}), where $n > 3$ was adopted as an effective approach. Sb- and Bi-doped O3′-$NaNiO_2$, with the chemical composition O3′-$Na_3Ni_2SbO_6$ and O3′-$Na_3Ni_2BiO_6$, were explored as possible cathodes for SIBs.[24,25] They are similar to layered structures, where Ni^{2+}/Sb^{5+} and Ni^{2+}/Bi^{5+} occupy the transition metal slabs and each SbO_6/BiO_6 octahedron are surrounded by six NiO_6 octahedrons forming a honeycomb-like structure.[24,25] The structure O3′-$Na_3Ni_2SbO_6$ showed ~117 mAh g^{-1} corresponding to the extraction of 2 Na^+ ions (based on a Ni^{2+}/Ni^{3+} redox reaction) at 3.3 and 3.7 V, respectively. Apart from the high capacity, it showed an excellent rate capability at 2 C rate for 500 cycles with 70% capacity retention.[24] A similar electrochemical performance was also reported for $Na_2Ni_2TeO_6$ and $Na_{1.6}Sr_{0.2}Ni_2TeO_6$ high voltage cathodes.[26] A specific capacity of ~110 mAh g^{-1} based on a Ni^{2+}/Ni^{3+} redox reaction was achieved for $Na_2Ni_2TeO_6$. It exhibited two bi-phasic plateaus at ~3.6 and ~4.4 V corresponding to two different ordering due to the extraction of 1/3 and 2/3 of Na^+ ions. The addition of Sr^{2+} to $Na_2Ni_2TeO_6$ further reduced the Na^+ ordering leading to the stabilization of capacity.[26] Ti-substituted O3′-$NaNiO_2$ with the composition O3′-$NaNi_{0.5}Ti_{0.5}O_2$ showed an excellent cycling performance, where it delivered ~121 mAh g^{-1} based on Ni^{2+}/Ni^{4+} at ~3.1 V.[27] The presence of Ti^{4+} helps in retaining the phase during cycling, leading to a longer cycle life.

Partial substitution of transition metals, such as Mn for Ni, has shown significant improvement in the electrochemical performance by suppressing the Jahn–Teller distortion in $NaNiO_2$. Komaba *et al.* reported an improved electrochemical performance of Mn-doped $NaNiO_2$ with the composition $Na(Mn_{0.5}Ni_{0.5})O_2$, where the Ni and Mn are in 2+ and 4+ oxidation states, respectively.[28] The presence of Mn^{4+} in the crystal structure makes Ni^{2+}. The reversible specific capacity of Mn-doped $NaNiO_2$ is mainly from a Ni^{2+}/Ni^{4+} redox reaction when cycled between 2.5–4.5 V, and corresponds to 185 mAh g^{-1}. Such an improved performance observed for Mn-doped $NaNiO_2$ is due to the highly reversible phase transfer of O3→O3′→P3→P3′ preferred at a low cut-off voltage of 3.8 instead 4.5.[28] Further improvement in the electrochemical performance of $Na(Mn_{0.5}Ni_{0.5})O_2$ was achieved with Fe doping. The improved performance was explained based on the highly reversible phase transformation of P3 → OP2 instead P3 → P3″, which was seen in $Na(Mn_{0.5}Ni_{0.5})O_2$. The formation the OP2 phase due to Fe substitution led to a smaller inter-slab distance (5.13 Å) compared to the P3″ phase (5.72 Å). Such a smaller inter-slab distance prohibits the insertion of solvents molecules/electrolyte anions, resulting in a stable cycle life.[29,30]

3.2.3 Layered Sodium Manganese Oxide and Its Derivatives as Cathodes for SIBs

Manganese-based layered materials are interesting because Mn is cheap and environmentally-friendly, and they have various oxidation states, which enables their use in different voltage ranges (Mn^{3+}/Mn^{4+} being the most common). Sodium manganese oxides with composition Na_xMnO_2 ($0.44 \leq x \leq 1$) are extensively used as cathode materials for SIBs due to their excellent electrochemical properties.[31] Na_xMnO_2 adopts two different crystal structures based on the sodium content: three-dimensional (3D), known as the tunnel structure at $x \leq 0.44$, and two-dimensional (2D), known as the layered structure at $0.5 \leq x \leq 1$.[31] Two different stable phases exist for the sodium-rich layered structure ($x = 1$): the first one being α-$NaMnO_2$, which has an O3′ structure with a space group of *C*2/*m* that is stable at low temperature, and the second being orthorhombic β-$NaMnO_2$ with a space group of *Pmnm* that is stable at high temperature.[31] α-$NaMnO_2$ with a monoclinic crystal structure is energetically more stable than β-$NaMnO_2$, unlike $LiMnO_2$, which favors an orthorhombic structure. The crystal structure arrangement in both α and β-$NaMnO_2$ differs in the MnO_6 octahedra arrangement. The MnO_2 sheet consists of edge sharing of the MnO_6 octrahedra in α-$NaMnO_2$; whereas double stacking of the MnO_6 octahedra is noticed in β-$NaMnO_2$.[1,2] In the case of $NaMnO_2$, it is worth discussing the distortion in the crystal structure caused by the Mn^{3+}. In Mn^{3+} (d^4-system), three electrons occupy t_{2g} orbitals (one each in d_{xy}, d_{yz} and d_{xz}), whereas the other one electron occupies the e_g orbital. The one electron occupancy in the energy degenerate e_g orbital (d_{z2} and $d_{x2\text{-}y2}$) leads to the crystal structure distortion.[32]

In 1985, Hagenmuller *et al.* explored the possibility of using O3′ α-$NaMnO_2$ as an electrode and evaluated its performance against sodium. They observed two different de-intercalated phases separated by a two-phase domain during a Na^+ ion extraction. The first de-intercalated phase was seen from $Na_{0.93}MnO_2$ to $Na_{0.77}MnO_2$; whereas the second phase continued up to $Na_{0.73}MnO_2$.[33] However, a poor cyclic performance with low Na^+ ion intercalation/de-intercalation was noticed. A different de-intercalation/intercalation behavior of O3′α-$NaMnO_2$ was reported by Ceder's group, where they reported a higher Na^+ ion de-intercalation/interaction from/into O3′α-$NaMnO_2$ when cycled between 2.0–3.8 V. During charge 0.85 Na^+ ions were extracted corresponding to 210 mAh g^{-1}; whereas only 0.8 Na^+ ions were able to be intercalated during discharge corresponding to 197 mAh g^{-1}.[34] The O3′α-$NaMnO_2$ showed various phase transformations during both charge (3.8 V)/discharge (2.0 V) as evident from the potentiostatic intermittent titration technique (PITT), where eight oxidation peaks and five reduction peaks were noticed. Each peak is associated with phase transformation. During Na^+ ion extraction, a large voltage plateau indicating two-phase was noticed at 2.66 V. $Na_{0.7}MnO_2$ with a different crystal structure compared to that of the orthorhombic and hexagonal $Na_{0.7}MnO_2$ was found at the end of the voltage plateau, confirmed from *ex situ* XRD.[34] In spite of the interesting electrochemical properties, O3′α-$NaMnO_2$ is associated with a major drawback of capacity degradation due to the drastic phase transformation and presence of Mn^{3+}. In addition, the dissolution of Mn^{2+} ions in electrolyte, which are formed due to the disproportion of Mn^{3+} ($2\,Mn^{3+} \rightarrow Mn^{4+} + Mn^{2+}$), is another reason for capacity degradation in O3′α-$NaMnO_2$ as reported by Ceder *et al.* They found the presence of Mn content in the electrolyte after the charge/discharge cycle using ICP-AES techniques.[34] The high temperature β-$NaMnO_2$ phase has a different crystal structure arrangement compared to that of conventional $NaMnO_2$ showing a different Na^+ ion intercalation/de-intercalation mechanism. Billaud *et al.* showed the Na cycling behavior of β-$NaMnO_2$ and investigated the structural transformation during charge/discharge using XRD, solid-state NMR and high-resolution TEM techniques.[35] Compared to α-$NaMnO_2$, it shows fewer numbers of oxidation and reduction voltage peaks, indicating a moderate phase transformation during cycling. A large voltage plateau was noticed around 2.7 V with a composition of $Na_{0.57}MnO_2$ followed by an increase in voltage in small voltage steps at $Na_{0.49}MnO_2$ and $Na_{0.39}MnO_2$, respectively.

Among all the polymorphs of Na_xMnO_2 at lower sodium content, the P2 phase and its derivatives show the best electrochemical performance. The Na_xMnO_2 crystallizes in hexagonal (P2, P3) and distorted orthorhombic (P2′) crystal structure when the sodium content is reduced.[36] In 2002, Caballero *et al.* prepared the P2-$Na_{0.6}MnO_2$ by a sol–gel route at 800 °C and showed a high specific capacity of 140–150 mAh g^{-1} when cycled between 2.0–3.8 V. They mentioned that P2-$Na_{0.6}MnO_2$ undergoes various structural changes during the charge/discharge reaction, reflected as redox peaks noticed in both differential capacity and galvanostatic cycling plots.[37] Though P2-$Na_{0.6}MnO_2$ is considered to be an interesting cathode material for SIBs, the major issue associated with P2-$Na_{0.6}MnO_2$ is the poor kinetics, which is the main reason for its capacity degradation. In addition, the intercalation of H_2O molecules into interstitial sites (Na^+ ion position) of P2-$Na_{0.6}MnO_2$, when exposed to air, causes an increase in the inter-slab distance, leading to a poor cycle performance.[2,38] It is necessary to improve the kinetic properties of P2-$Na_{0.6}MnO_2$ to enhance the electrochemical properties. A number of effective approaches were adopted to enhance the sodium storage capacity and life cycle of P2-$Na_{0.6}MnO_2$. Nickel-doped P2-$Na_{2/3}MnO_2$ has been studied extensively due to its various advantages: (i) It is stable when exposed to air and no intercalation of H_2O is seen. (ii) High specific capacity (theoretical value: 173 mAh g^{-1}) and high operating voltage (~3.5 V) corresponding to the $Ni^{2+/4+}$ redox reaction; (iii) Ni^{2+} having a similar ionic radius as Mn^{3+}

reduces the Mn^{3+} content and hence can minimize the Jahn–Teller distortion.[39–41] However, the contribution to specific capacity due to the $Ni^{2+/4+}$ redox reaction causes severe volume change due to P2→O2 transformation and exhibits poor cycle life.[39–41] Dahn *et al.* showed a high capacity of 161 mAh g^{-1} for P2-$Na_{2/3}[Ni_{1/3}Mn_{2/3}]O_2$, which is nearly the same as its theoretical specific capacity (173 mAh g^{-1}) when cycled between 2.0–4.5 V. *In situ* XRD measurement revealed the phase transformation of P2→O2 when cycled to higher charge potential and is detrimental in nature causing capacity degradation due to the large volume change.[39] Wang *et al.* also argued that the cyclic stability of P2-$Na_{2/3}[Ni_{1/3}Mn_{2/3}]O_2$ greatly depends on a higher cut-off voltage. They showed that restricting the higher cut-off voltage to 4.0 V enhanced the structural stability in P2-$Na_{2/3}[Ni_{1/3}Mn_{2/3}]O_2$ compared to 4.5 V, which resulted in capacity degradation due to the irreversible phase transformation from P2→O2.[42] No phase transformation from P2→O2 was noticed when cycled at a relatively lower cut-off voltage range, *i.e.* 1.5–3.75 V, supported by *in situ* XRD measurement from a similar report.[43] Reducing Ni content by partial substitution of different electrochemically active/inactive metal elements improved the capacity retention. Formation of P_2-$Na_{2/3}(Ni_{1/3-x}M_xMn_{2/3})O_2$ (where M = Li, Mg, Zn, Al, Co, Fe and Ti) showed improved electrochemical performance due to the suppression of P2→O2 phase transformation at the cost of low specific capacity. It was seen that the P2→O2 phase transformation was substantially minimized (no characteristic large voltage plateau at ~4.1 to 4.2 V due to the P2→O2 phase transformation even after charging to 4.4 V) after partial Li doping leading to an excellent electrochemical performance.[44] A similar improvement in cyclic stability was achieved while partially replacing Ni^{2+} with Mg^{2+}, which has a similar ionic radius (0.72 Å). The presence of inactive Mg^{2+} ions in the Ni^{2+}/Mn^{4+} metal layer helps to retain Na^{+} ions in trigonal prismatic sites during the charge/discharge process, leading to stabilization of the P2 structure even at high charge voltage. *Ex situ* XRD confirmed the retention of the P2 phase during the charge/discharge reaction even at the high charge voltage of ~4.35 V (a minor phase of OP4 was found along with P2); whereas the P2→O2 phase transformation was noticed in Mg-free P2-$Na_{0.67}(Ni_{0.33}Mn_{0.67})O_2$.[45]

3.2.4 Layered Sodium Iron Oxide and Its Derivatives as Cathodes for SIBs

$NaFeO_2$, which is analogous to $LiFeO_2$, is considered as a potential candidate for SIBs because of the high specific capacity, non-toxicity, low cost (due to the abundance of Fe resources) and high operating voltage. In addition, synthesis and functioning of $NaFeO_2$ is feasible compared to $LiFeO_2$, where the migration of Fe^{4+} to the Na^{+} site does not occur during Na^{+} ion extraction due to the difference in ionic radii of Fe^{4+} (0.72 Å) and Na^{+} (1.06 Å).[46,47] $NaFeO_2$ crystallizes in two different forms, such as α-$NaFeO_2$ and β-$NaFeO_2$. Synthesis and characterization of both phases have been performed over the years and it has been concluded that the electrochemical activity of α-$NaFeO_2$ is higher than that of β-$NaFeO_2$.[48] α-$NaFeO_2$ has an O3-type crystal structure, having a cubic closed packed oxygen array. The O3-α-$NaFeO_2$ has a theoretical capacity of ~242 mAh g^{-1} based on the Fe^{3+}/Fe^{4+} redox reaction, whereas it shows only 80–100 mAh g^{-1} corresponding to insertion/de-insertion of only 1/3 of Na^+ ions at a flat voltage of 3.3 V, possibly due to the chemical instability of Fe^{4+}.[48] The reversibility of the Na^+ ion insertion/de-insertion depends on the cut-off voltage, as an irreversible phase change due to the Jahn–Teller distortion is anticipated at higher operating voltages. To validate α-$NaFeO_2$ as a potential candidate, numerous experiments such as metal substitution, particle size reduction, control in morphology and surface coating were performed to improve its electrochemical performance.[15,49]

Despite the various types of development carried out on O3-type α-$NaFeO_2$, it has several drawbacks, including the phase transformation from O3 → P3′ *via* the intermediate phases of O3′ and P3. In addition, further de-intercalation of sodium ions leads to the formation of the P3″ phase. The enlarged interlayer spacing in the P3′ (~5.6 Å) and P3″ (~7.0 Å) phases are not preferred because of the intercalation of electrolytic molecules, which causes poor specific capacity.[50] P2-type layered oxide systems are more favorable compared to the O3-type, due to the structural stability discussed earlier. The attractive features of P2-type layered oxides motivated earlier research groups to synthesize P2-Na_xFeO_2. However, P2-Na_xFeO_2 cannot be considered as a viable electrode material because it is structurally unstable. This is due to the instability of the Fe^{4+} in the oxide structure. To overcome this drawback, Yabuuchi *et al.* introduced partial substitution of Mn to the Fe site to synthesize structurally stabilized P2-$Na_xFe_{0.5}Mn_{0.5}O_2$, which delivered a higher capacity of 190 mAh g^{-1}, compared to O3-$NaFe_{0.5}Mn_{0.5}O_2$, which showed only 100–110 mAh g^{-1}.[51] With Mn substitution, the reversible redox reaction of $Fe^{3+/4+}$ contributed to the specific capacity when charged to 3.8–4.2 V. However, the P2 phase was retained only when the potential window was restricted to 3.8 V during charging, corresponding to the oxidation of $Mn^{3+/4+}$. Beyond 3.8 V, the phase change from P2 → OP4 was seen as evident from an *in situ* XRD study.[51]

The cyclic stability of the P2-$Na_xFe_{0.5}Mn_{0.5}O_2$ material also strongly depends on the surface coating (*e.g.* carbon coating) and morphology. Park *et al.* performed uniform carbon coating onto $Na_{2/3}Fe_{1/2}Mn_{1/2}O_2$ prepared by the solid-state route, using 2,3-Dihydroxynaphthalene (DN) as a carbon precursor. They argued that even 0.2% DN-coated $Na_{2/3}Fe_{1/2}Mn_{1/2}O_2$ exhibited excellent cyclic stability due to improved reaction kinetics and minimized contact between the electrode and electrolyte. A higher initial capacity of 178 mAh g^{-1} at C/10 was reported, which is 20% higher than the uncoated electrode.[52] Morphology and particle size of the electrode materials also play an important role in achieving a better cycle life and higher specific capacity, while controlling the Na^+ ion diffusion along the electrode materials during the discharge/charge cycle. Kalluri *et al.* presented a comparison study on the electrochemical performance of $Na_{2/3}Fe_{1/2}Mn_{1/2}O_2$ with nanoparticle and nanofiber morphology. The $Na_{2/3}Fe_{1/2}Mn_{1/2}O_2$ with nanofiber morphology showed a higher capacity and better capacity retention compared to $Na_{2/3}Fe_{1/2}Mn_{1/2}O_2$ with nanoparticle.[53] The better connectivity among the fibers and enhanced Na^+ ion transfer are the reasons for the improved performance in the $Na_{2/3}Fe_{1/2}Mn_{1/2}O_2$ nanofiber. Although P2-$Na_{2/3}Fe_{1/2}Mn_{1/2}O_2$ demonstrated excellent performance in terms of high specific capacity and retention while adopting various modification approaches, still the use of P2-$Na_{2/3}Fe_{1/2}Mn_{1/2}O_2$ in a full cell remains a challenge. In full cell configurations, the Na^+ ion deficiency will lead to a great capacity reduction, especially when the anode doesn't contain sodium. To overcome this difficulty, Singh *et al.* proposed the addition of sacrificial salts, sodium azide (NaN_3) into $Na_{2/3}Fe_{1/2}Mn_{1/2}O_2$ cathode material during synthesis to decrease the first cycle's irreversible capacity. The addition of 5 wt.% of NaN_3 decreased the irreversible capacity of 59 mAh g^{-1} to 27 mAh g^{-1}. Such an improvement will be a great advantage for using $Na_{2/3}Fe_{1/2}Mn_{1/2}O_2$ as a cathode material in a full cell configuration.[54]

3.2.5 Layered Sodium Chromium Oxide and Its Derivatives as Cathodes for SIBs

$NaCrO_2$, which has a crystal structure that is similar to α-$NaFeO_2$, is attractive as a cathode for SIBs because of its improved electrochemical storage and thermal stability.[55] Hagenmuller *et al.* were the first to report the reversible Na^+ ion insertion/de-insertion into/from $NaCrO_2$, which shows intercalation/de-intercalation of 0.5 moles of sodium ion at an average voltage of ~3.0V, resulting in ~120 mAh g^{-1}.[56] $LiCrO_2$ is known as electrochemically inactive to lithium ion intercalation/de-intercalation because of the irreversible migration of the Cr^{6+} ($3Cr^{4+} \rightarrow 2Cr^{3+} + Cr^{6+}$) ions into the interstitial position during the charge/discharge process. However, such migration does not happen in $NaCrO_2$ due to the ionic radii mismatch of Na^+/Cr^{6+} and as a result Cr^{6+} stays in the metal layer even after Na^+ ion extraction, making it a suitable cathode material.[56] However, $NaCrO_2$ undergoes a phase transformation of O3→P3 during sodium-ion extraction, which causes capacity degradation upon cycling.[55,56] Ceder *et al.* also reported capacity degradation in $NaCrO_2$ due to the transformation of layered $Na_{0.4}CrO_2$ to rock salt CrO_2 during full Na^+ ion removal from the structure.[57] The oxidative decomposition of electrolyte in $NaCrO_2$ is also another reason for the capacity degradation, where the large inter-slab spacing in $NaCrO_2$ (~3.15 Å) enables penetration of propylene carbonate (PC) and /or PC solvated Na^+ into the layered electrode materials, causing electrolyte decomposition at a higher voltage (charged state).[58] This issue has been tackled by charging the $NaCrO_2$ electrode at a relatively higher current to provide a shorter duration for electrolyte decomposition. Tsuchiya *et al.* reported the preparation of nanosized $NaCrO_2$ to address the O3→P3 phase change, where the presence of grain boundaries and randomly oriented nano-crystalline structures suppress the phase change from O3 → P3, resulting in a better electrochemical performance in NaCrO2.[58]

Surface carbon coating is considered to be the most effective approach to enhance the electrochemical performance of $NaCrO_2$. The reducing atmosphere used for the synthesis of $NaCrO_2$ to prevent further oxidation of Cr^{3+} is beneficial for carbon coating. Citric acid was used as a source of carbon coating for $NaCrO_2$ as reported by Ding *et al.*[59] Higher capacity with better stability was achieved after carbon coating to $NaCrO_2$ due to an increase in electronic conductivity and decrease in lower polarization. However, the inhomogeneous (in the form of a composite) and thick carbon coating, evident from transmission electron microscopy (TEM) analysis, resulted in a poor rate performance at high current. Controlled and homogeneous carbon coating from pitch exhibited an excellent electrochemical performance as reported by Yu *et al.*[60] The carbon-coated $NaCrO_2$ showed an excellent capacity retention as well as rate performance at a high current rate of 150 C due to the structural stability achieved after carbon coating. In addition to an increase in electronic conductivity, it plays a vital role in delaying the exothermic decomposition by preventing oxygen loss from the crystal structure and prevents the moisture attack being hydrophobic in nature. The formation of electronically insulating NaOH and Na_2CO_3 on the surface of $NaCrO_2$ while exposed to air is minimized due to protective carbon coating.[59,60]

3.2.6 Layered Sodium Vanadium Oxide and Its Derivatives as Cathodes for SIBs

Vanadium-based oxide materials show rich redox chemistry due to various oxidation states of vanadium and hence result in different phase structures. In addition, good electrical conductivity of vanadium-based oxide from semiconducting to metallic makes them useful as electrode materials.[1,2,61] Na_xVO_2, which is analogous to $LiVO_2$, crystallizes in two different phases, such as O3 with monoclinic distortion (high Na^+ ion content phase, where $x = 1$) and P2 with hexagonal structure (low Na^+ ion content phase, where $x = 0.7$).[61–63] O3-$NaVO_2$ showed highly reversible Na^+ ion intercalation/deintercalation at $0.5 \leq x \leq 1$, where a large voltage plateau *i.e.*, a two-phase domain, was observed until the composition $Na_{0.66}VO_2$ was reached. With further Na^+ ion extraction, a complicated phase transformation was noticed with three different voltage plateaus at $0.5 \leq x \leq 0.66$. Such phase transformations are similar to that of Na_xCoO_2 reported elsewhere.[61–64] During the Na^+ ion extraction process, the phase transformation from O3 ↔ O3′ *via* monoclinic distortion was observed. The Na^+ ion extraction beyond 0.5 moles in $Na_{1-x}VO_2$ causes severe capacity degradation due to the migration of vanadium to the inter-slab vacancy sites. It showed reversible specific capacity of ~120 mAh g^{-1} when cycled between 1.2–2.4 V. However,

$NaVO_2$ is not stable in air and reacts when exposed for few seconds forming a de-sodiated O3-$Na_{1-x}VO_2$ phase.[59,63,65] The phase stability of $NaVO_2$ in a reducing atmosphere leads to effective carbon coating during synthesis, which could reduce the oxygen uptake from the air. Further, the carbon coating will increase the electronic conductivity and may facilitate better battery performance at higher current rates.

The P2-$Na_{0.7}VO_2$ phase shows interesting properties due to $Na^+/V^{3+/4+}$ and Na^+/Na^+ inter-slab repulsion, leading to a complex phase diagram during Na^+ ion insertion/de-insertion. De-intercalation to a higher voltage (>2.5 V) causes an irreversible process. During de-sodiation, it shows various voltage plateaus, which are the characteristic of two-phase (continuous voltage plateaus) and single-phase (abrupt voltage drops) reactions. The voltage drops indicate the presence of single phases, corresponding to specific Na/V ratios of 1/2, 5/8 and 2/3. These single phases are due to the special ordering of Na^+ and/or V^{3+}/V^{4+} ions between and/or within the VO_2 slabs. The *in situ* XRD during de-sodiation carried out on $Na_{0.92}VO_2$ (discharged state of $Na_{0.7}VO_2$) in the range of $0.5 \leq x \leq 0.92$ showed three different ordered phases at $x = 1/2$, 5/8 and 2/3. The formation of an extra peak noticed from the *in situ* XRD pattern confirms the existence of a superstructure.[66]

3.2.7 Layered Sodium Titanium Oxide and Its Derivatives as Cathodes for SIBs

Titanium-based layered electrode materials are considered as potential electrode materials due to their structural stability during the charge/discharge process. The structure remains stable to a deep discharge, *i.e.* 0 V, unlike other transition metal oxides, which undergo decomposition *via* a displacement or conversion reaction below to ~1.5 V.[1,2] $NaTiO_2$ consists of edge-sharing TiO_6 and NaO_6 octahedra, forming alternate slabs similar to $NaFeO_2$.[67,68] Heigenmuller *et al.* and Maazaz *et al.* were the first to report the electrochemical properties of $NaTiO_2$.[67,68] They showed O3→O3′ phase transformation in $NaTiO_2$ with the extraction of ~0.3 Na^+ ions corresponding to a specific capacity of ~75 mAh g^{-1}. The observed reversible capacity is from a Ti^{3+}/Ti^{4+} redox reaction occurring at ~1.0 V. The $NaTiO_2$ shows relatively low intercalation potential and can be used as a negative electrode for SIBs.[67,68] The solid solution prepared from $NaNiO_2$ and $NaTiO_2$ with a composition of $Na(Ni_{0.5}Ti_{0.5})O_2$ is a stable host as cathode materials for Na^+ ion intercalation/de-intercalation proposed by Yu *et al.*[27] The presence of Ti in the structure enabled the structural stability during cycling leading to long term cycling, whereas the high $Ni^{2+/4+}$ redox couple contributed to the specific capacity. The presence of both Ni and Ti in the structure can enable this material to be used as both an anode and cathode due to the different redox potentials. A similar study has been reported by Shanmugam *et al.*, where they proposed a P2-type layered bi-functional electrode material with the composition P2-$Na_{2/3}(Ni_{1/3}Ti_{2/3})O_2$.[69] However, it showed improved electrochemical performance as an anode possibly due to better structural stability.

The cobalt-doped P2-layered oxide with the composition $Na_{2/3}(Co_{1/3}Ti_{2/3})O_2$ showed excellent electrochemical properties as an anode material, 90 mAh g^{-1} with 84.84% capacity retention at the end of 3000 cycles. The improved cyclic stability is due to the prolonged structural stability, which is evident from the *in situ* XRD pattern (almost no volume change is associated during charge/discharge).[70] Similar cyclic behavior was noticed for lithium-substituted P2 layered titanium oxide with the composition $Na_{0.66}(Li_{0.22}Ti_{0.78})O_2$ used as an anode material.[71]

3.2.8 Na-rich Layered Oxides as Cathode Materials for SIBs

The alkali-rich layered metal oxide cathodes, A_2MO_3 (A = Li, Na and M = metal) are an interesting class of materials due to their promising electrochemical properties, where they deliver higher reversible capacity against their theoretical value, which is calculated based on the oxidation state of transition metals. The observed extra capacity is due to the high voltage redox reaction of the anions.[72–74] The Na-rich layered oxides are more stable compared to the Li-rich, as the ionic radius of the Na^+ ion is larger than that of M^+. The excess of sodium ions present in the structure (*i.e.* within the $Na_{1/3}M_{1/3}$ slab) leads to higher capacity due to two electron reactions ($Na_2MO_3 \leftrightarrow M^{6+}O_3 + 2\ Na^+ + 2e^-$).[72–74] Tamaru *et al.* reported the sodium ion storage behavior of Na_2RuO_3 prepared by solid-state reaction and showed a reversible specific capacity of 140–150 mAh g^{-1} due to a Ru^{4+}/Ru^{5+} redox reaction at 2.8 V.[73] Apart from the high specific capacity, an excellent rate performance was noticed for Na_2RuO_3 due to (i) faster sodium-ion diffusion, (ii) higher electronic conductivity and (iii) low desolvation energy at the electrode–electrolyte interphase. Na_2RuO_3 demonstrates a two-phase reaction such as hexagonal-I and hexagonal-II during sodium ion insertion/de-insertion, as evident from the *ex situ* XRD patterns obtained at various charge/discharge states. For $x < 0.5$, the existence of the hex-I phase was noticed, whereas both hex-I and hex-II were present for $0.5 \le x \le 0.6$. The hex-I phase completely disappeared when $x > 0.6$.[73] Further increase in specific capacity was noticed for Sn^{4+}-doped Na_2RuO_3 compared to Sn^{4+}-free Na_2RuO_3 as reported by Rozier *et al.* They showed that replacement of Ru^{4+} by an electrochemically inactive Sn^{4+} ion increased the specific capacity due to the contribution of both cationic (Ru^{4+}/Ru^{5+}) and anionic (O^{2-}/O_2^{n-}) redox process.[74] In a similar work reported by Song *et al.*, Zr-doping to Na-rich layered cathode Na_2RuO_3 was performed and showed better cyclability. Such improved cyclability in $Na_2(Ru_{0.95}Zr_{0.05})O_3$ resulted from a larger ionic radius of Zr^{4+} (0.72 Å) and strong affinity for octahedral coordination. The sodium-ion insertion/de-insertion in Zr-doped $Na_2(Ru_{0.95}Zr_{0.05})O_3$ shows a bi-phasic reaction mechanism.[75]

3.2.9 Layered $Na_2Ti_3O_7$ as an Anode Material for SIBs

Intercalation-based anode materials are largely used for battery applications owing to their excellent cycle life and high rate performance. However, limited intercalation-based anode materials are known for SIBs. $Na_2Ti_3O_7$ (NTO) is considered to be the most promising intercalation-based anode material for SIBs. It shows an ultra-low Na^+ ion intercalation/de-intercalation potential of ~0.3 V and a high theoretical capacity of 177 mAh g^{-1}.[76,77] NTO is the first intercalation-based oxide with the lowest Na^+ ion intercalation potential. It crystallizes in a monoclinic structure with a space group of $P2_1/m$ and consists of 2D sheets of $(Ti_3O_7)^{2-}$ with edge-sharing TiO_6 octahedral chains forming a layered structure, where Na^+ ions are accommodated in between these zig-zag sheets of $(Ti_3O_7)^{2-}$ (Figure 3.3).[77] For Na^+ ion occupancy, two different sites are available in this layered structure, one with nine-fold coordination and the other with seven-fold coordination. Apart from this, further accommodation of two Na^+ ions into this structure is possible to form a fully sodiated $Na_4Ti_3O_7$, where Na^+ ions occupy either octahedral or prismatic sites.[78] For the diffusion of Na^+ ions between the TiO_6 octahedron layers, a low activation energy of 0.186 eV is required as reported by Pan *et al.*[79] From first principle calculations, the intercalation of a maximum of 3.5 Na^+ ions per NTO corresponding to a specific capacity of 311 mAh g^{-1} (eqn (3.1)) has been reported by Zou *et al.* They proposed that the squeezing of Na^+ ions into the Ti–O structure during intercalation can lead to the accommodation of more Na^+ ions.[80]

$$Na_2Ti_3O_7 + 3.5Na^+ + 3.5e \rightarrow Na_{5.5}Ti_3O_7 \tag{3.1}$$

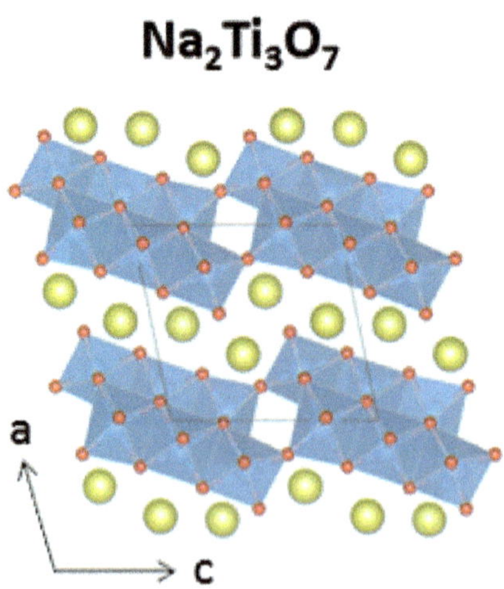

Figure 3.3 Layered structure of $Na_2Ti_3O_7$.
Reproduced from ref. 77 with permission from the Royal Society of Chemistry.

During the intercalation process, the existing Na^+ ion coordination in pristine NTO changes from nine and seven coordination to six coordination at a fully sodiated state of NTO (Figure 3.4a) as explained by Xu *et al.*[81] They also argued that shearing down Ti–O layers controls the change in lattice parameters of NTO leading to the accommodation of more Na^+ ions. Further, the sodiation caused the widening of the joint angle between the two neighboring Ti–O layers from 82.11 to 93.25°, suggesting the flexible nature of NTO. Such flexibility of NTO during Na^+ ion intercalation makes it unique among other layered transition metal oxides.[81]

Different synthesis approaches have been adopted for the synthesis of NTO, such as solid-state, hydrothermal, solvothermal, and sol–gel using a wide range of precursors. Irrespective of the synthesis procedure and precursors, the NTO material showed poor capacity retention, cyclic instability and limited rate capability.[82,83] An extensive study has been carried out to determine the reasons for the gradual capacity fade in NTO. Zerrabeitia *et al.* found that the effect of critical synthesis parameters has a great impact on the performance of NTO material. The small amounts of Na_2CO_3, either as an insulating film formed by the surface corrosion or from the unreacted precursor, strongly contribute to the capacity fade in NTO. They also reported that the atmosphere during the cooling process degrades the performance *i.e.*, the presence of H_2O and CO_2.[83] As NTO is an insulator with a band gap of 3.07 eV, most of the initial studies have been focused on the improvement of electrical conductivity. Carbon coating performed to NTO with nanotube morphology demonstrated excellent electrochemical performance as reported by Li *et al.*[82] The composite NTO@C showed a high reversible specific capacity of ~142 mAh g^{-1} at the end of 100 cycles

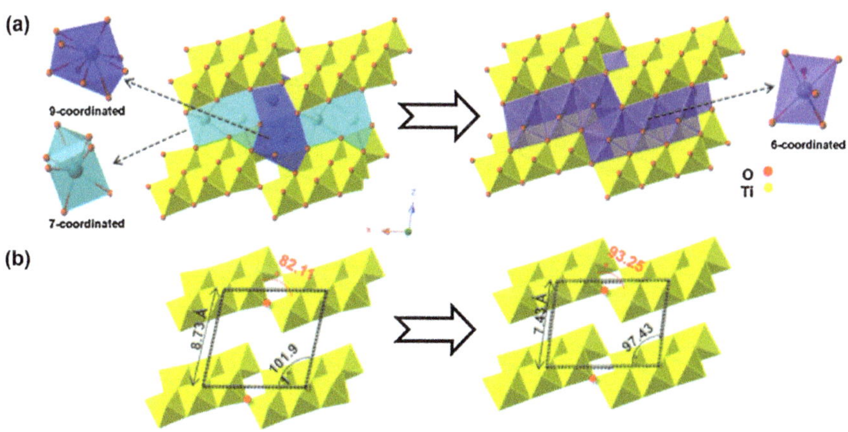

Figure 3.4 (a) Change in coordination of Na^+ ions during phase transformation. (b) Structural changes upon sodiation in the $Na_2Ti_3O_7$ structure. Reproduced from ref. 81 with permission from the Royal Society of Chemistry.

at 1 C rate, and with a good rate performance at 10 C where it delivered ~84 mAh g^{-1}. The reason for the improved performance is due to the increase in electronic conductivity of NTO by carbon coating as well as faster sodium-ion diffusion facilitated by the presence of nanotube morphology and a carbon layer.[82] Bharadwaj *et al.*, addressed the rapid capacity fade in NTO electrodes due to the continual growth of the passivation film from the side reactions with the electrolyte, by uniform wrapping of NTO particles with MWCNTs.[84] The poor electronic conductivity of NTO was addressed by partial replacement of Ti with Nb, which exhibited a lower band gap compared to that of pure NTO, resulting in higher reversible capacity, cyclic stability and an increase in rate capability due to the enhanced kinetics of the Na^+ ions.[85] Xia *et al.* showed an improvement in the electrochemical performance of NTO after lanthanides (Ln) doping, which introduces distortion and creation of oxygen vacancies leading to better electronic conductivity. The shrinkage of the lattice parameter in Ln-doped NTO observed due to strong Ln–O bonding is responsible for such distortion in NTO. The mechanism of oxygen vacancy formation is explained using the Kroger–Vink notation as follows:

$$Na_2Ti_3O_7 + xLn^{3+} \rightarrow Na_2Ti_{3-x}Ln_xO_{7-\frac{x}{2}} + xTi^{4+} + \frac{x}{2}V_O'' + Ti_{Ln} + \frac{x}{4}O_2 + xe^- \quad (3.2)$$

Among the Ln, Ytterbium (Yb^{3+})-doped NTO demonstrated the best cycling performance yielding a high reversible capacity of 71.6 mAh g^{-1} at 5 C after 1600 cycles (Figure 3.5).[76]

Wang *et al.* reported a better electrochemical performance for microspherical NTO material, which consists of tiny nanotubes and showed a

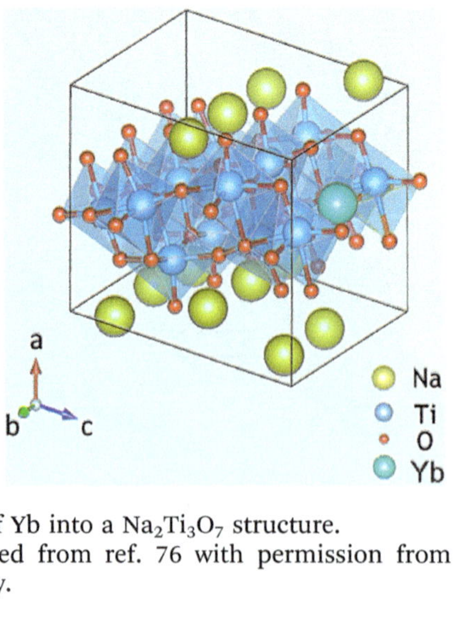

Figure 3.5 Doping of Yb into a $Na_2Ti_3O_7$ structure.
Reproduced from ref. 76 with permission from the Royal Society of Chemistry.

reversible specific capacity of 108 mAh g^{-1} at 354 mA g^{-1} over 100 cycles. The synthesized nanotubes showed a good cycle life and rate capabilities due to an increase in the specific surface area.[86] A similar improvement in performance was noticed for the 3D spider-web architecture assembled from NTO nanotubes prepared through the hydrothermal route as reported by Zhang *et al.*[87]

Another major reason for capacity degradation is due to the formation of the Ti-rich secondary phase, $Na_2Ti_6O_{13}$, during the synthesis of NTO. Thermodynamically, $Na_2Ti_6O_{13}$ is more stable than NTO as per the Gibbs free energy calculations mentioned in an earlier report.[88] The $Na_2Ti_6O_{13}$ phase was used previously as an anode for LIBs, which showed ~150 mAh g^{-1} at a high discharge/charge voltage of ~1.5 V. However, it showed a very poor performance when used as an anode for SIBs. In $Na_2Ti_6O_{13}$, three cubic sodium sites are available, in which two Na^+ ions are already filled and the reversible intercalation of the Na^+ ion into a third cubic site is limited as only 1/6th of Ti^{4+} can be reduced to Ti^{3+} during cycling.[78] Therefore, it exhibited only 49.5 mAh g^{-1} capacity at 1 C rate corresponding to the intercalation of <0.5 moles of Na + ions per formulae unit.[89] The reason for such low capacity is explained by Kataoka *et al.*, through neutron diffraction. They proposed that Na^+ ions are strongly bonded in $Na_2Ti_6O_{13}$, *i.e.*, they are located in the cubic sites with each ion having coordination of eight O^{2-} ions whereas Li^+ ions are bonded with only four O^{2-} ions with planar coordination.[90]

3.3 Sodium-ion Full Cell Based on Layered Cathode Materials

For practical applications, the electrochemical performance of various combinations of cathodes and anode materials in a full cell level have been demonstrated. Layered sodium transition metal oxides as cathodes were extensively used along with various anodes for their performance evaluation. The first sodium-ion full cell with a cathode layered P2-Na_xCoO_2 and an anode Na–Pb alloy composite has been proposed. The full cell demonstrated a good cycling performance, up to 300 cycles with a capacity retention of 50%.[14] Further improvement in performance, such as an increase in cell voltage and specific energy in the full cell has been achieved with the introduction of hard carbon, a potential anode material with low Na^+ ion intercalation potential (~0.2–0.25 V) and high specific capacity (~300 mAh g^{-1}).[91] Komaba *et al.* developed a 3 V sodium-ion full cell consisting of a layered O3-$Na(Ni_{0.5}Mn_{0.5})O_2$ cathode and hard carbon anode with different electrolytes that showed a high specific capacity of 200 mAh g^{-1}. The practical specific energy achieved in this full cell is 60% of LIBs.[92] A full cell with voltage >3 V has been developed by Kim *et al.* using a hard carbon anode and layered O3-$Na[Ni_{0.33}Mn_{0.33}Fe_{0.33}]O_2$ cathode, which exhibited a high specific capacity of 100 mAh g^{-1} that is stable up to

150 cycles.[93] A laminated battery developed by the Sumitomo Chemical Co. Ltd. using a O3-type $Na(Fe_{0.4}Mn_{0.3}Ni_{0.3})O_2$ cathode and hard carbon anode has showed good performance when charged in excess of 200% of the practical capacity with a voltage of 12 V.[94] A 2 Ah pouch cell fabricated using $Na_{0.9}[Cu_{0.22}Fe_{0.3}Mn_{0.48}]O_2$ as a cathode and soft carbon synthesized from anthracite as an anode, showed an excellent rate capability with an energy density of 100 Wh kg^{-1}.[95]

3.4 Summary

The low cost and high specific energy of sodium-ion batteries make them good alternative storage systems to lithium-ion batteries. Although extensive research and development activities are going on world-wide, its commercialization is far away due to the number of problems associated with its electrodes and electrolytes. Further research and development should be focused to identify suitable electrode materials, which would exhibit high specific capacity and prolonged life cycles. Layered sodium transition metal oxides are potential candidates as electrode materials for sodium-ion batteries owing to their high specific capacity. In this context, this chapter describes the electrochemical properties in terms of specific capacity, cycle life and the phase transformation of various layered sodium transition metal oxides and their derivatives. However, the cycle life still remains a challenge and needs improvement; hence, further modification of electrode materials and new concepts are needed to address this problem.

Acknowledgements

The authors would like to acknowledge financial support from the Technical Research Centre (TRC), Department of Science & Technology (TRC project: AI/1/65/ARCI/2014), Government of India. Authors also thank G. Padmanabham, Director, ARCI, for his encouragement and support for this work.

References

1. J. Y. Hwang, S. T. Myung and Y. K. Sun, *Chem. Soc. Rev.*, 2017, **46**, 3529–3614.
2. N. Yabuuchi, K. Kubota, M. Dahbi and S. Komaba, *Chem. Rev.*, 2014, **114**, 11636–11682.
3. C. Delmas, *Adv. Energy Mater.*, 2018, 1703137.
4. B. Dunn, H. Kamath and J. M. Tarascon, *Science*, 2011, **334**, 928–935.
5. M. D. Slater, D. Kim, E. Lee and C. S. Johnson, *Adv. Funct. Mater.*, 2013, **23**, 947–958.
6. M. Armand and J. Tarascon, *Nature*, 2008, **451**, 652–657.
7. H. Pan, Y. S. Hu and L. Chen, *Energy Environ. Sci.*, 2013, **6**, 2338.
8. Z. Dai, U. Mani, H. T. Tan and Q. Yan, *Small Methods*, 2017, 170098.
9. C. Delmas, C. Fouassier and P. Hagenmuller, *Phys. B + C*, 1908, **99**, 81–85.

10. M. H. Han, E. Gonzalo, G. Singh and T. Rojo, *Energy Environ. Sci.*, 2015, **8**, 81.
11. S. Guo, J. Yi, Y. Sun and H. Zhou, *Energy Environ. Sci.*, 2016, **9**, 2978–3006.
12. Y. Tsuchiya, K. Takanashi, T. Nishinobo, A. Hokura, M. Yonemura, T. Matsukawa, T. Ishigaki, K. Yamanaka, T. Ohta and N. Yabuuchi, *Chem. Mater.*, 2016, **28**, 19.
13. C. Delmas, J.-J. Braconnier, C. Fouassier and P. Hagenmuller, *Solid State Ionics*, 1981, **3/4**, 165–169.
14. L. W. Shacklette, T. R. Jow and L. Townsend, *J. Electrochem. Soc.*, 1988, **135**, 2669–2674.
15. H. Yoshida, N. Yabuuchi and S. Komaba, *Electrochem. Commun.*, 2013, **34**, 60–63.
16. Y. Ma, M. M. Doeff, J. Steven, J. De and C. Lutgard, *J. Electrochem. Soc.*, 1993, **140**, 2726–2733.
17. R. Berthelot, D. Carlier and C. Delmas, *Nat. Mater.*, 2011, **10**, 74–80.
18. J.-H. Cheng, C.-J. Pan, J.-F. Lee, J.-M. Chen, M. Guignard, C. Delmas, D. Carlier and B.-J. Hwang, *Chem. Mater.*, 2014, **26**, 1219–1225.
19. D. Carlier, J. H. Cheng, R. Berthelot, M. Guignard, M. Yoncheva, R. Stoyanova, B. J. Hwang and C. Delmas, *Dalton Trans.*, 2011, **40**, 9306–9312.
20. P. Yang, C. Zhang, M. Li, X. Yang, C. Wang, X. Bie, Y. Wei, G. Chen and F. Du, *ChemPhysChem*, 2015, **16**, 3408–3412.
21. X. Chen, X. Zhou, M. Hu, J. Liang, D. Wu, J. Wei and Z. Zhou, *J. Mater. Chem. A*, 2015, **3**, 20708–20714.
22. M. Matsui, F. Mizukoshi and N. Imanishi, *J. Power Sources*, 2015, **280**, 205–209.
23. P. Vassilaras, X. Ma, X. Li and G. Ceder, *J. Electrochem. Soc.*, 2013, **160**, A207–A211.
24. D. Yuan, X. Liang, L. Wu, Y. Cao, X. Ai, J. Feng and H. Yang, *Adv. Mater.*, 2014, **26**, 6301–6306.
25. D. S. Bhange, G. Ali, D.-H. Kim, D. A. Anang, T. J. Shin, M.-G. Kim, Y.-M. Kang, K. Y. Chung and K.-W. Nam, *J. Mater. Chem. A*, 2017, **5**, 1300–1310.
26. A. Gupta, C. B. Mullinsa and J. B. Goodenough, *J. Power Sources*, 2013, **243**, 817–821.
27. H. Yu, S. Guo, Y. Zhu, M. Ishida and H. Zhou, *Chem. Commun.*, 2014, **50**, 457–459.
28. S. Komaba, N. Yabuuchi, T. Nakayama, A. Ogata, T. Ishikawa and I. Nakai, *Inorg. Chem.*, 2012, **51**, 6211–6220.
29. D. D. Yuan, Y. X. Wang, Y. L. Cao, X. P. Ai and H. X. Yang, *ACS Appl. Mater. Interfaces*, 2015, **7**, 8585–8591.
30. J.-Y. Hwang, S.-T. Myung, D. Aurbach and Y.-K. Sun, *J. Power Sources*, 2016, **324**, 106–112.
31. J. P. Parant, R. Olazcuaga, M. Devalette, C. Fouassier and P. Hagenmuller, *J. Solid State Chem.*, 1971, **3**, 1–11.

32. C. Jiang, Z. Tang, S. Deng, Y. Hong, S. Wang and Z. Zhang, *RSC Adv.*, 2017, 7, 3746.
33. A. Mendiboure, C. Delmas and P. Hagenmuller, *J. Solid State Chem.*, 1985, **57**, 323–331.
34. X. Ma, H. Chen and G. Ceder, *J. Electrochem. Soc.*, 2011, **158**, A1307–A1312.
35. J. Billaud, R. J. Clément, A. R. Armstrong, J. Canales-Vázquez, P. Rozier, C. P. Grey and P. G. Bruce, *J. Am. Chem. Soc.*, 2014, **136**, 17243–117248.
36. R. Stoyanova, D. Carlier, M. Sendova-Vassileva, M. Yoncheva, E. Zhecheva, D. Nihtianova and C. Delmas, *J. Solid State Chem.*, 2010, **183**, 1372–1379.
37. A. Caballero, L. Hernan, J. Morales, L. Sanchez, J. Santos Pena and M. A. G. Aranda, *J. Mater. Chem.*, 2012, **12**, 1142–1147.
38. N. Yabuuchi and S. Komaba, *Sci. Technol. Adv. Mater.*, 2014, **15**, 043501.
39. Z. Lu and J. R. Dahn, *J. Electrochem. Soc.*, 2001, **148**, A1225–A1229.
40. K. Hemalatha, M. Jayakumar, P. Bera and A. S. Prakash, *J. Mater. Chem. A*, 2015, **3**, 20908–20912.
41. D. H. Lee, J. Xu and Y. S. Meng, *Phys. Chem. Chem. Phys.*, 2013, **15**, 3304–3312.
42. H. Wang, B. Yang, X.-Z. Liao and Z.-F. Ma, *Electrochim. Acta*, 2013, **113**, 200–2004.
43. Y. Wen, B. Wang, G. Zeng, K. Nogita, D. Ye and L. Wang, *Chem. – An Asian J.*, 2015, **10**, 661–666.
44. J. Xu, D. H. Lee, R. J. Clement, X. Yu, M. Leskes, A. J. Pell, G. Pintacuda, X.-Q. Yang, C. P. Grey and Y. S. Meng, *Chem. Mater.*, 2014, **26**, 1260–1269.
45. G. Singh, N. Tapia-Ruiz, J. Miguel Lopez del Amo, U. Maitra, J. W. Somerville, A. R. Armstrong, J. Martinez de Ilarduya, T. Rojo and P. G. Bruce, *Chem. Mater.*, 2016, **28**, 5087–5094.
46. S. Miyazaki, S. Kikkawa and M. Koizumi, *Synth. Met.*, 1983, **6**, 211–217.
47. S. Kikkawa, S. Miyazaki and M. Koizumi, *J. Power Sources*, 1984, **14**, 231–234.
48. N. Yabuuchi, H. Yoshida and S. Komaba, *Electrochem.*, 2012, **80**, 716–719.
49. X. Wang, G. Liu, T. Iwao, M. Okubo and A. Yamada, *J. Phys. Chem. C*, 2014, **118**, 2970.
50. S. Komaba, T. Nakayama, A. Ogata, T. Shimizu, C. Takei, S. Takada, A. Hokura and I. Nakai, *ECS Trans.*, 2009, **16**, 43–55.
51. N. Yabuuchi, M. Kajiyama, J. Iwatate, H. Mishikawa, S. Hitomi, R. Okuyama, R. Usui, Y. Yamada and S. Komaba, *Nat. Mater.*, 2012, **11**, 512–517.
52. K. Park, D. Han, J. K. Shon, S. G. Doo and S. Lee, *RSC Adv.*, 2015, **5**, 6340.
53. S. Kalluri, K. H. Seng, W. K. Pang, Z. Guo, Z. Chen, H.-K. Liu and S. X. Dou, *ACS Appl. Mater. Interfaces*, 2014, **6**, 8953–8958.
54. G. Singh, B. Acebedo, M. C. Cabanas, D. Shanmugaraj, M. Armand and T. Rojo, *Electrochem. Commun.*, 2013, **37**, 61–63.
55. X. Xia and J. R. Dahn, *Electrochem. Solid-State Lett.*, 2012, **15**, A1–A4.

56. J. J. Broconnier, C. Delmas and P. Hagenmuller, *Mater. Res. Bull.*, 1982, **17**, 993–1000.
57. S.-H. Bo, X. Li, A. J. Toumar and G. Ceder, *Chem. Mater.*, 2016, **28**, 1419–1429.
58. Y. Tsuchiya, A. M. Glushenkov and N. Yabuuchi, *ACS Appl. Nano Mater.*, 2018, **1**, 364–370.
59. J. J. Ding, Y. Zhou, Q. Sun and Z. Fu, *Electrochem. Commun.*, 2012, **22**, 85–88.
60. C.-Y. Yu, J.-S. Park, H.-G. Jung, K.-Y. Chung, D. Aurbach, Y.-K. Sun and S.-T. Myung, *Energy Environ. Sci.*, 2015, **8**, 2019–2026.
61. W. Luo, J.-J. Gaumet and L. Mai, *MRS Commun.*, 2017, 1–14.
62. M. Onoda, *J. Phys.: Condens. Matter*, 2008, **20**, 145205.
63. T. McQueen, P. Stephenes, Q. Huang, T. E. Klimczuk, F. Ronning and R. Cava, *Phys. Rev. Lett.*, 2008, **101**, 166402.
64. D. Hamani, M. Ati, J.-M. Tarascon and P. Rozier, *Electrochem. Commun.*, 2011, **13**, 938–941.
65. S. Komaba, C. Takei, T. Nakayama, A. Ogata and N. Yabuuchi, *Electrochem. Commun.*, 2010, **12**, 355–358.
66. M. Guignard, C. Didier, J. Darriet, P. Bordet, E. Elkaïm and C. Delmas, *Nat. Mater.*, 2013, **12**, 74–80.
67. P. Hagenmuller, A. Le Cerf and M. Onillon, *C. R. Acad. Sci.*, 1962, 255–928.
68. A. Maazaz, C. Delmas and P. Hagenmuller, *J. Inclusion Phenom.*, 1983, **1**, 45–51.
69. R. Shanmugam and W. Laiz, *ECS Electrochem. Lett.*, 2014, **3**, A23–A25.
70. H. Yu, Y. Ren, D. Xiao, S. Guo, Y. Zhu, Y. Qian, L. Gu and H. Zhou, *Angew. Chem., Int. Ed.*, 2014, **53**, 1–8.
71. Y. Wang, X. Yu, S. Xu, J. Bai, R. Xiao, Y.-S. Hu, H. Li, X.-Q. Yang, L. Chen and X. Huang, *Nat. Commun.*, 2013, **4**, 2365.
72. B. M. Shikano, C. Delmas and J. Darriet, *Inorg. Chem.*, 2004, **43**, 1214–1216.
73. M. Tamaru, X. Wang, M. Okubo and A. Yamada, *Electrochem. Commun.*, 2013, **33**, 23–26.
74. P. Rozier, M. Sathiya, A.-R. Paulraj, D. Foix, T. Desaunay, P.-L. Taberna, P. Simon and J.-M. Tarascon, *Electrochem. Commun.*, 2015, **53**, 29–32.
75. S. Song, M. Kotobuki, F. Zheng, C. Xu, N. Hu, L. Lu, Y. Wang and W.-D. Z. Li, *ACS Sustainable Chem. Eng.*, 2017, **5**, 4785–4792.
76. J. Xia, H. Zhao, W. K. Pang, Z. Yin, B. Zhou, G. He, Z. Guo and Y. Du, *Chem. Sci.*, 2018, **9**, 3421–3425.
77. J. Nava-Avendaño, A. Morales-García, A. Ponrouch, G. Rousse, C. Frontera, P. Senguttuvan, J. M. Tarascon, M. E. Arroyo-de Dompablo and M. R. Palacín, *J. Mater. Chem. A*, 2015, **3**, 22280–22286.
78. M. M. Doeff, J. Cabana and M. Shirpour, *J. Inorg. Organomet. Poly. Mater*, 2014, **24**, 5–14.
79. H. Pan, X. Lu, X. Yu, Y. S. Hu, H. Li, X. Q. Yang and L. Chen, *Adv. Energy Mater.*, 2013, **3**, 186–1194.

80. W. Zou, J. Li, Q. Deng, J. Xue, X. Dai, A. Zhou and J. Li, *Solid State Ionics*, 2014, **262**, 192–196.
81. J. Xu, C. Ma, M. Balasubramanian and Y. S. Meng, *Chem. Commun.*, 2014, **50**, 12564–12567.
82. M. Li, X. Xiao, X. Fan, X. Huang, Y. Liu and L. Chen, *J. Alloys Compds.*, 2017, **712**, 365–372.
83. M. Zarrabeitia, E. Castillo-Martínez, J. M. L. Del Amo, A. Eguía-Barrio, M. Á. Muñoz-Márquez, T. Rojo and M. Casas-Cabanas, *Acta Mater.*, 2016, **104**, 125–130.
84. H. S. Bhardwaj, T. Ramireddy, A. Pradeep, M. K. Jangid, V. Srihari, H. K. Poswal and A. Mukhopadhyay, *ChemElectroChem*, 2018, **5**, 1219–1229.
85. J. Chen, X. Zhou, C. Mei, J. Xu and C. P. Wong, *Electrochim. Acta*, 2017, **224**, 446–451.
86. W. Wang, C. Lin, Z. Yu, J. Hou, H. Zhu and S. Jiao, *Nanoscale*, 2013, **5**, 594–599.
87. Y. Zhang, L. Guo and S. Yang, *Chem. Commun.*, 2014, **50**, 14029–14032.
88. M. Holzinger, A. Benisek, W. Schnelle, E. Gmelin, J. Maier and W. Sitte, *J. Chem. Thermodyn.*, 2003, **35**, 1469–1487.
89. N. D. Trinh, O. Crosnier, S. B. Schougaard and T. Brousse, *ECS Trans.*, 2011, **35**, 91–98.
90. K. Kataoka, J. Awaka, N. Kijima, H. Hayakawa, K. I. Ohshima and J. Akimoto, *Chem. Mater.*, 2011, **23**, 2344–2352.
91. K. Wang, Y. Jin, S. Sun, Y. Huang, J. Peng, J. Luo, Q. Zhang, Y. Qiu, C. Fang and J. Han, *ACS Omega*, 2017, **2**, 1687–1695.
92. S. Komaba, W. Murata, T. Ishikawa, N. Yabuuchi, T. Ozeki, T. Nakayama, A. Ogata, K. Gotoh and K. Fujiwara, *Adv. Funct. Mater.*, 2011, **21**, 3859–3867.
93. D. Kim, E. Lee, M. Slater, W. Lu, S. Rood and C. S. Johnson, *Electrochem. Commun.*, 2012, **18**, 66–69.
94. S. Kuze, J. I. Kageura, S. Matsumoto, T. Nakayama, M. Makidera, M. Saka, T. Yamaguchi, T. Yamamoto and K. Nakane, *Sumitomo Kagaku*, 2013, **2013**, 1–13.
95. Y. Li, Y. S. Hu, X. Qi, X. Rong, H. Li, X. Huang and L. Chen, *Energy Storage Mater.*, 2016, **5**, 191–197.

CHAPTER 4

Ionic Liquid Electrolytes for Graphene-based Supercapacitors with an Ultrahigh Energy Density

CHUNBAO DU,*[a,b] YUAN CHENG,*[c] RONG XUE[a] AND XIAOLING HU[a]

[a] School of Natural and Applied Science, Northwestern Polytechnical University, Xi'an 710072, PR China; [b] College of Food Science and Engineering, Northwest A&F University, Yangling, Shaanxi 712100, PR China; [c] Institute of High Performance Computing, A*STAR, 138632, Singapore
*Email: duchunbao218@126.com; chengy@ihpc.a-star.edu.sg

4.1 Introduction

Energy is the driving force for the development of human society. Nowadays, the growing shortage of traditional fossil energy resources and the deterioration of the environment have vigorously promoted the development of other energies. In order to meet the growing energy demand, many devices have thus been designed, explored and fabricated for energy storage applications. A supercapacitor, different from conventional capacitors and secondary batteries, is a special device for electrochemical energy storage. Due to its high power density, long cycle life, high safety and excellent performance, the supercapacitor has shown great potential in the area of energy recycling and storage.[1–8] Moreover, the supercapacitor is a particular

Smart Materials No. 34
Layered Materials for Energy Storage and Conversion
Edited by Dongsheng Geng, Yuan Cheng and Gang Zhang

Published by the Royal Society of Chemistry, www.rsc.org

energy device with various properties of efficiency, cleanability and sustainability, which meet the up-to-date concept of green energy development. Differing from pseudocapacitors with a reversible chemical adsorption/desorption or redox reaction, the supercapacitor is based on the adsorption/desorption of the electrode and electrolyte. With the aforementioned advantages, supercapacitors have been widely used in many fields such as consumer electronics, green energy, military, transportation and industry, which are summarized as follows (see Figure 4.1). (1) In particular places such as offshore islands, remote mountain areas and sparsely populated grassland farms, wind and solar energies are the reliable way to deal with production and living energy.[9–11] However, these energies cannot be provided steadily. With the combination of a supercapacitor and wind power plant or solar cell to form hybrid power, the energy source can be stored by a supercapacitor through electric energy in full sun or strong wind. Then, discharging is conducted at night or in weak wind conditions to ensure a sustained energy supply. (2) An electron device always needs an instantaneous current that conventional batteries cannot meet.[12] Combined with a supercapacitor, the power supply could provide instantaneous power and long life. In particular, it is very economical and convenient due to its low cost and light weight. For example, by using a supercapacitor as the energy element of a flashlight, it only takes a few seconds to recharge with hundreds

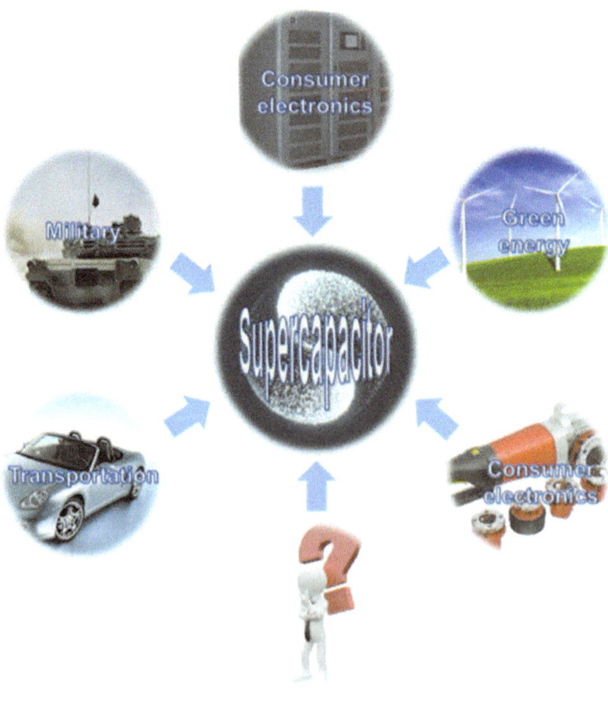

Figure 4.1 Application areas of supercapacitors.

of thousands of cycles. (3) The supercapacitor has also gained the attention of militarists due to its high power density, fast charging rate and long cycle life.[13] When applying supercapacitors to heavy trucks, tanks and armoured vehicles, these mechanisms can be started quickly. Also, electromagnetic catapults of aircraft carriers could launch warcraft within several seconds and the catapulted energy would reach hundreds of megajoules with the shortest cycle time of about forty seconds. Supercapacitors are a very good choice. (4) In order to improve the starting speed of an automobile, a supercapacitor can provide a large starting current for the running of the engine in combination with the accumulator.[14–17] Meanwhile, the lifetime of the accumulator can be prolonged greatly. Besides, the consumed energy in the braking process could be recycled by the supercapacitor for further utilization. (5) The use of the supercapacitor in industry mainly involves an uninterruptible power supply, safety protection equipment, instruments and apparatus.[18–20]

There are usually four parts in a supercapacitor including the electrode active material, electrolyte, diaphragm and current collector. As the major parts of a supercapacitor, the electrode material and electrolyte determine the performance, which can be clearly exhibited by the following equation.

The energy density (E) of the supercapacitor is calculated as follows:

$$E = \frac{1}{2}CV^2 \tag{4.1}$$

where C is the capacitance and V is the nominal voltage. It can be found that the energy density E of a supercapacitor depends on its capacitance C and nominal voltage V. Therein, voltage V is generally controlled by the stability of the electrolyte, while capacitance C is mainly determined by the electrode materials and electrode/electrolyte interface.

Common electrode materials include carbonaceous material, metal oxide (metal hydroxide), conductive polymer and composite material.[21–32] In comparison to the other electrode materials, carbonaceous materials have been widely used in supercapacitors due to their high specific surface area, good electron conductivity and excellent chemical stability. Various carbon materials such as activated carbon, carbon nanotubes (CNTs), nano-onion-like carbon and graphene, have been prepared and used as the electrodes of supercapacitors to improve energy density and retain high power density. It has been found that there are a significant number of micropores in conventional activated carbon. However, these micropores are not easily accessible to electrolyte molecules, even very small molecules such as KOH. Unlike activated carbon materials with a distribution of pores, graphene depends on layers, which grant it a high effective specific surface. So, as a promising electrode material, graphene, with its unique two-dimensional structure, possesses the specific surface area of 2630 $m^2 g^{-1}$, which is twice as large as single-walled carbon nanotubes (SWNTs). Stoller and co-workers pioneered supercapacitors based on chemically modified graphene.[33] Thereafter, graphene has attracted a lot of attention for application as a supercapacitor electrode material. Moreover, graphene has shown excellent conductive thermal conductivity and high electron mobility, and the surface of graphene could form a double electrode layer, which is useful to diffuse electrolytes for good capacitive performance.

Besides the electrode materials, the electrolyte also exhibits a great effect on the performance of supercapacitors, which includes the following three aspects (see Figure 4.2). First, the steric effect of ions in the electrolyte hinders the diffusion of ions. Second, the interactions between the cation and anion hinder the diffusion of ions. Third, although ions' solvation could weaken the interactions between the cation and anion, solvated ions do not easily enter into the micropores of electrodes.

Generally speaking, the electrolyte of a supercapacitor should have the properties of a wide working voltage window, high electrochemical stability, large ionic concentration, small solvated ionic radius, low specific resistance and high purity. Moreover, low viscosity, volatility, toxicity and cost are also required. However, there is no electrolyte that could meet all of the above properties simultaneously. Common electrolytes used in supercapacitors include the aqueous electrolyte, organic electrolyte and ionic liquid electrolyte.

Aqueous electrolytes (Table 4.1, Figure 4.3) have the merits of high ion concentration, small ion radius and low internal resistance. Aqueous electrolytes include three categories, namely acid electrolytes, alkaline electrolytes and neutral electrolytes. With high ionic conductivity, H_2SO_4 is the most commonly used acid electrolyte, which shows an ionic conductivity of 0.8 $S\,cm^{-1}$ at 25 °C with 1 $mol\,L^{-1}$.[34] However, due to strong causticity, H_2SO_4 can easily corrode the electrode materials and shell, which would lead to the poor performance of a supercapacitor. As for the alkaline electrolyte, KOH is the most commonly used electrolyte, which shows an ionic conductivity of 0.6 $S\,cm^{-1}$ with a concentration of 6 $mol\,L^{-1}$ at 25 °C.[35] Certainly, NaOH and LiOH aqueous solution also could be used as an alkaline electrolyte in supercapacitors.[36] Nevertheless, an alkaline electrolyte leaks easily,

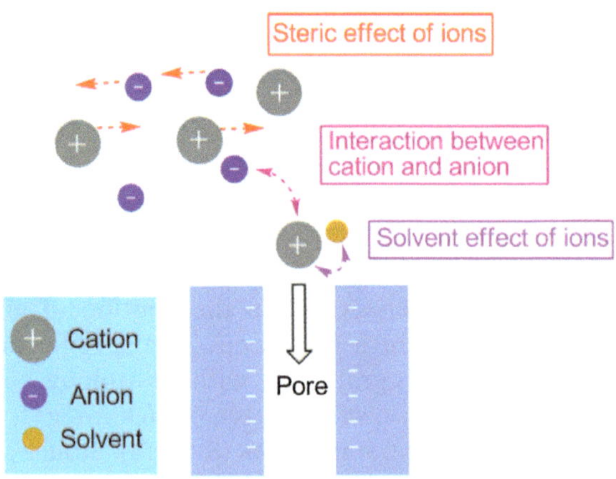

Figure 4.2 Electrolyte interactions scheme of a supercapacitor.

Table 4.1 Comparison of three types electrolytes used in supercapacitors.

Electrolytes	Electrochemical stable window (V)	Viscosity	Toxicity	Cost
Aqueous electrolyte	≤1.2	Low	Low	Low
Organic electrolyte	2.5–2.8	Medium/high	Medium/high	Medium/high
Ionic liquid electrolyte	2–6	High	Low	High

which has limited its extensive use. Neutral electrolytes mainly include sodium, sylvite and lithium salt, and Na_2SO_4 is the most commonly used neutral electrolyte.[37] Compared with the acid and alkaline electrolyte, the operating voltage of a neutral electrolyte is relatively high. Generally speaking, the gross defect of the aqueous electrolyte is the low operating voltage (<1.2 V), making it very difficult to improve the energy density and power density.

Although most academic research about supercapacitors is focused on the aqueous electrolyte, the leading electrolyte in the commercial market is the organic electrolyte with an operating voltage of 2.5–2.8 V (Table 4.1). An improvement of the operating voltage could increase the energy density and power density of supercapacitors. As shown in Figure 4.3, commonly used organic electrolytes are a mixture of conducting salts (tetraethylammonium tetrafluoroborate, triethylmethylammonium tetrafluoroborate, *etc.*) and organic solvents (acetonitrile, tetrahydrofuran, cyclohexane, *etc.*).[38–40] Until now, the most critical question for organic electrolytes in supercapacitors is the problem of safety, such as flammability, volatility and toxicity.

Compared with aqueous and organic electrolytes, ionic liquid-based electrolytes have gained more attention because their wide voltage window can be extended to over 3 V.[41] Ionic liquids are commonly defined as substances that are composed of cations and anions with a melting point below 100 °C.[42] Despite this definition, the family of ionic liquids is extremely broad. Now, the working voltage, temperature range, equivalent series resistance and ionic conductivity of ionic liquids can meet the requirements of supercapacitors because many potential cations and anions can be designed and prepared. It was found that ionic liquids can be synthesized from different inorganic and organic species. Figure 4.3 shows the most commonly used cations (imidazolium, pyridinium, quaternary ammonium and phosphonium *etc.*) and anions (hexafluorophosphate, tetrafluoroborate, chloridion *etc.*) with different hydrophobicity and hydrophilicity.[43]

Ionic liquids have aroused increasing interest for use in supercapacitors due to their unique properties such as a wide electrochemical potential window, high ionic conductivity, wide liquid range, and nonflammability, *etc.* Compared with the other two types of electrolytes, ionic liquids have been widely studied as an electrolyte in supercapacitors to take advantage of their wider electrochemical window and high ionic conductivity. As shown in Table 4.1, the electrochemical window of ionic liquids can reach up to 6 V,

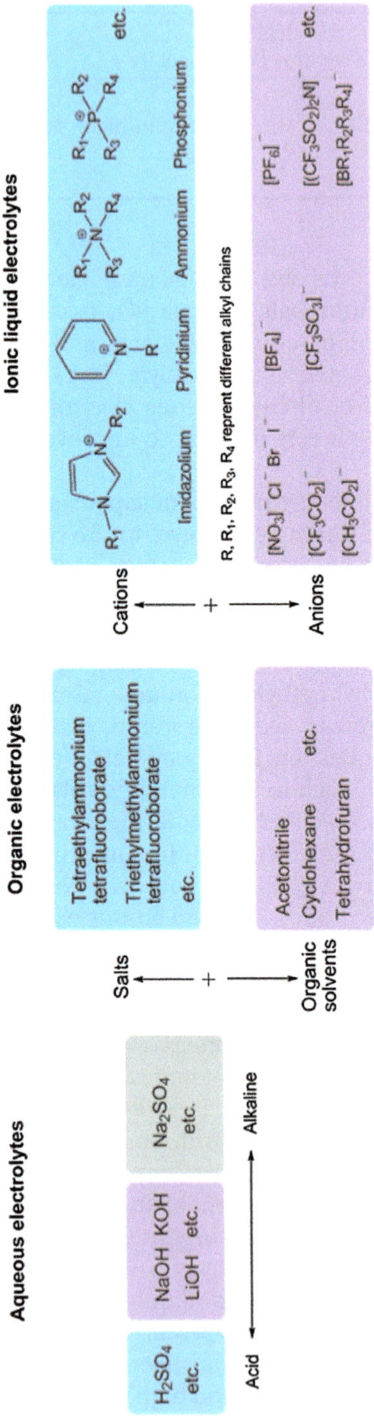

Figure 4.3 The chemical structures of typical electrolytes.

which is much higher than that of aqueous and organic electrolytes.[44] Most importantly, compared with the other electrolytes, ionic liquids electrolytes have low toxicity and are therefore very easy and safe to use.

In this chapter, the development of graphene-based electrodes with ionic liquid as an electrolyte in supercapacitors will be highlighted and discussed. We first review the functionalization and modification methods of graphene. In the second part, the development of ionic liquid electrolytes in graphene-based supercapacitors is summarized, namely single ionic liquid, mixed ionic liquid and ionic liquid/polymer as electrolytes. The chemical structures of cations and anions of ionic liquids used in graphene-based supercapacitors are presented in Tables 4.2 and 4.3. Moreover, the interface investigation of a graphene electrode and ionic liquid electrolyte is also analyzed. Finally, ionic liquid electrolytes for other 2D-based supercapacitors, and supercapacitors for the next generation of wearable and portable electronic devices are summarized and future prospects are discussed.

Table 4.2 Cations of ionic liquids used in graphene-based supercapacitors.

Name	Abbreviation	Chemical formula	References
1-Butyl-1-methylpyrrolidinium bis(fluorosulfonyl)imide	P_{14}	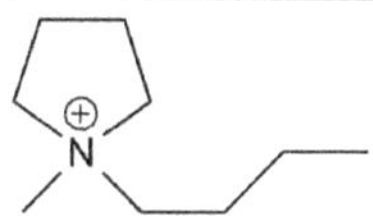	45, 55
1-Butyl-3-methylimidazolium	BMIM	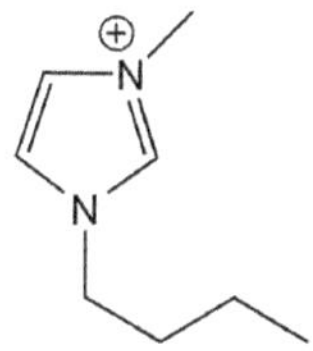	45, 56, 68, 75, 80, 82–84, 87
1-Ethyl-3-methylimidazolium	EMIM	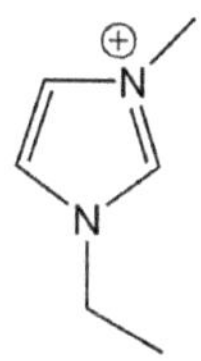	50–52, 54, 57, 59, 61, 67, 72, 76, 77, 78, 88, 89
1-Methyl-1-propylpiperidinium	PP_{13}	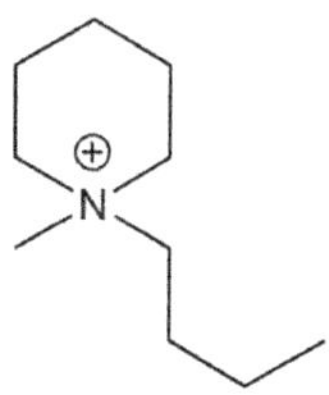	79, 89
1-Methyl-1-propylpiperidinium	PP_{14}	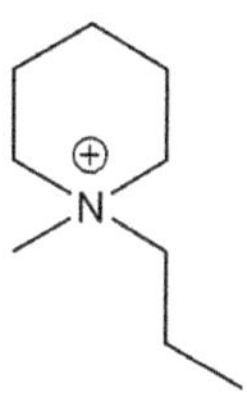	63, 79

Table 4.2 (*Continued*)

Name	Abbreviation	Chemical formula	References
Tetraethylammonium	N_{2222}	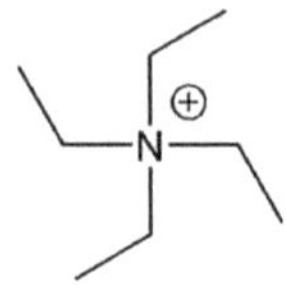	64, 72
Tetramethylammonium	N_{1111}		77
Octyl bis(3-methylimidazolium)	[C_8H_{16}BMIM]	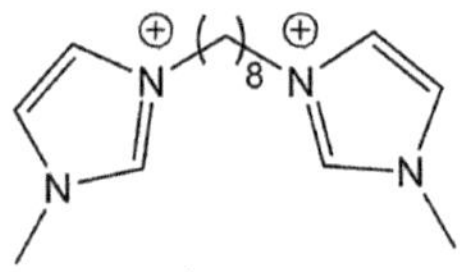	69, 73
1-Aminopropyl-3-methylimidazolium	[APMIM]	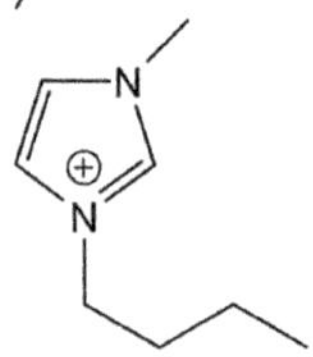	71, 119
Trihexyl (tetradecyl) phosphonium	P_{66614}	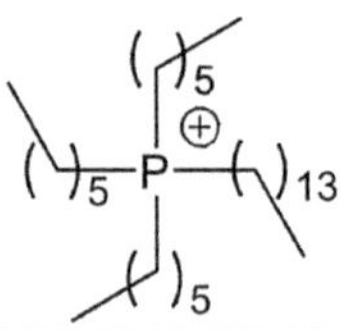	93

Table 4.3 Anions of ionic liquids used in graphene-based supercapacitors.

Name	Abbreviation	Chemical formula	References
Bis(trifluoromethylsulfonyl)-imide	NTf_2	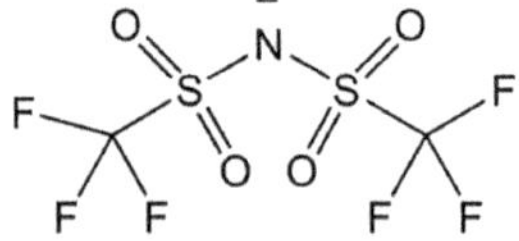	45, 56, 58, 67, 72, 76, 88, 89, 93, 117
Hexafluorophosphate	PF_6		49, 70, 75
Tetrafluoroborate	BF_4	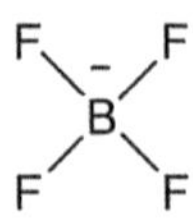	51, 52, 54, 57, 59, 60–62, 64, 68, 72, 77, 80, 81, 83, 85, 87, 89

Table 4.3 (*Continued*)

Name	Abbreviation	Chemical formula	References
Dicyanamide	$N(CN)_2$	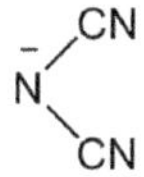	55
Iodide	I	I^-	69, 73
Bromide	Br	Br^-	71, 119
Bis(fluorosulfonyl)imide	FSI	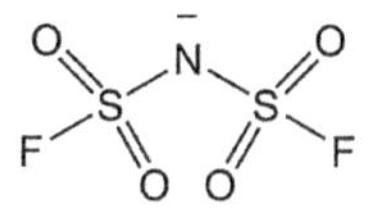	79
Thiocyanate	SCN	$N{\equiv}C{-}S^-$	116

4.2 Graphene-based Electrode Materials

In 2008, Vivekchand and co-workers pioneered graphene-based supercapacitors with an ionic liquid 1-butyl-1-methylpyrrolidinium bis-(trifluoromethanesulfonyl)imide ($[P_{14}][NTf_2]$) electrolyte.[45] They synthesized three types of graphene with different porosities by using three different methods, namely the thermal exfoliation of graphitic oxide, heating a nanodiamond at high temperature in a helium atmosphere, and decomposition of camphor over nickel nanoparticles.[46–48] The former two methods could grant graphene a higher specific capacitance of 75 F g^{-1} than the third method, which was also superior to single-wall or multi-walled carbon nanotubes (MWCNTs). Therefore, the performance of the graphene electrode has a great relationship with the quality, in terms of specific surface area. Compared with the H_2SO_4 electrolyte (1 V), the operating voltage of supercapacitors with $[P_{14}][NTf_2]$ as an electrolyte could be extended to 3.5 V. Moreover, the value of the maximum energy density of a graphene-$[P_{14}][NTf_2]$ supercapacitor was 31.9 Wh kg^{-1}, while it was only 17.0 Wh kg^{-1} for a graphene-H_2SO_4 supercapacitor. Therefore, it is clear that ionic liquids as electrolytes of graphene-based supercapacitors have a significant advantage. Henceforth, this work has opened the way for graphene and ionic liquids in the applications of supercapacitors.

However, current graphene-based supercapacitors with ionic liquids as electrolytes still face problems with graphene sheets' stacking, lower porosity, high ionic migration resistance, and low effective specific surface area (see Figure 4.4). At this time, it is not easy to obtain large-batch graphene sheets for use in supercapacitors. Compared with graphene, reduced graphene oxide (rGO) is easily obtained, so rGO is often fabricated with ionic liquid into supercapacitors. However, due to the high oxygen content in rGO, the performance of rGO-based supercapacitors is heavily discounted. In order to enhance the charge transfer and wetting capacity of electrode surfaces, an efficient route to fabricate electrochemically rGO films with lower

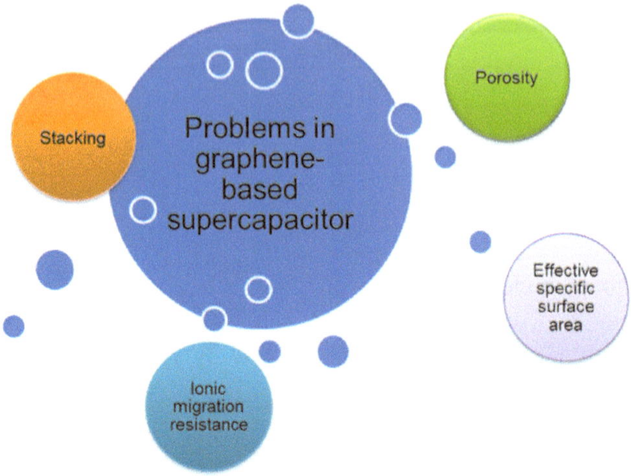

Figure 4.4 Problems that graphene needs to resolve.

oxygen content is still a challenge. Besides, other methods including functionalized graphene and hybrid graphene electrodes with heteroatoms, carbon and polymer are the current research trend.

4.2.1 Functionalized/Functional Graphene Electrodes

Owing to the defection of graphene itself, rGO is a relatively good substitute. To improve the dispersion of rGO in water, Chen and co-workers have fabricated partial rGO using an acid weak reductant, hydrobromic acid, with GO colloids.[49] The results showed that 2–3 layers of graphene could be clearly observed by using high-resolution transmission electron microscopy (HRTEM). The authors found that some residual oxygen groups of rGO can give the rGO electrode additional pseudocapacitance. With 1-butyl-3-methylimidazolium hexafluorophosphate ([BMIM][PF_6]) as the electrolyte, the maximum capacitance value of the supercapacitor was 158 $F\,g^{-1}$ at a current density of 0.2 $A\,g^{-1}$. Moreover, the residual oxygen groups of rGO not only had a significant impact on the capacitance but also on the life time performance.

When using chemical reagent reduction to obtain rGO, it usually involves dangerous and toxic reduction reagents such as hydrazine (N_2H_4), sodium borohydride ($NaBH_4$) and hydroiodic acid (HA). So, it is critical to find facile and green reduction reagents. Jeon and co-workers reported a facile synthesis of rGO and found that Zn could reduce graphene oxide (GO) to rGO, which also showed a mesoporous structure.[50] In their research, GO was dispersed in the aqueous phase and then acidized by hydrochloric acid (HCl). After added Zn power, the Zn/GO mixture was given ultrasonic treatment. After being filtrated, washed and dried, the obtained mesoporous rGO exhibited the specific capacitance of 104.3 F g^{-1} at 1 A g^{-1} with 1-ethyl-3-methylimidazolium bis(trifluoromethylsulfonyl)imide ([EMIM][NTf_2]) as the electrolyte. After 5000 cycles, the specific capacitance only decreased 3% and the improved performance was attributed to the significant mesopores, which were helpful for the mass transport of [EMIM][NTf_2].

Because of the special structure of ionic liquid, the performance of ionic liquid electrolyte-based supercapacitors is often limited, caused by the slow diffusion of big ions in the narrow pores. Thus, it is necessary to develop mesoporous (2–50 nm) or macroporous (>50 nm) graphene electrodes to improve the performance. Liu *et al.* developed a graphene-based supercapacitor using 1-ethyl-3-methylimidazolium tetrafluoroborate ([EMIM][BF_4]) as the electrolyte to achieve a high energy density with 85.6 Wh kg^{-1} at room temperature.[51] They developed curved graphene sheets with a pore size of 2–25 nm that could avoid restacking once the curved graphene sheets were compressed into the electrode. Thus, when fabricated with [EMIM][BF_4] into a supercapacitor, the high tolerable voltage of [EMIM][BF_4] (>4 V) could be fully utilized. Although the viscosity of the ionic liquid [EMIM][BF_4] was much higher than the organic and aqueous electrolyte, the mesoporous structure of the curved graphene sheets still exhibited a high power density. The curved morphology of graphene can ensure the accessibility and wettability of [EMIM][BF_4] with mesoporosity.

In another study, Pope and co-workers designed and prepared a rGO/ionic liquid composite as the electrode of a supercapacitor that showed a gravimetric capacitance of 156 F g^{-1}, energy density of 17.5 Wh kg^{-1} and operating voltage of 3 V, respectively.[52] To obtain the rGO/ionic liquid composite, GO was dispersed in a water-ethanol-ionic liquid ([EMIM][BF_4]) mixture to form colloidal gels. Then the colloidal gel was drop-cast onto the Pt disks. Due to the non-volatile property of [EMIM][BF_4], the volatile components (water and ethanol) can be evaporated at room temperature overnight. By controlling the reduction temperatures between 200 and 350 °C, GO was reduced to improve its electrical conductivity. [EMIM][BF_4] in the composite electrode, was not only used as a working electrolyte, but also as the spacer to prevent the stacking of graphene sheets. So, the [EMIM][BF_4] electrolyte-accessible surface area can be increased greatly. Moreover, the authors found that the capacitance of the supercapacitor can still be kept above 90% at a scan of 500 mV s^{-1} by the increase of the degree of thermal reduction and the content of ionic liquid. In addition, it was also found that if the energy density of the supercapacitor was not the primary goal, the electrodes could be tailored toward higher power.

Recently, Li and co-workers obtained a facile graphene-like active carbon, which has rational pore structures, a large surface area and good conductivity by using chemically exfoliating biomass.[53] They first "drilled" holes of biomass with H_2O_2 and then graphene-like nanosheets were obtained after a treatment of acetic acid. After that, the graphene-like nanosheets were carbonized at a high temperature for highly activated carbon, which possesses a unique pore structure and large surface area. The results showed that the activated graphene-like carbon (specific capacitance with 340 $F g^{-1}$ at 0.5 $A g^{-1}$) exhibited a specific energy density with 23.33 to 16.67 $W h kg^{-1}$. It is worth mentioning that exploring carbons from bio-wastes is low cost, which possesses the potential for large-scale production.

In summary, functionalized graphene and functional graphene are the easiest ways to change the properties of a graphene electrode. Combined with ionic liquid as the electrolyte, more effort is needed to improve the accessibility between graphene and the ionic liquid.

4.2.2 Graphene/Heteroatom Electrodes

To improve the capacitance of graphene-based supercapacitors, many scholars have employed the heteroatom strategy. This can be achieved by minimizing the adverse impact of graphene's high level of porosity and low packing density, so many materials can be applied for further improving the electrochemical performance of graphene. For example, the doping of nitrogen atoms, namely the introduction of nitrogen atoms into the graphene skeletons, can change the surface functionality of a graphene-based electrode, such as the surface wettability, electron distribution and electroactive surface area of graphene. For example, Wang and co-workers utilized a facial approach with confined polymerization and a carbonization strategy for a highly crumpled nitrogen-doped graphene-based supercapacitor with gravimetric capacitances and energy density of 128 $F g^{-1}$ and 56 $Wh kg^{-1}$, respectively.[54] They found that this strategy can avoid the restacking of graphene nanosheets and thus increase the ion-accessible surface area, ion transfer rate and packing density, which leads to the realization of high energy density. The nitrogen functional groups on the surface of two-dimensional graphene nanosheets can facilitate the wetting of a [EMIM][BF_4] electrolyte on the graphene nanosheets, which then further increases the accessible surface area. In addition, nitrogen-doping also endows the graphene nanosheets' surface with hydrophilic properties that allows [EMIM][BF_4] to wet the previously inaccessible pores.

Although the specific power of a supercapacitor can be effectively improved, the specific energy is still a problem. To deal with this, Ma and co-workers prepared a N-doped graphene electrode mixed with mixed-valence manganese oxide (MnO_x, $x = 2/3$, 3/4, 1, 2) to improve the performance of the supercapacitor. MnO_x was synthesized by adding $Mn(CH_3COO)_2 \cdot 4H_2O$ into ethylene glycol and polyvinyl-pyrrolidone followed by a series of simple post-processing.[55] Then an aerogel synthesis method was used to obtain a MnO_x/N-doped reduced graphene oxide (MnO_x/N-rGO) electrode. Typically, GO was dispersed in the MnO_x solution with ethylene glycol as the solvent. Thereafter, hydrazine hydrate was added and the mixed solution was conducted at 80 °C for three days to obtain a hydrogel. Thus, the MnO_x/N-rGO electrode with a diameter of <20 nm was obtained after the freeze-drying of hydrogel. In contrast to other research using pure ionic liquid as an electrolyte, they prepared a hybrid electrolyte by adding potassium ferrocyanide ($K_4[Fe(CN)_6]$) into 1-butyl-1-methylpyrrolidinium dicyanamide ($[P_{14}][N(CN)_2]$). The as-fabricated supercapacitor showed a high specific energy of 44.7 Wh kg^{-1}. The author clarified the charge storage mechanism of MnO_x in the MnO_x/N-rGO electrode by using *in situ* electrochemical X-ray absorption spectroscopy (XAS). The results showed that the notable increase in surface capacitive activity of this hybrid electrolyte was attributed to an additional reversible faradaic capacitive, which was caused by the presence of $K_4[Fe(CN)_6]$ on the surface of MnO_x/N-rGO.

4.2.3 Graphene/Carbon Hybrid Material Electrodes

Although the introduction of the heteroatom effect could effectively change the surface wettability and electrical conductivity of graphene-based electrodes in some way, a detailed investigation of the pore and heteroatom effect of graphene-based electrodes is somewhat difficult because it is hard to control the pore and heteroatom effect. However, the hybridization of graphene and carbonaceous material is a good strategy to optimize the pore effect.

Tamailarasan and Ramaprabhu first studied CNT/graphene/ionic liquid ternary electrode-based supercapacitors by using functionalized multiwalled carbon nanotubes (f-MWNTs), hydrogen exfoliated graphene (HEG) and 1-butyl-3-methylimidazolium bis(trifluoromethylsulfonyl)imide ($[BMIM][NTf_2]$).[56] The f-MWNTs were used as a spacer for the transfer of the electron between the graphene sheets. $[BMIM][NTf_2]$ with hydrophobic property ensured the wide potential window of 3.5 V, thus greatly increasing the electrolyte accessibility and enlarging the directional ion transport paths. With the same ionic liquid $[BMIM][NTf_2]$ as the electrolyte, the supercapacitors showed a specific capacitance of 201 F g^{-1}, energy density of 171 Wh kg^{-1} and power density of 148.43 kW kg^{-1}. The f-MWNTs can be seen as a bridge as well as a spacer for effective electron transfer between graphene. Similarly, hydrophobic ionic liquid in the electrode was not only utilized as the continuity of the ion diffusion path but also had a positive role in enhancing the electrolyte's accessibility.

Lei and co-workers have designed and prepared a graphene/carbon-based hybrid electrode by using graphene oxide (GO) and mesoporous carbon CMK-5 platelets through electrostatic interactions.[57] After thermal annealing, GO was easily reduced to reduced GO (RGO), and a hierarchical carbon nanostructure can be obtained in RGO-CMK-5. There was more exposed surface area on the separated RGO, which was favorable for the formation of electrical double layer capacitance. Moreover, the introduction of CMK-5 platelets served as a highway for the fast transport of the $[EMIM][BF_4]$ ionic liquid electrolyte. The RGO-CMK-5 based supercapacitor showed a specific capacitance of 144.4 F g^{-1}, energy density of 60.7 W h kg^{-1} and power density of 174 W kg^{-1} with a charged/discharged operating voltage of 3.5 V.

The supercritical CO_2 method is an ideal way to prepare nanomaterials because CO_2 is non-toxic, low cost and inflammable. To prevent stacking of the graphene nanosheets, Lee and co-workers used ($SCCO_2$) to prepare a graphene/α-MnO_2 hybrid electrode.[58] Due to the strong penetration ability and extremely low viscosity, $SCCO_2$ can disperse oxide nanorods and debundle the graphene nanosheets effectively. With the optimization of the electronic and ionic conducting routes in the hierarchical structure, they found that the graphene/α-MnO_2-based supercapacitor exhibited a specific capacitance of 230 $F\,g^{-1}$ with [EMIM][NTf_2] as the electrolyte.

Aerogel has a special structure such as additional macropores that are very promising for fabricating supercapacitors. As might be imagined, these macropores are helpful for the wettability between the aerogel electrode and electrolyte. Graphene aerogel has a three-dimensional macroscopic structure and is easy to assemble into macroscopic devices. In that sense, Shao and co-workers prepared a graphene-based aerogel electrode to enhance the capacitance of a supercapacitor.[59] Graphene was spaced by CNTs to form an aerogel electrode through a hydrothermal method. The added CNTs can improve the specific surface area and capacitance of the aerogel electrode. They found that the macropores created by the CNTs not only improved the wettability between the aerogel electrode and [EMIM][BF_4] electrolyte, but also facilitated the ion adsorption. Compared to graphene-based aerogels without CNTs, the CNT/graphene aerogels exhibited a higher specific capacitance of 183.3 $F\,g^{-1}$ and energy density of 80 $Wh\,kg^{-1}$.

Duan and co-workers advanced the field by the integration of 1D CNTs with 2D graphene nanosheets for a 3D sandwich-structured nanohybrid paper with abundant interconnected pores.[60] Here, CNTs with graphene were functionalized by hydrophilic (1-butyl-3methylimidazolium tetrafluoroborate) ([BMIM][BF_4]), which offered a high specific surface area for efficient and stable loading of polyaniline (PANI). In the 3D electrode, the integration of multiple components led to a synergic effect, which endows high capacitance and good cycling stability.

To lower the cost and enlarge the application range, the use of natural materials to fabricate graphene/carbon hybrid electrodes is very promising. With this in mind, Bozym and co-workers designed and prepared a graphene/carbon hybrid electrode with the addition of dehydrated sucrose (DS) and [EMIM][BF_4].[61] They first prepared DS particles *via* the dehydration of sucrose. Then, DS particles were mixed with graphene oxide with tip sonication. After addition of [EMIM][BF_4], the suspensions were stirred and drop-cast onto platinum pucks to obtain DS-GO-[EMIM][BF_4] electrodes. The authors found that DS nanoparticles were absorbed on the surface of GO, which provided a physical barrier to inter-sheet collapse. Thus, DS nanoparticles can prevent the restacking of GO due to van der Waals forces and improve the gravimetric capacitance.

Even under a high current, the ion transportation of ionic liquid might be confined by the ions' large size. Zhang and co-workers prepared hierarchical porous carbon spheres/graphene (CSG) and investigated the electrochemical performance in a [BMIM][BF_4] electrolyte.[62] They first prepared CS by using a hydrothermal treatment and then CS and poly(dimethyl diallyl ammonium chloride) (PDDA) were added in a GO dispersion. After the mixture was maintained in a hydrothermal reactor for 4 hours at 180 °C, a CSG electrode was obtained. They found that CS had a submicron size so it was used as the spacer of GO. The improved specific surface area of GO in CSG was attributed to the reduction of CS and there were more pores for the big ions of [BMIM][BF_4] to transport. CSG exhibited a specific capacity of 280.0 F g^{-1} and the supercapacitor showed an energy density of 87.5 Wh kg^{-1} in [BMIM][BF_4]. After 5000 cycles, CSG still maintained a specific capacity of 90.1% and good stability, which was due to the novel hierarchical porous structure.

If the electrochemical window of an ionic liquid electrolyte in a symmetric supercapacitor cannot be fully utilized, the performance of the supercapacitor also might not be displayed sufficiently. Li and co-workers have found that unequal electrodes could make full use of the electrochemical window of the electrolyte.[63] They optimized the voltage distribution on two SWNT/rGO (SWNT: rGO = 1 : 4, w/w) electrodes through adjusting the mass ratio of SWNT/rGO on the positive and negative electrode. The results showed that when the mass ratio of the positive and negative electrode was set at 1.5, the energy density of the supercapacitors was enlarged to 118 Wh kg^{-1} and the operating voltage could reach 4.7 V with ionic liquid 1-methyl-1-propylpiperidinium bis (trifluoromethyl sulfonyl) imide ([PP_{14}][NTf_2]) as the electrolyte. In contrast, the symmetric supercapacitor only showed the largest working voltage of 4.0 V. Here, the SWNT/rGO electrode was easy to obtain by dispersing SWNTs and GO in an aqueous solution of sodium dodecyl benzenesulfonate (SDBS) and reducing by hydrazine.

4.2.4 Graphene/Polymer Hybrid Material Electrodes

Due to the strong π–π interaction between graphene interlayers, the effective surface of graphene would be lost with the inevitable aggregation and restacking. Fu and co-workers developed a facile and low-cost strategy by using aniline monomer intercalating into the interlayers of expanded expandable graphite.[64] After *in-situ* polymerization of the aniline monomer, 2D quasi-ordered nitrogen-enriched porous carbon nanohybrids were obtained and showed a high energy density of 95.7 Wh kg^{-1} at a power density of 449.7 W kg^{-1} in ionic liquid electrolyte [N_{2222}][BF_4]-propylene borate ([N_{2222}][BF_4]-PC).

Another facile strategy is to utilize the conductivity of some polymers. Zhang and Zhao reported a method to fabricate the high performance of conductive polymer-reduced graphene oxide-based supercapacitor electrodes.[65] The results indicated that a superior performance can be observed by direct coating PEDOT, PANI, or PPY on reduced graphene oxide sheets because of the synergic effect between the two components.

Because of the advantages of ionic liquids as electrolytes, there is still one challenge to achieve graphene-based electrolyte materials, which could be wetted with a suitable ionic liquid. In recent years, it was found that polymeric/polymerized ionic liquids or poly(ionic liquid)s (PILs) exhibited potential application in supercapacitors. PILs not only have the unique properties of ionic liquids, but also possess the flexibility of macromolecular architectures, which revealed novel properties and functions.

In order to ensure compatibility of the ionic liquid electrolyte with the electrode, it is an effective strategy to modify the electrode with poly(ionic liquid)s (PILs). PILs are usually obtained by the polymerization of ionic liquid monomers, which include five types with unsaturated double bonds on cations or anions. Figure 4.5 illustrates the general synthetic route to PILs from ionic liquid monomers, which is similar to common ionic liquids.[66]

Kim *et al.* developed poly(1-vinyl-3-ethylimidazoliumbis(trifluoromethyl-sulfonyl)amide) modified rGO *via* electrostatic and cation–π interactions.[67] They found that PIL-modified reduced graphene oxide (PIL:rGO) could provide enhanced compatibility with a [EMIM][NTf_2] electrolyte and improved accessibility of ionic liquid electrolyte ions into the graphene electrodes. The supercapacitor assembled using the PIL:RGO electrode and [EMIM][NTf_2] electrolyte exhibited a stable electrochemical response. The experiment results showed that the operating voltage of the supercapacitor could reach up to 3.5 V and a maximum energy density of 6.5 Wh kg^{-1} can be obtained at a power density of 2.4 kW kg^{-1}.

Mao and co-workers have reported, for the first time, a new concept of graphene/polyfluorene imidazolium ionic liquid hybrid electrodes for a high-performance supercapacitor.[68] As shown in Figure 4.6, they designed and prepared two conjugated polyfluorene imidazolium homo-polymer (hoPILs) and co-polymer (co-PILs). Then, hoPILs and co-PILs were mixed with graphene oxide (GO) aqueous solution to intercalate GO. After this step,

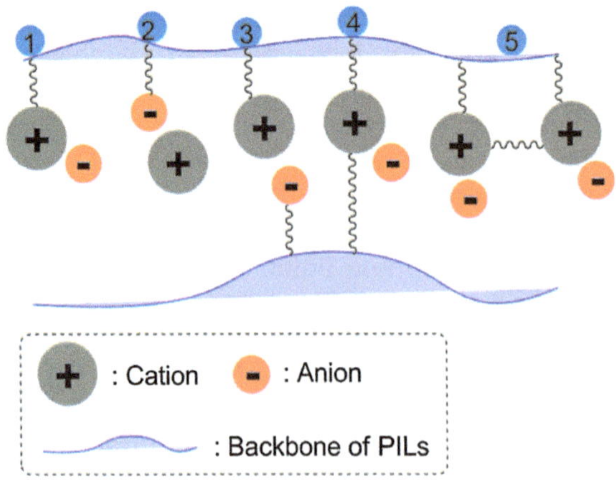

Figure 4.5 Basic ionic liquid monomers in PILs that have been found.

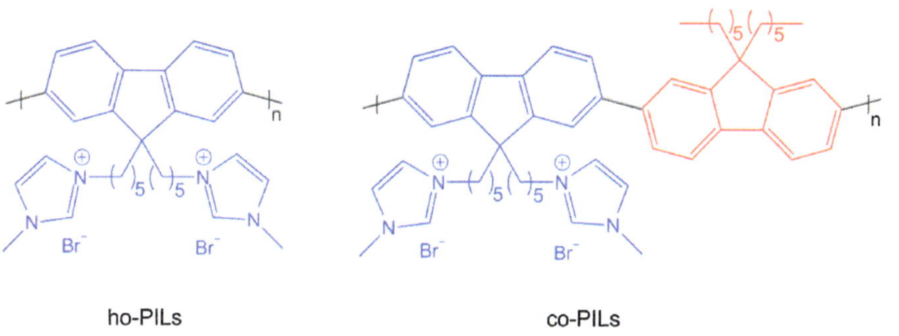

Figure 4.6 The chemical structures of ho-PILs and co-PILs.

hydrazine monohydrate was added to get PILs-intercalated rGO at 80 °C in a vacuum oven. Here, conjugated PILs were rigid and had a planar backbone and methylimidazolium bromide moieties. Compared with the other molecules, both ho-PILs and co-PILs had strong interaction with the RGO sheets because of their electron-delocalized planar backbone. Moreover, PILs were easily adsorbed on the surface of rGO sheets due the existence of strong π–π interactions. Meanwhile, the wettability of RGO was greatly improved by the presence of hydrophilic imidazolium groups. When assembled with 1-butyl-3-methylimidazolium tetrafluoroborate ([BMIM][BF_4]) as a supercapacitor, the specific capacitances of hoPILs-rGO and coPILs-rGO can be obtained as high as 223 and 222 F g^{-1} with good cycling stability. In addition, hoPILs-rGO and coPILs-rGO exhibited maximum power densities about 340 kW kg^{-1} at a current density of 5 A g^{-1}. It is worth mentioning that two kinds of conjugated PILs modified graphene electrodes can be produced on the gram scale, which has wide practicability.

Because of the diversity of polymer, the hybrid of graphene and polymer could come in a variety of forms, which assuredly grants supercapacitors with a different performance.

4.2.5 Conclusion

The above methods to modify graphene have greatly extended the application of graphene in supercapacitors. In addition, a graphene-based hybrid supercapacitor is a new strategy because each component could play a key role in the achievement of a high capacitance value. For example, Genc and co-workers developed α-manganese oxide-polypyrrole (PPy) nanorods decorated with C-dots as anode materials and reduced graphene oxide for the cathode.[69] By using octyl bis(3-methylimidazolium) diiodide ([C_8H_{16}BMIM]I_2) as an electrolyte at a working voltage of 3 V, the constructed hybrid supercapacitor exhibited an area capacitance value of 17.3 μF cm^{-2}.

Despite the fact that graphene electrodes are functional with different functional groups or can be hybridized with various functional materials such as heteroatoms, carbon materials or polymers, the ionic liquid electrolyte has proved to be a very capable target to hybridize with graphene. For example, Jieun and Soak reported graphene modified by [BMIM][PF_6] as an electrode for a supercapacitor, which exhibited a higher specific capacitance than nonmodified graphene.[70] Moreover, [BMIM][PF_6] modified graphene showed a stable structure and high electron transfer, and [BMIM][PF_6] was an effective exfoliation agent to prevent the restacking of graphene nanosheets. In another study, Raj and co-workers used hydrophilic imidazolium-based ionic liquid 1-aminopropyl-3-methylimidazolium bromide ([APMIM]Br) to modify rGO with covalent functionalization.[71] This strategy was designed to increase the interlayer distance of graphene and alter its wettability. After functionalization, the water contact angle was significantly decreased from 94 to 37°, and the graphene nanosheets were effectively prevented from restacking. Moreover, the covalent functionalization with [APMIM]Br also improved the energy density of the rGO-based supercapacitor with 36.67 Wh kg^{-1}. The results showed that the excellent performance of the rGO-based supercapacitor was not only due to the intrinsic property of high electronic conductivity and large surface area, but also to the enhanced wettability and increased interlayer distance.

This section has summarized several main strategies to improve the properties of graphene under the premise of ionic liquid as an electrolyte in supercapacitors. Finding other, better methods, or hybrid materials for graphene is still a good strategy and well worth exploring to discover a supercapacitor with excellent performance.

4.3 Ionic Liquid Electrolytes

Using non-aqueous electrolytes can broaden the voltage window and increase the energy density. Due to the special structures and properties, ionic liquids, the "green" solvents, have been widely used in supercapacitors. In general, ionic liquids have the properties of non-volatility, inflammability, thermal stability and electrochemical stability and a wide working voltage window (2–6 V). For example, Lewandowski *et al.* have used activated carbon cloth as the carbon electrode of a supercapacitor, and compared the performance of aqueous, organic and ionic liquids (tetraethylammonium tetrafluoroborate ([N2222][BF_4]), [EMIM][BF_4] and [EMIM][NTf_2]) as electrolytes.[72] The results showed that the operating voltage could reach 3.5 V by using ionic liquids as electrolytes, while the operating voltages of the classical organic system (acetonitrile/polycarbonate) and aqueous system were only 2.5 V and 1 V, respectively.

Up to now, cations used in supercapacitors only contain imidazolium, pyrrolidinium, alkylammonium and alkylphosphonium. Similarly, anions used in supercapacitors only contain [BF_4], [PF_6], NTf_2 and FSI. Although ionic liquids have showed great potential in graphene-based supercapacitors, some defects such as high viscosity, low ionic conductivity and the high cost of ionic liquids have restrained their practical application. Therefore, more efforts have been taken to deal with these problems.

4.3.1 Single Ionic Liquid Electrolytes

Because there is no ion-pairing and over screening effect for single ionic liquid as electrolytes in supercapacitors, more electrode pores could be utilized effectively for higher energy density. However, due to the strong interactions between the cation and anion, the ions' diffusion is still slow and the power energy is reduced. In contrast to conventional ionic liquid, Genc and co-workers have synthesized a novel dicationic-bis-imidazolium-based ionic liquid as an electrolyte using a *N*-alkylation reaction.[73] This dicationic ionic liquid, $[C_8H_{16}BMIM]I_2$ was found to enhance the wetting capacity and charge transfer of reduced graphene oxide electrodes' surfaces.

The other main research focus is on preventing the restacking of graphene sheets, including incorporating spacers between graphene sheets or creating 3D porous networks. Sun *et al.* reported a new structure of an asymmetric supercapacitor fabricated using ionic liquid–CNT–graphene gel and MnO_2-graphene gel as electrodes.[74] Superior performance can be obtained owing to a 3D porous network and spacers, which can effectively keep graphene sheets from restacking. Furthermore, hydrophilic ionic liquid can also ensure the accessibility of the ion onto the electrodes.

4.3.2 Mixed Ionic Liquid Electrolytes

It is widely believed that viscosity has a great relationship with conductivity. So, it was the same for the ionic liquid electrolyte. To decrease the viscosity and increase the ionic conductivity of the ionic liquid electrolyte, mixing ionic liquid with other agents such as an organic solvent is a convenient method. For example, Fu and co-workers prepared a new kind of supercapacitor with chemically reduced graphene as the electrode material and $[BMIM][PF_6]$/acetonitrile mixed solvent as the electrolyte.[75] First, chemically reduced graphene with a high surface area of 617 $m^2 g^{-1}$ was obtained by chemical reduction of graphene oxide. Because the electrode material is a mixture of single-layer graphene or few-layers graphene, its surface area was much lower than that of the theoretical value (2630 $m^2 g^{-1}$). Interestingly, this structure can store much more charge with a specific capacitance of 132 $F g^{-1}$. The addition of acetonitrile (25% volume ratio) in $[BMIM][PF_6]$ can decrease the viscosity of the mixed electrolyte and increase the ionic conductivity. Compared with pure ionic liquid as an electrolyte, the performance of this new supercapacitor with a mixed electrolyte was much higher, and the energy density and specific power density were 143.7 $Wh kg^{-1}$ and 2.8 $kW kg^{-1}$, respectively. Moreover, the electrochemical window still remained at about 2.8 V, which was much wider than that of the aqueous electrolyte (~1.0 V). Thus, it can be seen that diluting ionic liquid using other solvents such as acetonitrile is a useful method to reduce the viscosity of the ionic liquid. In another study, Aksay *et al.* used a low-dielectric constant solvent, 1,2-dichloroethane, to dilute a hydrophobic ionic liquid, $[EMIM][NTf_2]$, which can yield a great gain in capacitance near the open circuit potential for glassy carbon-based supercapacitors.[76]

The merit of this method, namely diluting ionic liquids with other solvents, can decrease the interactions between the cation and anion due to the solvated effect of ions. Thus, the ions' diffusion and power density can be increased. However, the solvated ions are relatively large and it is not easy for them to enter into the pores of electrodes, leading to a decrease of power energy. It seems as if this is an irreconcilable contradiction for energy density and power density. Therefore, it is particularly important to increase energy density and power energy simultaneously.

Recently, Wang and co-workers reported a mixed ionic liquid electrolytes-based supercapacitor with $[EMIM][BF_4]$ and tetramethylammonium tetrafluoroborate ($[N_{1111}][BF_4]$) for improving the energy density of a supercapacitor while keeping its high power density.[77] It was found that the strong ionic interactions between the $[EMIM]^+$ and $[N_{1111}]^+$ ions could induce their entrance into the mesopores of the carbon-based electrode materials. Here, there was attraction between the $[EMIM]^+$ ions with an electron-rich π-conjugate structure and positively charged $[EMIM]^+$ ions, thus the exclusion between the $[EMIM]^+$ and $[N_{1111}]^+$ ions could be partially offset. The strength of the π–cation interaction was of the same order as the hydrogen bonds.[78] This denser arrangement of $[EMIM]^+$ and $[N_{1111}]^+$ ions in the pores of electrodes can improve the energy density. Although this work is based on mesopore-rich carbon-based electrode materials, it might be promising in graphene-based supercapacitors. Regretfully, so far there are no related reports with graphene-based supercapacitors using mixed ionic liquid/ionic liquid electrolytes.

The current research efforts in graphene-based supercapacitors with ionic liquid electrolytes are devoted to enhance the performance of graphene-based supercapacitors, while there are fewer studies based on their operating temperature range. Due to the low melting point near room temperature, it is hard for neat ionic liquids to be used at sub-zero temperatures.

Negre *et al.* proposed a strategy by using an ionic liquid eutectic mixture (1-methyl-1-propylpiperidinium bis(fluorosulfonyl)imide ($[PP_{13}][FSI]$) and 1-butyl-1-methylpyrrolidinium bis(fluorosulfonyl)imide ($[PP_{14}][FSI]$)) as an electrolyte for the improvement of the operating temperature range and voltage window (*i.e.*, energy density) simultaneously.[79] At room temperature and below, they found that the ionic liquid eutectic mixture electrolyte still could work effectively with a large potential window of 3.5 V. At an operating temperature of −20 and −30 °C, a maximum gravimetric capacitance of a supercapacitor can be obtained with 130 and 100 $F\,g^{-1}$, respectively.

It is suggested that the interaction mechanism between the ionic liquid and the other solvents should be given effective statistics and analysis, and promising candidates may carry out this research.

4.3.3 Ionic Liquid/Polymer Electrolytes

Single or mixed ionic liquid electrolytes belong to liquid electrolytes. When considering the stretchability of supercapacitors, liquid electrolytes might not work in this situation. The use of ionic liquid electrolytes in flexible supercapacitors is currently rarely reported. This is mainly due to the fact that it is more complex to fabricate flexible supercapacitors with ionic liquid electrolytes than with aqueous/hydrogel electrolyte devices. In this case, the development of ionic liquid/polymer-based stretchable electrolytes is very important for supercapacitors that are usually called all-solid-state supercapacitors. Ionic liquid/polymer electrolytes are generally packed by the imbibition of ionic liquid into the host polymers where ionic liquids (mobile phase) endow ionic conductivity and the polymers provide mechanical stability.

Recently, many ionic liquid/polymer electrolytes, named "ion gels" have been exploited. Pandey and Rastogi have fabricated graphene-based all-solid-state supercapacitors by using ionic liquid/gel polymer as the electrolyte.[80] They chose ionic liquid [BMIM][BF_4] to immobilize poly(vinylidene fluoride-hexafluoropropylene) (PVDF-HFP). Impedance analysis showed that resistance contributions were related with nominal charge transfer and ion diffusion at the pores. The graphene-based solid-state supercapacitors exhibited a specific capacitance of 76 F g^{-1}, specific energy of 7.4 Wh kg^{-1} and specific power of 4.5 kW kg^{-1}. The advantage of solid-state supercapacitors is that the leakage of electrolytes can be effectively avoided.

Primarily, PEO, PVA, PMMA, PAN, PVDF have been used with ionic liquid to fabricate a high-performance supercapacitor. Among these polymers, P(VDF-HFP) has been demonstrated to show high performance due to the mechanical property of the crystalline parts (PVDF). Notably, ionic liquid can be trapped by the amorphous parts (PHFP).

To improve the ionic conductivity of supercapacitors, Yang and co-workers have further used GO to dope [EMIM][BF_4]/P(VDF-HFP) electrolytes.[81] First, GO was dispersed in DMF to obtain a clear solution. Then, P(VDF-HFP) and [EMIM][BF_4] were added in the above solution to form a homogeneous viscous mixture. After being cast onto the substrate and dried, the obtained P(VDF-HFP)- [EMIM][BF_4]-GO gel film was fabricated into an all-solid-state supercapacitor. Here, P(VDF-HFP), [EMIM][BF_4] and GO were the matrix, supporting the electrolyte and ionic conducting promoter, respectively. The results showed that the homogeneous-dispersed GO was doped in the 3D P(VDF-HFP) gel and facilitated the ion transport. Vividly, the doping GO acted like an ion "highway", and only 1 wt% GO could improve the ionic conductivity up to 260% compared with pure ion gel. This excellent performance was attributed to the compatibility of [EMIM][BF_4] and the polymer matrix, and GO can endow the supercapacitor with an electrochemical property and long-term stability.

Tamilarasan and co-workers have incorporated [BMIM][NTf_2] with polyacrylonitrile (PAN) as an electrolyte to from a solid-like layered phase.[82]

The ionic conductivity of a PAN/[BMIM] [NTf_2] electrolyte can reach up to 2.42 mS cm^{-1} with good thermal stability because the weight of the PAN/[BMIM][NTf_2] electrolyte only lost 3% at 400 °C. The phase that formed in the PAN/[BMIM][NTf_2] electrolyte was caused by the ionic interaction between C≡N of the PAN and ion of [BMIM][NTf_2]. When fabricated with rGO, a mechanically stable and flexible supercapacitor can be obtained, which exhibits a maximum energy storage capacity of 32.3 Wh kg^{-1}.

In later research, Ujjain and co-workers reported a supercapacitor with ionic liquid functionalized graphene as the electrode and ionic liquid incorporated polymer gel as the electrolyte.[83] They first used [BMIM][BF_4] to modify the reduced graphene oxide for the improvement of the wettability of graphene by using the imidazolium groups of [BMIM][BF_4]. Then, [BMIM][BF_4] was assembled with [PVDF-HFP] poly(vinylidenefluoride) hexafluoropropylene as a solid electrolyte, which also showed accessibility with the electrodes. The obtained supercapacitor exhibited a high capacitance with 242 F g^{-1} at 5 mV s^{-1}. Compared to any existing EDL system, the relaxation time constant of 2.4 ms was very low and the supercapacitor showed fast energy delivery at relatively high power.

In ionic liquid/polymer electrolytes, the interaction between the ionic liquids and host polymers determines the mechanical property, ionic conductivity and electrochemical stability. To improve the mechanical strength of solid-state ionic liquid/polymer electrolytes with good electrochemical properties, Ramaprabhu and co-workers first prepared a [BMIM][NTf_2] incorporated poly(methyl methacrylate) (PMMA) electrolyte (PMMA/[BMIM]-[NTf_2]) with a mass ratio of 1 : 2, which showed high stretchability, good ionic conductivity (0.78 mS cm^{-1} at 28 °C) and excellent transparency (>98% in a visible range).[84] When the PMMA/[BMIM][NTf_2] electrolyte was fabricated with hydrogen-exfoliated graphene, the supercapacitor exhibited an energy density of 25.7 Wh kg^{-1} and power density of 35.2 kW kg^{-1}, respectively. Moreover, it also showed a four-fold stretchability and low equivalent series resistance of 16 Ω, which was mainly caused by the compatibility of the interface of graphene and PMMA/[BMIM][NTf_2].

Because the ionic liquids used in supercapacitors are not stable against the moisture of the air, this will result in the spontaneous incorporation of water with ionic liquid, which changes the original properties of the ionic liquid. In addition, the potential leakage problem for the ionic liquid electrolyte is also unavoidable during consecutive bend cycles. To deal with this problem, Ma and co-workers developed a flexible solid-state supercapacitor by using ionic liquid/polymer as the electrolyte.[85] Here, ionic liquid [EMIM][BF_4] was combined with gel polymer polyvinyl fluoride-*co*-hexafluoropropylene (PVDF-HEP), and this ionic liquid/polymer electrolyte showed good compatibility. Based on hierarchically porous graphene nanocomposites (GNCs), with polyaniline (PANI) as the electrode, this flexible solid-state supercapacitor with a [EMIM][BF_4]/PVDF-HEP electrolyte exhibited a high specific capacitance of 180 F g^{-1} at 1 A g^{-1} and large energy density of 75 Wh Kg^{-1}. Moreover, these meso/microporous GNCs showed high ionic conductivity, a wide range of operating voltage, nonflammability, as well as superior flexibility under consecutive bending conditions.

4.3.4 Conclusion

The general conclusion is that ionic liquid as an electrolyte in a graphene-based supercapacitor is very promising. Besides the above methods to reduce the defects of ionic liquid, other attempts have also been reported. In Rennie's research, they compared ionic liquids with an ether bond and non-ether bond as an electrolyte in supercapacitors, and the results showed that the specific capacitance of ionic liquids with the ether bond electrolyte was double that of the non-ether bond ionic liquids.[86] The viscosity and melting point of ionic liquids with an ether bond were much lower than that of non-ether bond ionic liquids, which indicated that the performance of a supercapacitor could be improved by introducing an oxygen atom. Apart from ionic liquid as an electrolyte in supercapacitors, electrode materials can be functionalized by ionic liquid to give rise to abundant interconnected pores and improve the electrical conductivity of electrode materials. For example, Sun and co-workers developed a new type of flexible nanohybrid electrode with a three-dimensional structure through the hybrid of polyaniline with [BMIM][BF_4] functionalized graphene and CNTs.[87] Through cation–π interaction, the [BMIM][BF_4] molecules can interact strongly with and graphene/CNTs, which is helpful for the formation of interconnected pores. The large interlayer spacing and interconnected pores can facilitate the entrance of polyaniline, leading to high capacitance.

Recently, intensive research about supercapacitors has focused on high temperature application. In a practical application, it is very important for the electromobile and hybrid electric vehicle to work at high temperatures. Working with a battery, a supercapacitor is a good substitute to provide power and store energy for the improvement of battery life. In this system, the temperature of the combustor near the supercapacitor could easily reach more than 60 °C, which simultaneously influences the behavior of the supercapacitor. Due to good thermal stability, ionic liquid is very suitable to be used in this extreme condition.

4.4 Graphene/Ionic Liquid Supercapacitors: Investigation of the Interface of Graphene and Ionic Liquid

In the previous discussion, some research that involved functionalization and modification of graphene focused on solving the interfacial compatibility between graphene-based electrodes and ionic liquid electrolytes. This means that by employing ionic liquids as electrolytes, the relatively high viscosity of ionic liquids might limit their ionic conductivity and also restrain pore wetting.

Pope *et al.* pioneered a high energy density graphene-based supercapacitor with a high ion-accessible surface area by using a [EMIM][NTf_2]-surfactant (Tween 20) microemulsion system.[88] This system was helpful for the spontaneous adsorption of ionic liquid into GO. Through a heat treatment at 360 °C, the surfactant and GO were removed and reduced, respectively. In this process, [EMIM][NTf_2] played a double-acting role of spacer and electrolyte. The content of [EMIM][NTf_2] exhibited an important effect on the gravimetric capacitance of the supercapacitor and an excellent gravimetric capacitance (302 F g^{-1} at 1 A g^{-1}) with 80 wt% [EMIM][NTf_2] composites can be obtained.

As previously mentioned, the interactions between the graphene-based electrode and the ionic liquid electrolyte have determined the electrochemical performance of supercapacitors. Commonly, there were residual oxygenic functional groups on the graphene by using the chemical reduction method. It is widely believed that these functional groups have resulted in the interactions between the electrolyte and graphene electrode in supercapacitors. Qin *et al.* studied the existential form of interactions between the residual groups of graphene and three kinds of ionic liquid, [EMIM][NTf_2], [EMIM][BF_4], and 1-methyl-1-propylpiperidinium bis(trifluoromethyl sulfonyl) imide ([PP_{13}][NTf_2]).[89] The results showed that the residual hydroxyl groups of graphene were mainly transferred to $[EMIM]^+$ and $[NTf_2]^-$. In contrast, there was almost no reaction taking place in $[BF_4]^-$ or $[PP_{13}]^+$, during the process of charging. Therefore, they concluded that the device capacitance was derived from the chemical reactions, and the highest energy density was 169 Wh kg^{-1} in the [PP_{13}][NTf_2] electrolyte.

Most research about graphene/ionic liquid-based supercapacitors also involves the interface interactions between graphene and ionic liquid. However, the interface mechanism has not received enough in-depth study. Certainly, the interface mechanism should deserve more attention for the more scientific design of supercapacitors with high performance.

4.5 Supercapacitors for the Next Generation of Wearable and Portable Electronic Devices

Flexible and wearable electronics, as the name implies, is an electronic device or piece of equipment that exhibits mechanical flexibility and is compatible with human skin. As one of the excellent potential energy storage devices available, a supercapacitor with flexible components is always a great choice for smart portable electronic devices. Related research is progressing towards meeting the requirements of future potential markets. Recently, the exploitation of flexible supercapacitors has become the focus of intense research owing to their application in flexible devices such as bendable mobile phones, wearable sensors, and smart clothing.[90–92] Supercapacitors are highly desirable for an energy storage device with excellent flexibility, high energy density and long cycle life. Because one of the crucial features for solid-state devices is the physical flexibility of supercapacitors, the exploitation of supercapacitors for flexible and wearable electronic devices still lies in the appropriate electrode materials and electrolytes. Considering the required flexibility and wearability, the electrode materials and electrolytes of supercapacitors in electronic devices should be very special. For electrode materials, excellent mechanical strength is needed for practical applications. The commercial active electrodes for supercapacitors are usually activated carbons, but their rigid microstructure restricts flexibility. When the electrodes are reused with twisting, bending or stretching, it results in breakage of the electrode materials and it is very difficult for them to go back to their original structure. Moreover, the low cost of the materials for the fabrication of supercapacitors is very important to achieve mass production and application. Except for the requirements of electronic performance and wearability, the introduced pseudocapacitive materials for modifying the surface of carbon materials also need to be compatible with other components and retain superior performance. Recently, Shabeeba and co-workers prepared an attractive low-cost flexible supercapacitor with a graphene-based electrode and ionic liquid-based electrolyte by using a hot lamination method.[93] First, graphene, a binder and solvent 1-methyl-2-pyrrolidinone (NMP) were mixed together to form a slurry. Thereafter, the working electrode was obtained after the slurry was heated to remove NMP under vacuum. When fabricated with trihexyl (tetradecyl) phosphonium bis (trifluoromethanesulfonyl) imide ([P_{66614}][NTf_2]) in acetonitrile (25%) as the electrolyte, the supercapacitor showed a high rate scalability with a specific capacitance of 495 $F g^{-1}$. Moreover, the operating voltage could reach 4 V. Considering the complexity of flexible supercapacitors for the fabrication of wearable and portable electronic devices, some other issues still need to be summarized and solved such as wearing comfort, working environment and so on. For wearable supercapacitors, applying flexible electrodes is obviously necessary in addition to solid-state electrolytes or quasi-solid-state electrolytes. For the former, mixing individual graphene with other materials is a facile way to fabricate flexible electrodes that can retain the necessary mechanical properties with relatively high electrical conductivity. For example, Choi *et al.* reported a way to fabricate solid-state flexible asymmetric supercapacitors. GO was chemically functionalized by using [BMIM][BF_4]. Then, hydrous RuO_2 was modified on [BMIM][BF_4]-GO and used as the positive electrode with [BMIM][BF_4]-GO as the negative electrode of the supercapacitor.[94] In contrast to previous works that focused on reconstructing the surface configuration of electrode materials, this paper concentrated on cell voltage, which is also an important factor for supercapacitors. The results showed that a 1.8 V voltage range could be obtained with 175 $F g^{-1}$ specific capacitance. Furthermore, high energy of 19.7 $W h kg^{-1}$ and power density of 6.8 $kW g^{-1}$ could also be achieved. As for solid-state electrolytes or quasi-solid-state electrolytes, the purpose of fabricating such components is to make sure that wearable supercapacitors are free from the dangerous leakage of electrolytes.

Fiber-shaped supercapacitors with superior electrochemical performance may be the most widely studied in the wearable supercapacitors field. Such functional fiber devices have great potential for the next generation of portable and wearable electronics. Considerable efforts have been made in this area. Among them, graphene fiber (GF) may be a good choice because of its larger specific surface area than other common carbonaceous fibers, and it can be obtained by facile wet spinning or hydrothermal methods. Reduction treatment is unavoidable during the fabrication of GF to obtain higher conductivity. Then, pseudocapacitive materials can be applied on the surface of GF for high electrochemical performance. For example, Cai *et al.* proposed a facile way to solve the limited run-time of electronics when using fiber-shaped supercapacitors.[95] In this work, a 1.5 V high cell voltage and high volumetric energy densities asymmetric fiber-based supercapacitors could be obtained. These excellent properties were achieved by coating GF with $NiCo_2S_4$ nanoparticles. This flexible supercapacitor exhibited the high volumetric capacitances of 388 F cm^{-3} and 300 F cm^{-3} in a three-electrode cell and two-electrode cell, respectively. Furthermore, great enhancement of the electrochemical performance could also be achieved.

A wearable supercapacitor combines the excellent performance of capacitors with high energy densities and could be a big advantage in the portable energy storage field. Furthermore, wearable supercapacitors still work when deformed, this means high electrochemical performance and wearability can both be achieved.

4.6 Ionic Liquids for Other 2D-based Supercapacitors

The discoveries and applications of graphene have opened a window for 2D materials, and over more than twenty 2D materials have been found up to now including graphdiyne,[96] black phosphorus (BP),[97] borophene,[98] germanene,[99] silylene,[100] bismuthene,[101] antimonene,[102] *h*-BN,[103] *g*-C_3N_4,[104,105] transitional metal dichalcogenides (TMDs),[106] layered double hydroxides (LDHs),[107,108] transition metal oxide (TMOs),[109] MXenes,[110,111] APX_3,[112] layered metal–organic framework (MOFs),[113,114] layered covalent organic framework (COFs),[115] and so on. Recently, an increasing number of studies have been carried out to investigate the energy storage and conversion of these 2D materials. Compared with graphene as electrodes in ionic liquid electrolyte-based supercapacitors, other 2D materials have also shown a unique performance in supercapacitors. Zhang and co-workers investigated the electrochemical performance of a MnO_2-based electrode with a mixture of [BMIM][PF_6] and *N,N*-dimethylformamide (DMF) as the electrolyte.[116] They first prepared a MnO_2 electrode through an electrodeposition experiment in $Mn(CH_3COO)_2$ aqueous solution. After washing and annealing, MnO_2 was assembled with CNT and nickel foam to a prepared nanocomposite electrode. They found that the electrochemical properties of the MnO_2 nanocomposite electrode was affected by the ratio of [BMIM][PF_6] and DMF. Adjusting the volume ratio of [BMIM][PF_6] and DMF with 1 : 1, the MnO_2

nanocomposite electrode exhibited the best electrochemical performance with an operation potential window of 2.1 V and specific capacitance of 523.3 F g^{-1} at 3 A g^{-1}. When fabricated into an asymmetric supercapacitor by using a MnO_2 nanocomposite electrode and activated carbon as asymmetric electrodes, the supercapacitor showed a maximum specific energy of 67.5 W h kg^{-1} and maximum specific power of 20.4 kW kg^{-1}.

Besides being an electrolyte, ionic liquid also can be used to prepare a hollow electrode. For example, Xu *et al.* designed and prepared a g-C_3N_4 hybridized a-Fe_2O_3 electrode. g-C_3N_4 was prepared through the heating of dicyandiamide and then composited with a-Fe_2O_3 to manufacture hollow microspheres by an eco-friendly solvothermal method.[117] Differing from other studies using ionic liquid as an electrolyte, they developed a special ionic liquid [BMIM]-tetrachlorideferrate(III) system, which was called [BMIM][$FeCl_4$] containing iron ions. It was found that [BMIM][$FeCl_4$] played triple roles simultaneously, namely reactant, dispersing media and template. The results of the electrochemical experiments showed that the g-C_3N_4/a-Fe_2O_3 hollow microspheres exhibited an excellent electrochemical performance. In another study, Luo and co-workers developed a Ni_3S_4-MoS_2 heterojunction electrode with the assistance of [BMIM][SCN] through a one-step extended strategy.[118] [BMIM][SCN] can provide an *S* source in the sulfidation processing and induce the formation of a Ni_3S_4-MoS_2 heterojunction. So, ionic liquid can act as an attractive medium for the synthesis of different distinctive electrodes. In another example, Sun and co-workers developed a sandwich-structured flexible supercapacitor electrode. First, Go was modified by [APMIM]Br ionic liquid in the presence of KOH.[119] Then, [APMIM]Br functionalized GO ([APMIM]Br-GO) was filtrated by using a cellulose acetate membrane to obtain nanohybrid papers. Thereafter, MnO_2 with a nanostructure was electrodeposited on the surface of nanohybrid paper through a controllable template-free ultrasonic electrodeposition. The excellent pseudocapacity of MnO_2 was inherited by the obtained electrode material, leading to an enhanced supercapacitor performance. The electrochemical measurement showed that this electrode exhibited excellent mechanical stability and flexibility with distinguished rate capability and cycling stability.

As discussed earlier, the devices would be exposed at very low temperature for special applications. Although ionic liquid is a very safe electrolyte in the fabrication of electrochemical devices, its conductivity would decrease sharply at the low temperature of 253 K or below, due to the strong interactions arising between the component ions. Therefore, it is necessary to pursue some effective methods to tune the intermolecular interactions of ionic liquid for controlling its phase behavior and ionic conductivity. To deal with this problem, Fujie and co-workers, for the first time, loaded [EMIM][TFSA] (*i.e.*, [EMIM][NTf_2]) in MOF of ZIF-8 ($Zn(MeIM)_2$, H(MeIM) = 2-methylimidazole).[120] It is well known that MOFs have a highly porous and flexible structure with tunable properties and have been applied to various domains.[111–124] In this work, it was found that the dynamics of [EMIM][TFSA] can be controlled *via* the tunable interactions between [EMIM][TFSA] and MOF. At low temperature, the ionic conductivity of [EMIM][TFSA]@ZIF-8 had no marked decrease while the bulk [EMIM][TFSA] decreased markedly. This strategy has exhibited the great potential to design electrolytes for an all-solid-state and flexible supercapacitor. As a type of emerging nanoporous material, MOF has tunable pore sizes, distinguished architectures and topologies. Because of its thermal stability, MOF can be used as a template to prepare various electrodes.[125,126]

It is clear at this point, after a survey of the literature, that ionic liquid has great potential as an electrolyte or medium in other 2D-based supercapacitors. Undoubtedly, more in-depth studies and research are required in this field.

4.7 Conclusion

Graphene and ionic liquid have opened a new window for developing energy storage in supercapacitors. This chapter has summarized the research progress of graphene/ionic liquid-based supercapacitors. On the one hand, the special lamellar structure of graphene-based electrodes renders them good electron conductivity and excellent chemical stability differing from blocky graphite. On the other hand, ionic liquid-based electrolytes show extraordinary properties differing from conventional aqueous or organic phase electrolytes. These unique properties give them an edge in supercapacitors.

According to the progress in the aforementioned research, the development of graphene/ionic liquid-based supercapacitors still focuses on the electrode materials, electrolytes and interfacial compatibility. As for future research, there are several directions for graphene-based electrode materials, such as novel preparation methods for a high specific surface area, proper pore effect for the diffusion of ionic liquid, a composite with low resistance by doping and good surface structure for excellent compatibility with ionic liquid to improve the cycle life of supercapacitors. As for ionic liquid electrolytes, this challenge could be addressed by increasing the electrochemical window of ionic liquid for high energy density, increasing the purity of ionic liquid and reducing the negative influence of impurity to the working voltage window and cycle life, exploiting novel ionic liquids with low viscosity and high ionic conductivity, and developing new-style ionic liquids without fluorine to avoid health and environmental hazards. It is worth mentioning that interfacial compatibility of graphene-based electrodes and ionic liquid electrolytes is very important for supercapacitors, which lies in the matching ability between the pore structures and surface chemistry of a graphene-based electrode and the size of ionic liquid. Currently, these studies are still in the early stages, and there are few theoretical calculation studies about the interfacial compatibility of graphene and ionic liquid. The revelation of mechanism still needs more *in situ* characterization techniques to demonstrate the functionality of electrodes and electrolytes. In addition to the issues discussed above, the cost of supercapacitors also determines their development and extensive application. Although challenges still exist, it is predicted that graphene-based supercapacitors with ionic liquid as electrolytes will herald the coming of a new era in energy storage.

Acknowledgements

The authors thank Dr Qinzhi Wang at Northwest A&F University for support.

References

1. M. Winter and R. J. Brodd, *Chem. Rev.*, 2004, **35**(50), 4245.
2. D. S. Su and R. Schlögl, *ChemSusChem*, 2010, **3**(2), 136.
3. D. P. Dubal, O. Ayyad, V. Ruiz and P. Gomez-Romero, *Chem. Soc. Rev.*, 2015, **44**(7), 1777.
4. B. A. Zhang, F. Y. Kang, J. M. Tarascon and J. K. Kim, *Prog. Mater. Sci.*, 2016, **76**, 319.
5. L. Liu, Z. Niu and J. Chen, *Chem. Soc. Rev.*, 2016, **45**(15), 4340.
6. F. Wang, X. Wu, X. Yuan, Z. C. Liu, Y. Zhang, L. J. Fu, Y. S. Zhu, Q. M. Zhou, Y. P. Wu and W. Huang, *Chem. Soc. Rev.*, 2017, **46**(22), 6816.
7. B. Bayatsarmadi, Y. Zheng, A. Vasileff and S. Z. Qiao, *Small*, 2017, **13**(21), 1700191.
8. X. Peng, L. Peng, C. Wu and Y. Xie, *Chem. Soc. Rev.*, 2014, **43**(10), 3303.
9. W. Li, G. Joós and J. Bélanger, *IEEE Trans. Ind. Electron.*, 2010, **57**(4), 1137.
10. L. Qu and W. Qiao, *IEEE Trans. Ind. Appl.*, 2011, **47**(1), 359.
11. A. H. M. A. Rahim and E. P. Nowicki, *Energy Convers. Manage.*, 2012, **59**(3), 96.
12. A. S. Aricò, P. Bruce, B. Scrosati, J. M. Tarascon and W. Van Schalkwijk, *Nat. Mater.*, 2005, **4**(5), 366.
13. K. S. Kumar, N. Choudhary, Y. Jung and J. Thomas, *ACS Energy Lett.*, 2018, **3**(2), 482.
14. P. Rodatz, G. Paganelli, A. Sciarretta and L. Guzzella, *Control Eng. Pract.*, 2005, **13**(1), 41.
15. J. N. Marie-Francoise, H. Gualous, R. Outbib and A. Berthon, *J. Power Sources*, 2005, **143**(1), 275.
16. T. Mishima, E. Hiraki and M. Nakaoka, *J. Power. Electron.*, 2010, **10**(1), 27.
17. H. Gualous, R. Gallay, M. A. Sakka, A. Oukaour, B. Tala-Ighil and B. Boudart, *Microelectron. Reliab.*, 2012, **52**(9–10), 2477.
18. X. Zhang, H. Xue, Y. Xu, H. S. Chen and C. Q. Tan, *Energ. Convers. Manage.*, 2014, **85**(9), 784.
19. W. Choi, J. W. Howze and P. Enjeti, *J. Power Sources*, 2006, **157**(1), 311.
20. Y. Zhan, Y. Guo, J. Zhu and L. Li, *Int. J. Elec. Power Energy Syst.*, 2015, **67**, 598.

21. Y. F. Li, Y. Z. Liu, W. K. Zhang, C. Y. Guo and C. M. Chen, *Mater. Lett.*, 2015, **157**, 273.
22. C. C. Hu, K. H. Chang, M. C. Lin and Y. T. Wu, *Nano Lett.*, 2006, **6**(12), 2690.
23. K. H. Chang, C. C. Hu and C. Y. Chou, *Electrochim. Acta*, 2009, **54**(3), 978.
24. Y. T. Wang, A. H. Lu, H. L. Zhang and W. C. Li, *J. Phys. Chem. C*, 2011, **115**(13), 5413.
25. S. I. Kim, J. S. Lee, H. J. Ahn, H. K. Song and J. H. Jang, *ACS Appl. Mater. Interfaces*, 2013, **5**(5), 1596.
26. H. R. Ghenaatian, M. F. Mousavi, S. H. Kazemi and M. Shamsipur, *Synth. Met.*, 2009, **159**(17–18), 1717.
27. R. K. Sharma, A. C. Rastogi and S. B. Desu, *Electrochem. Commun.*, 2008, **10**(2), 268.
28. C. Zhou, Y. Zhang, Y. Li and J. P. Liu, *Nano Lett.*, 2013, **13**(5), 2078.
29. M. E. Plonskabrzezinska, J. Mazurczyk, B. Palys, J. Breczko, A. Lapinski, A. T. Dubis and L. Echegoyen, *Chem. – Eur. J.*, 2012, **18**(9), 2600.
30. M. M. Huq, C. T. Hsieh and C. Y. Ho, *Diamond Relat. Mater.*, 2016, **62**, 58.
31. L. Xie, F. Su, L. Xie, X. Li, Z. Liu, Q. Kong, X. Guo, Y. Zhang, L. Wan, K. Li, C. Lv and C. Chen, *ChemSusChem*, 2015, **8**(17), 2917.
32. Y. Xu, L. Wang, P. Cao, C. Cai, Y. Fu and X. Ma, *J. Power Sources*, 2016, **306**, 742.
33. M. D. Stoller, S. Park, Y. Zhu, J. An and R. S. Ruoff, *Nano Lett.*, 2008, **8**(10), 3498.
34. S. A. Ali and M. A. Sami, *Pak. J. Pharm. Sci.*, 2005, **18**(1), 6.
35. C. C. Yang, S. T. Hsu and W. C. Chien, *J. Power Sources*, 2005, **152**(1), 303.
36. A. Kacprzak, R. Kobylecki and Z. Bis, *J. Power Sources*, 2013, **239**(10), 409.
37. J. M. Macak, K. Sirotna and P. Schmuki, *Electrochim. Acta*, 2005, **50**(18), 3679.
38. J. Read, K. Mutolo, W. Behl, J. Wolfenstine and A. Driedger, *J. Electrochem. Soc.*, 2003, **150**(10), 1985.
39. J. J. Read, *J. Electrochem. Soc.*, 2006, **153**(1), A96.
40. E. M. Benbow, S. P. Kelly, L. Zhao, J. W. Reutenauer and S. L. Suib, *J. Phys. Chem. C*, 2011, **115**(44), 22009.
41. T. Welton, *Chem. Rev.*, 1999, **99**, 2071.
42. M. Yoshizawa and H. Ohno, *Ionics*, 2002, **8**, 267.
43. P. Hapiot and C. Lagrost, *Chem. Rev.*, 2008, **108**, 2238.
44. J. Chen, W. K. Zhang, S. U. Fang-Yuan, H. Y. Yang, R. X. Liu and C. M. Chen, *New Carbon Mater.*, 2017, **32**(2), 106.
45. S. R. C. Vivekchand, C. S. Rout, K. S. Subrahmanyam, A. Govindaraj and C. N. R. Rao, *J. Chem. Sci.*, 2008, **120**, 9.
46. H. C. Schniepp, J. L. Li, M. J. McAllister, H. Sai, M. Herrera-Alonso, D. H. Adamson, R. K. Prud'homme, R. Car, D. A. Saville and I. A. Aksay, *J. Phys. Chem. B*, 2006, **110**(17), 8535.

47. O. E. Andersson, B. L. V. Prasad, H. Sato, T. Enoki, Y. Hishiyama, Y. Kaburagi, M. Yoshikawa and S. Bandow, *Phys. Rev B.*, 1998, **58**, 16387.
48. P. R. Somani, S. P. Somani and M. Umeno, *Chem. Phys. Lett.*, 2006, **430**(1–3), 56.
49. Y. Chen, X. Zhang, D. Zhang, P. Yu and Y. Ma, *Carbon*, 2011, **49**(2), 573.
50. H. Jeon, J. H. Han, D. M. Yu, J. Y. Lee, T. H. Kim and Y. T. Hong, *J. Ind. Eng. Chem.*, 2016, **45**, 105.
51. C. Liu, Z. Yu, D. Neff, A. Zhamu and B. Z. Jang, *Nano Lett.*, 2014, **10**(12), 4863.
52. M. A. Pope, S. Korkut, C. Punckt and I. A. Aksay, *J. Electrochem. Soc.*, 2013, **160**(10), A1.
53. S. Y. Lu, M. Jin, Y. Zhang, Y. B. Niu, J. C. Gao and C. M. Li, *Adv. Energy Mater.*, 2017, 1702545.
54. J. Wang, B. Ding, Y. Xu, L. Shen, H. Dou and X. Zhang, *ACS. Appl. Mater. Interfaces*, 2015, 7(40), 22284.
55. N. Ma, N. Phattharasupakun, J. Wutthiprom, T. Chan, P. Wuanprakhon and P. Kidkhunthod, *Electrochim. Acta*, 2018, **271**, 110.
56. P. Tamailarasan and S. Ramaprabhu, *J. Phys. Chem. C*, 2012, **116**(27), 14179.
57. Z. Lei, Z. Liu, H. Wang, X. Sun, L. Lu and X. S. Zhao, *J. Mater. Chem. A*, 2013, **1**(6), 2313.
58. M. T. Lee, C. Y. Fan, Y. C. Wang, H. Y. Li, J. K. Chang and C. M. Tseng, *J. Mater. Chem. A*, 2013, **1**(10), 3395.
59. Q. Shao, T. Jie, Y. Lin, L. Jing, F. Qin and J. Yuan, *J. Power Sources*, 2015, **278**, 751.
60. Y. Sun, Z. Fang, C. Wang, A. Zhou and H. Duan, *Nanotechnology*, 2015, **26**(37), 374002.
61. D. J. Bozym, S. Korkut, M. A. Pope and I. A. Aksay, *ACS Sustainable Chem. Eng.*, 2016, **4**(12), 7167.
62. N. Zhang, N. Gao, C. Fu, D. Liu, S. Li, L. Jiang, H. Zhou and Y. Kuang, *Electrochim. Acta*, 2017, **235**, 340.
63. J. Li, J. Tang, J. Yuan, K. Zhang, Y. Sun, H. Zhang and L. Qin, *Electrochim. Acta*, 2017, **258**, 1053.
64. T. Kuila, A. K. Mishra, P. Khanra, N. H. Kim and J. H. Lee, *Nanoscale*, 2013, **5**(1), 52.
65. J. Zhang and X. S. Zhao, *J. Phys. Chem. C*, 2012, **116**(9), 5420.
66. J. Yuan and M. Antonietti, *Polymer*, 2011, **52**(7), 1469.
67. T. Y. Kim, H. W. Lee, M. Stoller, D. R. Dreyer, C. W. Bielawski, R. S. Ruoff and K. S. Suh, *ACS Nano*, 2011, **5**(1), 436.
68. L. Mao, Y. Li, C. Chi, H. S. Chan and J. Wu, *Nano Energy*, 2014, **6**(10), 119.
69. R. Genc, M. O. Alas, E. Harputlu, S. Repp, N. Kremer, M. Castellano, S. G. Colak, K. Ocakoglu and E. Erdem, *Sci. Rep.*, 2017, 7(1), 11222.
70. J. Kim and S. Kim, *Electrochim. Acta*, 2014, **119**(6), 11.
71. S. Bag, A. Samanta, P. Bhunia and C. R. Raj, *Int. J. Hydrogen Energy*, 2016, **41**(47), 22134.

72. A. Lewandowski, A. Olejniczak, M. Galinski and I. Stepniak, *J. Power Sources*, 2010, **195**(17), 5814.
73. R. Genc, M. O. Alas, E. Harputlu, S. Repp, N. Kremer, M. Castellano, S. G. Colak, K. Ocakoglu and E. Erdem, *Sci. Rep.*, 2017, 7(1), 11222.
74. Y. Sun, Y. Cheng, K. He, A. Zhou and H. Duan, *RSC Adv.*, 2015, **5**(14), 10178.
75. C. Fu, Y. Kuang, Z. Huang, X. Wang, Yi. Yin, J. Chen and H. Zhou, *J. Solid. State. Electrochem.*, 2011, **15**(11–12), 2581.
76. D. J. Bozym, B. Uralcan, D. T. Limmer, M. A. Pope, N. J. Szamreta, P. G. Debenedetti and I. A. Aksay, *J. Phys. Chem. Lett.*, 2015, **6**(13), 2644.
77. X. Wang, A. Y. Mehandzhiyski, B. Arstad, K. L. Van Aken, T. S. Mathis, A. Gallegos, Z. Tian, D. Ren, E. Sheridan, B. A. Grimes, D. Jiang, J. Wu, Y. Gogotsi and D. Chen, *J. Am. Chem. Soc.*, 2017, **139**(51), 18681.
78. J. P. Gallivan and D. A. Dougherty, *J. Am. Chem. Soc.*, 2000, **122**(5), 870.
79. L. Negre, B. Daffos, V. Turq, P. L. Taberna and P. Simon, *Electrochim. Acta*, 2016, **206**, 490.
80. G. P. Pandey and A. C. Rastogi, *Mater. Res. Soc. Symp. Proc.*, 2012, 1440.
81. X. Yang, F. Zhang, L. Zhang, T. Zhang, Y. Huang and Y. Chen, *Adv. Funct. Mater.*, 2013, **23**(26), 3353.
82. P. Tamilarasan and S. Ramaprabhu, *Energy*, 2013, **51**(1), 374.
83. S. K. Ujjain, V. Sahu, R. K. Sharma and G. Singh, *Electrochim. Acta*, 2015, **157**, 245.
84. P. Tamilarasan and S. Ramaprabhu, *Mater. Chem. Phys.*, 2014, **148**(1–2), 48.
85. L. Feng, K. Wang, X. Zhang, X. Sun, C. Li, X. Ge and Y. Ma, *Adv. Funct. Mater.*, 2018, **28**(4), 1704463.
86. A. J. Rennie, N. Sanchezramirez, R. M. Torresi and P. J. Hall, *J. Phys. Chem. Lett.*, 2013, **4**(17), 2970.
87. Y. Sun, Z. Fang, C. Wang, A. Zhou and H. Duan, *Nanotechnology*, 2015, **26**(37), 374002.
88. Z. She, D. Ghosh and M. A. Pope, *ACS Nano*, 2018, **11**(10), 10077.
89. J. Li, J. Tang, J. Yuan, K. Zhang, Q. Shao, Y. Sun and L. Qin, *Electrochim. Acta*, 2016, **197**, 84.
90. N. Cui, J. Liu, L. Gu, S. Bai, X. Chen and Y. Qin, *ACS Appl. Mater. Interfaces*, 2015, 7(33), 18225.
91. T. P. J. Knowles and R. Mezzenga, *Adv. Mater.*, 2016, **28**(31), 6546.
92. F. Yi, H. Ren, J. Shan, X. Sun, D. Wei and Z. Liu, *Chem. Soc. Rev.*, 2018, **47**(9), 3152.
93. P. Shabeeba, K. K. Thasneema, M. Shahin Thayyil, M. P. Pillai and C. V. Niveditha, *Mater. Res. Express*, 2017, **4**(8), 085501.
94. B. G. Choi, S. J. Chang, H. W. Kang, C. P. Park, H. J. Kim, W. H. Hong, S. Lee and Y. S. Huh, *Nanoscale*, 2012, **4**(16), 4983.
95. W. Cai, T. Lai, J. Lai, H. Xie, L. Ouyang, J. Ye and C. Yu, *Sci. Rep.*, 2016, **6**, 26890.
96. H. Du, H. Yang, C. Huang, J. He, H. Liu and Y. Li, *Nano Energy*, 2016, **22**, 615.

97. X. Ren, P. Lian, D. Xie, Y. Yang, Y. Mei and X. Huang, *J. Mater. Sci.*, 2017, **52**(17), 10364.
98. H. R. Jiang, Z. Lu, M. C. Wu, F. Ciucci and T. S. Zhao, *Nano Energy*, 2016, **23**, 97.
99. L. Li and M. Zhao, *Phys. Chem. Chem. Phys.*, 2013, **15**(39), 16853.
100. H. Wang, M. Y. Yeh, C. Chen, T. S. Lim, W. Fann and T. Y. Luh, *Macromolecules*, 2008, **41**(8), 2762.
101. F. Reis, G. Li, L. Dudy, M. Bauernfeind, S. Glass and W. Hanke, *Science*, 2016, **357**, 287.
102. W. Tian, S. Zhang, C. Huo, Daming Zhu, Q. Li, L. Wang, X. Ren, L. Xie, S. Guo, P. K. Chu, H. Zeng and K. Huo, *ACS Nano*, 2018, **12**(2), 1887.
103. H. Beomyong, H. Jeongwoon, Y. J. Keon, L. Sungjun, K. Sungmin, L. Minjun, H. K. Jeong, B. Hongwoo, S. Dongchul, K. Gunn, H. Suklyun, I. Jisoon, A. S. Joseph and K. Young, *Sci. Rep.*, 2016, **6**, 31160.
104. Q. Weng, X. Wang, X. Wang, Y. Bando and D. Golberg, *Chem. Soc. Rev.*, 2016, **45**(14), 3989.
105. S. Patnaik, D. P. Sahoo and K. Parida, *Renewable Sustainable Energy Rev.*, 2018, **82**, 1297.
106. Q. Peng and S. De, *Phys. Chem. Chem. Phys.*, 2013, **15**(44), 19427.
107. S. Samuei, Z. Rezvani and A. R. Amani-Ghadim, *Environ. Prog. Sustainable Energy*, 2016, **36**, 4046.
108. H. Wang, X. Liu, Y. Wu, C. Hou, Y. Qiu and K. Guo, *Energy Fuels*, 2017, **31**(9), 9898.
109. X. Guo, G. Zhang, Q. Li, H. Xue and H. Pang, *Energy Storage Mater.*, 2018, **15**, 171.
110. Y. Wu, P. Nie, J. Wang, H. Dou and X. Zhang, *ACS Appl. Mater. Interfaces*, 2017, **9**(45), 39610.
111. B. Anasori, M. R. Lukatskaya and Y. Gogotsi, *Nat. Rev. Mater.*, 2017, **2**(2), 16098.
112. J. Liu, X. Li, D. Wang, W. Lau, P. Peng and L. Liu, *J. Chem. Phys.*, 2014, **140**(5), 37.
113. L. Wang, Y. Han, X. Feng, J. Zhou, P. Qi and B. Wang, *Coord. Chem. Rev.*, 2016, **307**, 361.
114. Y. R. Miao, Z. Su and K. S. Suslick, *J. Am. Chem. Soc.*, 2017, **139**(13), 4667.
115. X. Liu, C. Guan, D. Wang and L. Wan, *Adv. Mater.*, 2014, **26**(40), 6912.
116. X. Zhang, D. Zhao, Y. Zhao, P. Tang, Y. Shen, C. Xu, H. Li and Y. Xiao, *J. Mater. Chem. A*, 2013, **1**(11), 3706.
117. L. Xu, J. Xia, H. Xu, S. Yin, K. Wang, L. Huang, L. Wang and H. Li, *J. Power Sources*, 2014, **245**, 866.
118. W. Luo, G. Zhang, Y. Cui, Y. Sun, Q. Qin, J. Zhang and W. Zheng, *J. Mater. Chem. A*, 2017, **5**, 11278.
119. Y. Sun, Z. Fang, C. Wang, K. R. Rakhitha Malinga Ariyawansha, A. Zhou and H. Duan, *Nanoscale*, 2015, 7, 7790.
120. K. Fujie, K. Otsubo, R. Ikeda, T. Yamada and H. Kitagawa, *Chem. Sci.*, 2015, **6**(7), 4306.

121. E. M. Dias and C. Petit, *J. Mater. Chem. A*, 2015, **3**(45), 22484.
122. F. Yi, J. Li, D. Wu and Z. Sun, *Chemistry*, 2015, **21**(32), 11475.
123. J. Zhu, P. M. Usov, W. Xu, P. J. Celis-Salazar, S. Lin, M. C. Kessinger, C. Landaverde-Alvarado, M. Cai, A. M. May, C. Slebodnick, D. Zhu, S. D. Senanayake and A. J. Morris, *J. Am. Chem. Soc.*, 2018, **140**, 993.
124. W. P. Lustig, S. Mukherjee, N. D. Rudd, A. V. Desai, J. Li and S. K. Ghosh, *Chem. Soc. Rev.*, 2017, **46**(11), 3242.
125. B. Liu, H. Shioyama, H. Jiang, X. Zhang and Q. Xu, *Carbon*, 2010, **48**(2), 456.
126. I. A. Khan, A. Badshah, I. Khan, D. Zhao and M. A. Nadeem, *Microporous Mesoporous Mater.*, 2017, **253**, 169.

CHAPTER 5

Properties and Applications of Layered Thermoelectric Materials

YAJIE HUANG, BINGKUN HU AND LIANGLIANG LI*

School of Materials Science and Engineering, Tsinghua University, Beijing 100084, China
*Email: liliangliang@mail.tsinghua.edu.cn

5.1 Introduction

Thermoelectric materials are capable of transferring heat energy directly to electricity based on the Seebeck effect.[1] Due to the simplicity and reliability of thermoelectric conversion, thermoelectric materials have been used in various applications and are expected to play an important role in energy conversion in the future. Up until now, radioisotope thermoelectric generators (TEGs) have been used as power sources in spacecrafts such as satellites and deep-space probes.[2] Another attractive application of thermoelectric materials is to extract thermal energy from automobile exhausts thereby increasing the efficiency of fuel.[3] In addition, TEGs can be applied to scalable power generation converting solar heat or the Earth's internal heat into electricity,[3] or they can serve as long-term power supplies for implantable medical devices and wearable electronics.[4] Aside from power generation, thermoelectric materials are also used in refrigerators based on the Peltier effect.[5] Thermoelectric coolers are smaller and less noisy in comparison with traditional compressor-based refrigerators, they do not

Smart Materials No. 34
Layered Materials for Energy Storage and Conversion
Edited by Dongsheng Geng, Yuan Cheng and Gang Zhang

Published by the Royal Society of Chemistry, www.rsc.org

contain moving parts, and do not use chloroflurocarbons that can cause ozone-layer depletion.

To generate power efficiently, it is required that thermoelectric materials have the ability to produce a large voltage difference and a high current at a fixed temperature difference ΔT, which means that the materials have a large Seebeck coefficient S and electrical conductivity σ. The materials also need to have a low thermal conductivity κ to maintain the ΔT. A dimensionless thermoelectric figure of merit zT is used to quantify the performance of thermoelectric materials and is defined as

$$zT = \frac{S^2 \sigma T}{\kappa_l + \kappa_e}, \tag{5.1}$$

where T, κ_l, and κ_e are the absolute temperature, lattice thermal conductivity, and the thermal conductivity of electronic carriers, respectively.[1] It is considered that $zT > 3$ is necessary to make the TEGs possible for wide commercialization.[6] Much research effort has been paid to achieve this goal. Thermoelectric materials such as skutterudites, clathrates, bismuth and bismuth-antimony, bismuth telluride-based alloys, lead telluride analogues, and oxides have been developed, and many of them possess a $zT > 1$. Among them, it is interesting to notice that many thermoelectric materials with high zT are layered materials. A representative material is Bi_2Te_3, which has been intensively explored and is regarded as the most promising thermoelectric material near room temperature (RT).[7,8] SnSe, an IV–VI compound with a layered structure, has also been developed, and a remarkable zT of 2.8 has been achieved.[9] Recently, liquid-like thermal conduction in $AgCrSe_2$ has been observed, suggesting that layered materials with heavy intercalants between layers have ultralow thermal conductivity.[10] Here we summarize several layered thermoelectric material systems including Bi_2Te_3, SnSe, BiCuSeO, and intercalated layered materials, and introduce some TEGs based on these layered materials.

5.2 Bi_2Te_3

5.2.1 Materials Development

Bi_2Te_3 is a narrow bandgap semiconductor with a rhombohedral crystal structure (space group $R\bar{3}m$).[11] Figure 5.1 shows a Bi_2Te_3 unit cell.[12] Five atomic layers bonded with strong covalent-ionic bands in the order of Te_1–Bi–Te_2–Bi–Te_1 form a unit, named a quintuple layer. Two adjacent quintuple layers are bonded by van der Waals force.[13] The electrical conductivity and thermal conductivity along the a axis of the Bi_2Te_3 crystal are higher than those along the c axis due to the anisotropic structure.

Bi_2Te_3 is a prototypical thermoelectric material that has been thoroughly studied over decades. It was used as a thermoelectric material for the first time in 1954, when Goldsmid and coworkers used Bi_2Te_3 as the p-type

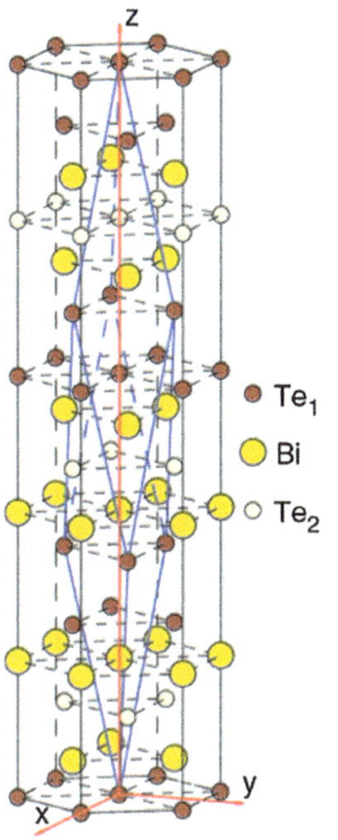

Figure 5.1 A Bi_2Te_3 unit cell with a rhombohedral crystal structure. Reproduced from ref. 12, http://dx.doi.org/10.1103/PhysRevB.84.165208, with permission from American Physical Society, Copyright 2011.

material in their thermoelectric refrigeration.[8] In their initial experiment, the Bi_2Te_3 sample was produced with a zone melting method, which caused a surplus of Bi atoms and vacancies in Te sites, resulting in a p-type electrical conduction. The S, σ, and κ of the material was 220 $\mu V\,K^{-1}$, 4.0×10^2 $S\,cm^{-1}$, and 2.1 $W\,m^{-1}\,K^{-1}$, respectively. Since comparable n-type materials were not available at that time, pure bismuth was used to construct the thermocouple. The thermocouple generated a temperature difference of 26 K because of the Peltier effect. In 1955, n-type Bi_2Te_3 was obtained through iodine doping. A thermoelectric refrigeration based on p-type and n-type Bi_2Te_3 was fabricated and an improved cooling performance up to 40 K was achieved.[14] These results have already shown that Bi_2Te_3-based materials are very promising for thermoelectric applications.

One challenge for Bi_2Te_3 thermoelectric material is its relatively high thermal conductivity. Several approaches can be adopted to suppress κ. One useful strategy is alloying two semiconductors with the same crystal structure, since the substituted atoms increase phonon scattering. This strategy has been applied successfully in the Bi_2Te_3 system. The representative examples are $(Bi–Sb)_2Te_3$ and $Bi_2(Te–Se)_3$, which are the best-known commercial p-type and n-type thermoelectric materials, respectively. With optimized concentrations, κ can be significantly reduced while the reduction in electrical conductivity is marginal, causing an enhanced zT. For p-type

$(Bi–Sb)_2Te_3$, a high zT of 1.4 has been achieved.[15] Meanwhile, several groups have reported zT values of around 1.2 in n-type $Bi_2(Te–Se)_3$.[16,17]

The concept of nanostructure was introduced to further improve the thermoelectric performance of Bi_2Te_3-based materials. In the last few decades, much research effort has been paid to create nanostructures in this material system. Figure 5.2 summarizes various nanostructures including nanoparticles, nanowires, and nanosheets in thermoelectric materials.[18] These research works were initially inspired by the idea that nanostructures enabled a significant enhancement of $S^2\sigma$ because of the quantum confinement effect.[19] While this idea remains in doubt, it is reasonable to expect an enhanced zT in nanostructured Bi_2Te_3, because defects, dislocations, grain boundaries, and boundaries can effectively increase phonon scattering, leading to a considerable reduction of κ. Indeed, most research works have confirmed that nanostructured materials have a reduced κ. Poudel *et al.* synthesized nanocrystalline BiSbTe alloys with a facile two-step method combining ball milling and hot press (Figure 5.3).[15] Due to a low κ of around 1.1 W m^{-1} K^{-1} from 25 to 250 °C, the alloys showed a significantly

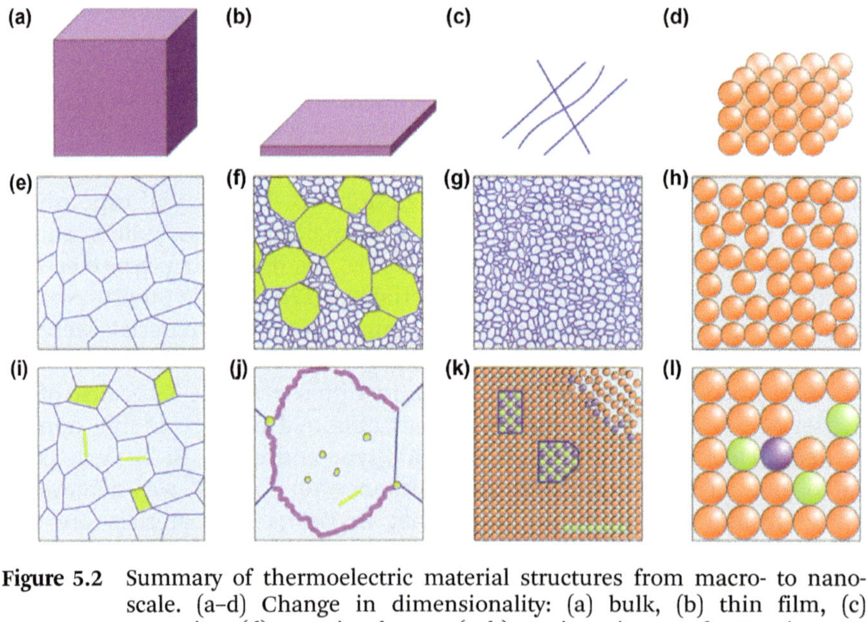

Figure 5.2 Summary of thermoelectric material structures from macro- to nanoscale. (a–d) Change in dimensionality: (a) bulk, (b) thin film, (c) nanowire, (d) atomic cluster. (e–h) Grain mixtures from micro- to nanoscale: (e) normal micro-grained bulk, (f) mixture of coarse and fine grains, (g) nano-grained bulk, (h) amorphous. (i–l) Size evolution of isolated distinct phases or atoms in the composite: (i) normal composite, (j) nano-dispersions located inside grains or at grain boundaries, (k) nano-inclusions or nanodots, boundary modification, (l) atomic doping or alloying, and vacancies.
Reproduced from ref. 18 with permission from Springer Nature, Copyright 2010.

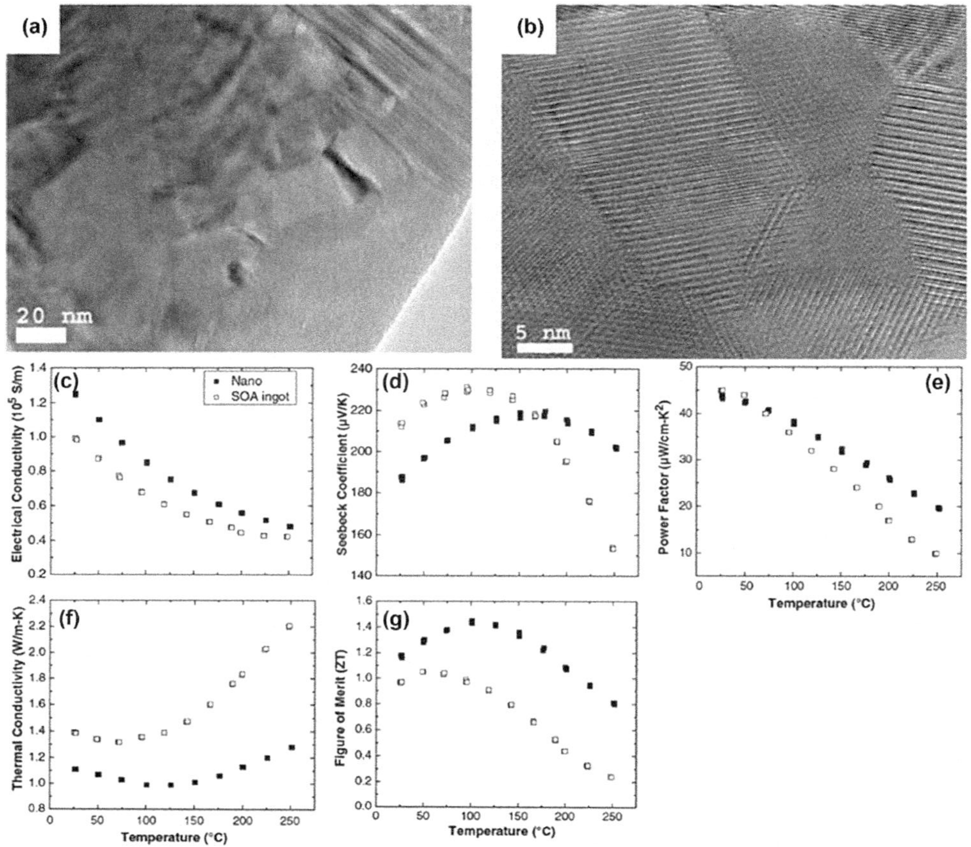

Figure 5.3 (a) Low-magnification TEM image of a hot-pressed nanocrystalline bulk sample showing the nanograins. (b) High-magnification TEM image of the hot-pressed bulk sample showing the nanosize, high crystallinity, random orientation, and clean grain boundaries. Temperature dependence of σ (c), S (d), $S^2\sigma$ (e), κ (f), and zT (g) of a hot-pressed nanocrystalline bulk sample (black squares) as compared with that of a state-of-the-art ingot (white squares). Reproduced from ref. 15 with permission from AAAS.

enhanced thermoelectric performance. The maximum *zT* was about 1.4 at 100 °C, which was superior to that of BiSbTe ingots prepared with the zone melting method.[15] The nanocrystalline BiSbTe alloys showed a nearly isotropic thermoelectric performance, which is preferable in device fabrication. Moreover, the fabrication process of nanocrystalline BiSbTe alloys is simple and promising for the scalable production of thermoelectric materials.

Another strategy is to make Bi_2Te_3-based nanocomposites. It has been reported that *zT* can be improved by 20% by adding SiC nanoparticles into a Bi_2Te_3 matrix, and the mechanical properties of the Bi_2Te_3–SiC nanocomposites are improved at the same time.[20] In addition, adding other second phases such as In_2Te_3[21] and Ag[22] nanoparticles in Bi_2Te_3 is also effective for the enhancement of *zT* due to the decrease of κ. Recently, bulk pellets produced by compressing Bi_2Te_3 nanowires[23] and nanoplates[24] have been synthesized. Similarly, enhanced *zT* values are observed in these materials because of low κ.

The above-mentioned strategies indeed significantly reduce the κ of Bi_2Te_3-based thermoelectric materials. Sometimes, the κ of Bi_2Te_3 can be very close to its amorphous limit;[6] however, the deterioration of electrical properties is often observed in nanostructured Bi_2Te_3-based thermoelectric materials. To achieve a net enhancement of *zT*, it is necessary to realize independent optimization in thermoelectric parameters including S, σ, and κ. Venkatasubramanian *et al.* reported a Bi_2Te_3/Sb_2Te_3 superlattice structure, which can block phonon transport but enable electron transmission.[25] The thickness of the layers were adjusted in order to achieve a significantly reduced κ as well as enhanced power factor $S^2\sigma$. A p-type thermoelectric material with a Bi_2Te_3(10 Å)/Sb_2Te_3(50 Å) superlattice structure was fabricated and a remarkable high *zT* of 2.4 was obtained at 300 K. Meanwhile, an n-type superlattice material with a Bi_2Te_3(10 Å)/Sb_2Te_3(50 Å) superlattice structure was synthesized with a *zT* of 1.46 at 300 K. Despite the excellent thermoelectric performance of superlattice devices, it is difficult to commercialize them because of the complex and high-cost fabrication processes. Research effort is still needed to realize low-cost production of high-performance Bi_2Te_3-based materials.

Table 5.1 summarizes the recent research progress on Bi_2Te_3-based thermoelectric materials. At elevated temperatures, the *zT* values are in the range of 0.7–2.4.

5.2.2 Device Performance

It is necessary to fabricate thermoelectric devices to realize thermoelectric power generation or cooling. Thermoelectric devices consist of a number of thermoelectric thermocouples. Figure 5.4 shows schematic figures of a TEG and a thermoelectric cooler (TEC) containing one thermoelectric thermocouple, where n-type and p-type thermoelectric legs are connected "thermally in parallel and electrically in series".[1]

Table 5.1 The *zT* values of Bi_2Te_3-based thermoelectric materials.

Reference	Year	Material	Maxiumum *zT*	Temperature (K)
26	2017	n-Type $Bi_2Te_{2.5}Se_{0.5}$	1.18	463
27	2017	p-Type $Ag_{0.002}Bi_{0.5}Sb_{1.498}Te_3$	1.25	400
24	2016	n-Type $Bi_2Te_{3-x}Se_x$	1.23	480
28	2016	p-Type Bi_2Te_3-based materials	1.4	453
29	2016	p-Type Bi_2Te_3-based materials	1.15	373
16	2016	n-Type $Bi_2Te_{3-x}Se_x$	1.1	473
17	2015	n-Type $Bi_2Te_{3-x}Se_x$	1.2	375
30	2015	p-Type $Bi_{0.5}Sb_{1.5}Te_3$	1.86	320
31	2014	n-Type $Bi_2Te_{2.3}Se_{0.7}$	1.2	445
31	2014	p-Type $Bi_{0.3}Sb_{1.7}Te_3$	1.3	380
32	2013	Bi_2Te_3/Sb_2Te_3 composites	0.7	400
33	2011	n-Type $Bi_2Se_{0.3}Te_{2.7}$	0.99	400
34	2008	n-Type Bi_2Te_3	1.18	423
35	2005	p-Type $(Bi,Sb)_2Te_3$	1.15	350
25	2001	p-Type Bi_2Te_3/Sb_2Te_3	2.4	300

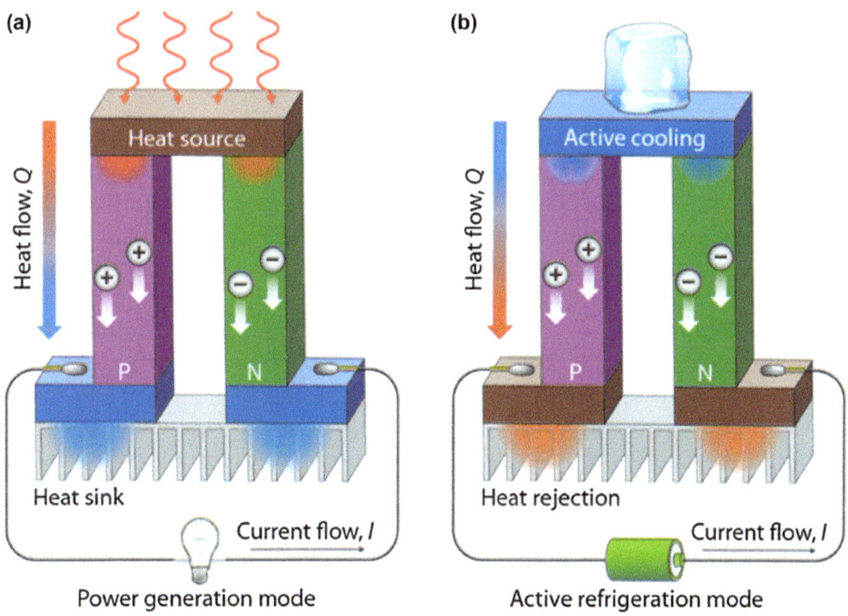

Figure 5.4 Schematic figures of a TEG and TEC.
Reproduced from ref. 18 with permission from Springer Nature, Copyright 2010.

The TEG efficiency is measured by a parameter η, which is defined as the ratio of generated electricity to consumed thermal energy. The maximum η can be written as[1]

$$\eta_{\max} = \frac{T_H - T_C}{T_H} \cdot \frac{\sqrt{1+ZT} - 1}{\sqrt{1+ZT} + \frac{T_C}{T_H}}, \tag{5.2}$$

where T_H and T_C are the temperatures of the hot end and cold end, respectively. Here, ZT is defined as[1]

$$ZT = \frac{(S_p - S_n)^2}{((\kappa_p \cdot \rho_p)^{\frac{1}{2}} + (\kappa_n \cdot \rho_n)^{\frac{1}{2}})^2} T, \tag{5.3}$$

where S_p and S_n are the Seebeck coefficients of p- and n-type materials, respectively, ρ_p and ρ_n are the electrical resistivity of p- and n-type materials, respectively, κ_p and κ_n are thermal conductivity of p- and n-type materials, respectively.

The performance of a TEC is evaluated by the coefficient of the performance (COP). Similarly, the maximum COP COP_{max} can be estimated by the following equation:[1]

$$COP_{\max} = \frac{T_C}{T_H - T_C} \cdot \frac{\sqrt{1+ZT} - \frac{T_H}{T_C}}{\sqrt{1+ZT} + 1}. \tag{5.4}$$

Bi_2Te_3-based materials are the most widely-used materials in commercial thermoelectric devices because of their high zT at RT. For example, the state-of-the-art thermoelectric elements of KELK Ltd., are fabricated with Bi_2Te_3-based alloys. Devices with a size of 50 mm×50 mm×4.2 mm can generate a maximum output power of 24 W and exhibit a very high conversion efficiency η of 7.2%.[36]

Several factors such as material properties, device structure, and fabrication processes can affect the energy conversion efficiency of thermoelectric devices. At first, the zT values of thermoelectric materials play a key role as shown in eqn (5.2). Particularly, high zT values across the whole operating temperature range are favorable. Feng *et al.* synthesized doped p-type $Bi_{0.5}Sb_{1.5-x}M_xTe_3$ (M = Cd, Cu, and Ag), exhibiting a high averaged zT value of 1.2 over a broad temperature range from 100 to 300 °C. Thermoelectric devices were fabricated with doped p-type $Bi_{0.5}Sb_{1.5-x}M_xTe_3$ and commercial n-type Bi_2Te_3-based material, and an efficiency up to 6.0% was shown at a temperature difference of 217 K, without any intentional optimization of device fabrication.[28]

Besides raising zT, designing a segmented structure with more than one kind of thermoelectric material is a promising method to enhance the energy conversion efficiency. Hu *et al.* reported that the maximum efficiency of their segmented Bi_2Te_3/PbTe module was as high as ~11% at a temperature difference of 590 K.[37] Zhang *et al.* reported Bi_2Te_3/skutterudite segmented modules that exhibited an efficiency of 12% at a temperature difference of 541 K.[38] Li *et al.* demonstrated that a high-performance TEG with segmented legs (n-type PbTe/BiTe and p-type $Bi_{0.3}Sb_{1.7}Te_3/Zn_4Sb_3$) can be fabricated by a one-step spark plasma sintering (SPS) method, which eliminated the soldering process between different thermoelectric materials.[39] These segmented TEGs are shown in Figure 5.5.

The properties of the thermoelectric devices including efficiency, output power, and output voltage are also influenced by fabrication methods. Much research effort has been focused on device fabrication. Methods including SPS,[39] electrodeposition,[40] sputtering,[41] screen-printing,[42] and 3D-printing[43,44] have been developed to make thermoelectric devices. Roth *et al.* reported cross-plane TEGs fabricated with electrochemically deposited Bi_2Te_3 and Sb_xTe_y.[40] They also demonstrated a new reflow soldering method using $Bi_{57}Sn_{42}Ag_1$ soldering paste to connect Cu electrodes with n-type Bi_2Te_3 and p-type Sb_xTe_y. Tan *et al.* fabricated thin film devices by magnetron sputtering.[41] In their device, densely columnar Cu film electrodes were used to reduce the electrical resistance of the device, resulting in an enhancement of output power. Navone *et al.* manufactured Bi_2Te_3 thick-film-based devices using a printing method, which was beneficial for scalable production.[42] More recently, attempts of using advanced 3D-printing technology to fabricate Bi_2Te_3-based thermoelectric devices were made.[43,44]

Another important direction for research into thermoelectric devices is microminiaturization in order to expand their applications in microelectronics. Snyder *et al.* used a scalable microelectromechanical-systems (MEMS) process to fabricate a thermoelectric device that consists of 126 n-type and p-type $(Bi–Sb)_2Te_3$ thermoelectric elements with a thickness of 20 μm and a diameter 60 μm.[45] By repeatedly developing and removing photoresist layers, these thermoelectric elements were connected electrically with bridging metal interconnects and formed a MEMS-like three-dimensional structure. The author also demonstrated that the fabricated thermoelectric devices can be used for both cooling and power generation. Zhang *et al.* reported a cross-plane micro-thermoelectric generator containing 127 pairs of n-type Bi_2Te_3 and p-type Sb_2Te_3 thermoelectric thermocouples deposited by electroplating.[46] Due to the low electrical resistance, this micro-TEG showed a high power density up to 9.2 $mW\,cm^{-2}$ (see Figure 5.6).

In summary, Bi_2Te_3-based materials are the best thermoelectric materials used at around RT so far; therefore, it is necessary to optimize their performance. In the last twenty years, significant research progress has been made and ideas involving nanostructure, metallic alloying, and nanocomposite have raised the zT values of Bi_2Te_3-based materials. The best p-type and n-type Bi_2Te_3-based bulk materials possess zT values of 1.86 and 1.23, respectively. The zT of the superlattice Bi_2Te_3 film is as high as 2.4, which is the largest value in the Bi_2Te_3 material system. However, there is a bottleneck to further raise the zT of Bi_2Te_3 with existing research strategies.

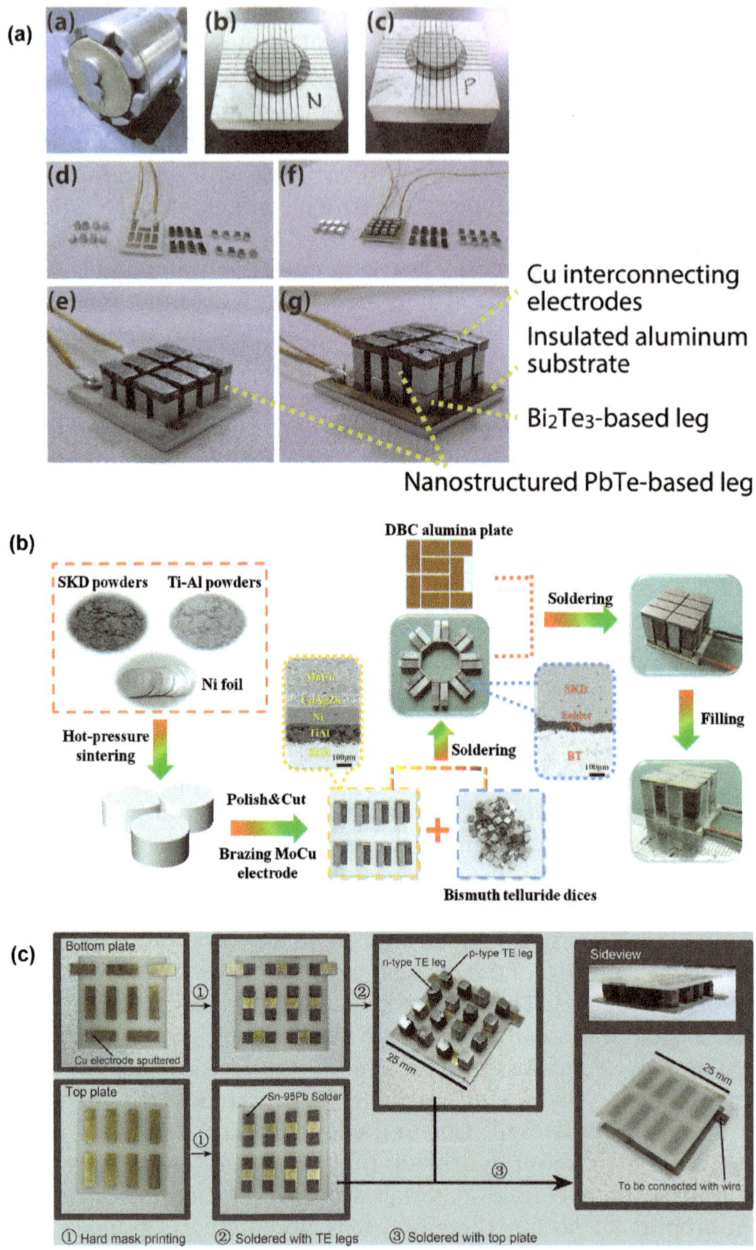

Figure 5.5 Fabrication processes of (a) segmented Bi_2Te_3/PbTe modules. Reproduced from ref. 37 with permission from the Royal Society of Chemistry. (b) Bi_2Te_3/skutterudite segmented modules. Reproduced from ref. 38 with permission from the Royal Society of Chemistry. (c) n-Type PbTe/BiTe and p-type $Bi_{0.3}Sb_{1.7}Te_3/Zn_4Sb_3$ segmented modules. Reproduced from ref. 39 with permission from Elsevier, Copyright 2016.

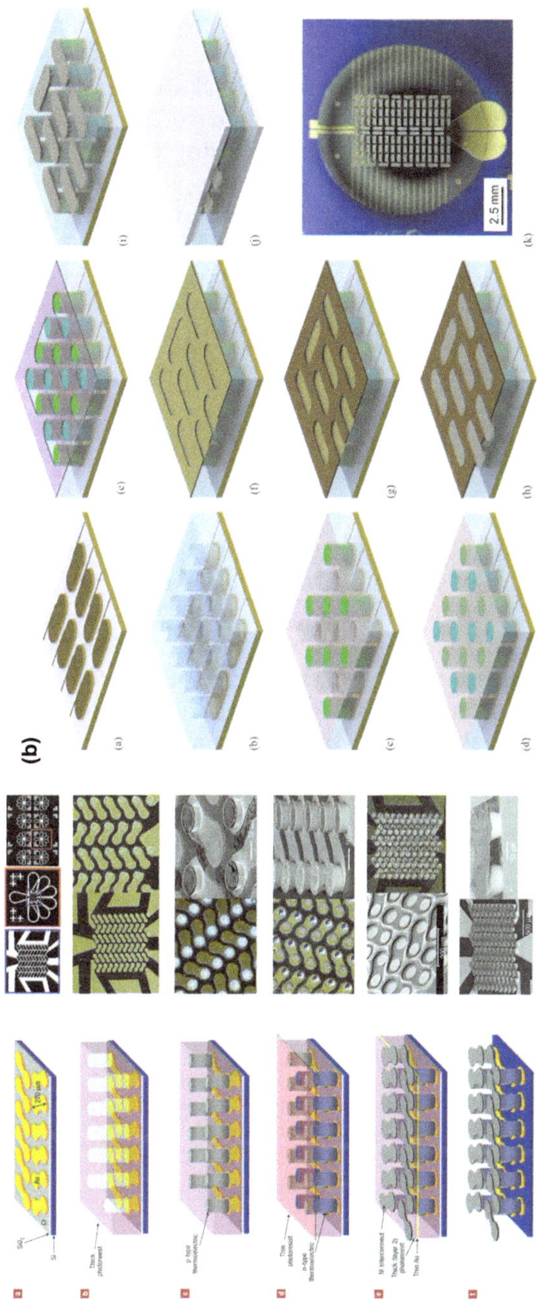

Figure 5.6 Micro-thermoelectric devices. (a) MEMS-like fabrication processes of thermoelectric devices. Reproduced from ref. 45 with permission from Springer Nature, Copyright 2003. (b) Fabrication steps of electroplated micro-thermoelectric generators. Reproduced from ref. 46 with permission from IEEE, Copyright 2016.

Hence, it is still an important topic to further enhance the properties of Bi_2Te_3.

Many companies such as KELK, Thermonamic Electronics, Marlow, HiZ, and Lairdtech have commercialized Bi_2Te_3-based thermoelectric devices.[47] The current thermoelectric devices have several shortcomings such as low efficiency, high cost, and poor reliability. Much work is required to optimize the material synthesis and device manufacturing to realize the mass production of high-performance, low-cost, and long lifespan of thermoelectric devices. In addition, the market of thermoelectric devices needs to be expanded in order to promote their commercialization. Current commercial thermoelectric devices are mostly used for cooling, and they are very limited for large-scale power generation due to low efficiency. Recently, the rapid development of wearable electronics provides a good opportunity for Bi_2Te_3-based TEGs since they have the advantages of small volume, low noise, and no moving parts.

5.3 SnSe

SnSe is an orthorhombic layered material with a *Pnma* space group under room temperature. A SnSe unit cell consists of two monolayers stacking along the *a* axis, as shown in Figure 5.7.[48] There is an armchair structure along the *b* axis and a zigzag structure along the *c* axis in a monolayer. Each Sn atom is bonded with seven Se atoms to form a $SnSe_7$ coordination polyhedral, which contains three short Sn–Se bonds within the *bc*-plane and four long Sn–Se bonds along the *a* axis. At a temperature above 750 K, SnSe transforms to a high-symmetry orthorhombic structure with a *Cmcm* space group.[48]

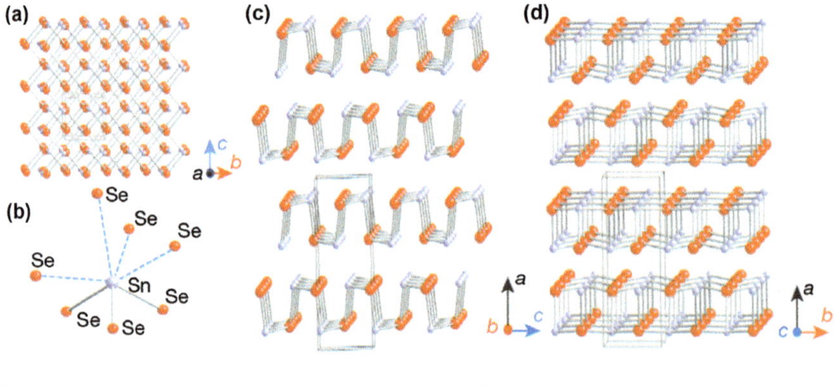

Figure 5.7 Crystal structure of SnSe. (a) Crystal structure along the *a* axis: gray, Sn atoms; red, Se atoms. (b) Highly distorted $SnSe_7$ coordination polyhedron with three short and four long Sn–Se bonds. (c) Structure along the *b* axis. (d) Structure along the *c* axis.
Reproduced from ref. 48 with permission from Springer Nature, Copyright 2014.

SnSe has attracted much research interest since Zhao *et al.* achieved a remarkable zT of 2.6 in single crystal SnSe in 2014.[48] The single crystal SnSe shows anisotropic thermoelectric properties because of its low-symmetry structure. Figure 5.8 shows the σ, S, power factor (PF), and κ along a, b, and c directions. The highest power factor of 10 $\mu W\,cm^{-1}\,K^{-2}$ and the highest zT of 2.6 is obtained along the b axis at 923 K. The electrical and thermal properties along the c axis are close to those along the b axis, leading to a zT of 2.3 at 923 K. Although the thermal conductivity along the a axis is the lowest, the zT along this direction is only 0.8 at 923 K, which is attributed to the significantly reduced electrical conductivity.[48]

For p-type SnSe, Na doping is usually adopted. The σ along the b axis can be increased from 10–12 $S\,cm^{-1}$ to 1300–1500 $S\,cm^{-1}$ in single crystal SnSe through Na doping.[49,50] The S of a Na-doped single crystal SnSe along the b axis is about 160 $\mu V\,K^{-1}$ at 300 K, which is comparable to that of commercial $Bi_{2-x}Sb_xTe_3$ with similar carrier concentrations. Both the enhanced σ and the high S lead to an ultra-high PF along the b axis, which is ~40 $\mu W\,cm^{-1}\,K^{-2}$ at 300 K. Due to the ultra-high PF, the Na-doped single crystal SnSe can possess high zT values of 0.7–2.0 in a temperature range of 300–773 K, leading to a considerably large average ZT of 1.34 from 300 to 773 K.

N-type single crystal SnSe was also successfully synthesized by Bi doping and a remarkable zT of 2.2 was achieved with a carrier concentration of $2.1\times10^{19}\,cm^{-3}$ at 773 K.[51] Atomic resolution scanning tunneling microscopy (STM) was used to study the microstructure of Bi-doped SnSn. It was observed that Bi dopants occupied the Sn sites. The STM image in Figure 5.9(a) shows a defect in Bi-doped SnSe with a unique topographic feature indicated by a dotted ellipse.[51] It is assumed that this defect is generated by Bi substitution. To verify the assumption, the STM image of the defect (Figure 5.9(b)) was simulated by density functional theory (DFT) calculations with a $5\times5\times1$ supercell. In the simulation, a Sn atom is removed from the lattice and a Bi atom is placed in the same location. Figure 5.9(c) shows the simulated STM image in which the defect with similar features as those in Figure 5.9(b) is reproduced; therefore, it is confirmed that Bi dopants are located in the Sn sites in Bi-doped SnSe.[51]

Recently, Chang *et al.* reported that with Br doping in single crystal SnSe, an impressively high zT of 2.8 was achieved in the out-of-plane direction, rather than the in-plane direction, which provided a novel method to enhance zT in thermoelectric materials with layered structures.[9]

Besides single crystal SnSe, polycrystalline SnSe is being extensively studied since polycrystals have better mechanical properties in comparison with single crystals. Doping is a widely-used method to enhance the electrical conductivity of polycrystalline SnSe. Various acceptors (Na,[52] Ag,[53] Ge,[54] *etc.*) and donors (Br,[55] I,[56] *etc.*) have been used in SnSe polycrystals. Wei *et al.* performed a systematic study on the thermoelectric properties of alkali metals (Li, K, and Na) doped polycrystalline SnSe. The result showed that the electrical resistivity ρ can be effectively reduced from 1.2 to 0.18, 0.036, and 0.011 Ωcm by doping 1% Li, K, and Na, respectively. Na was found to be the

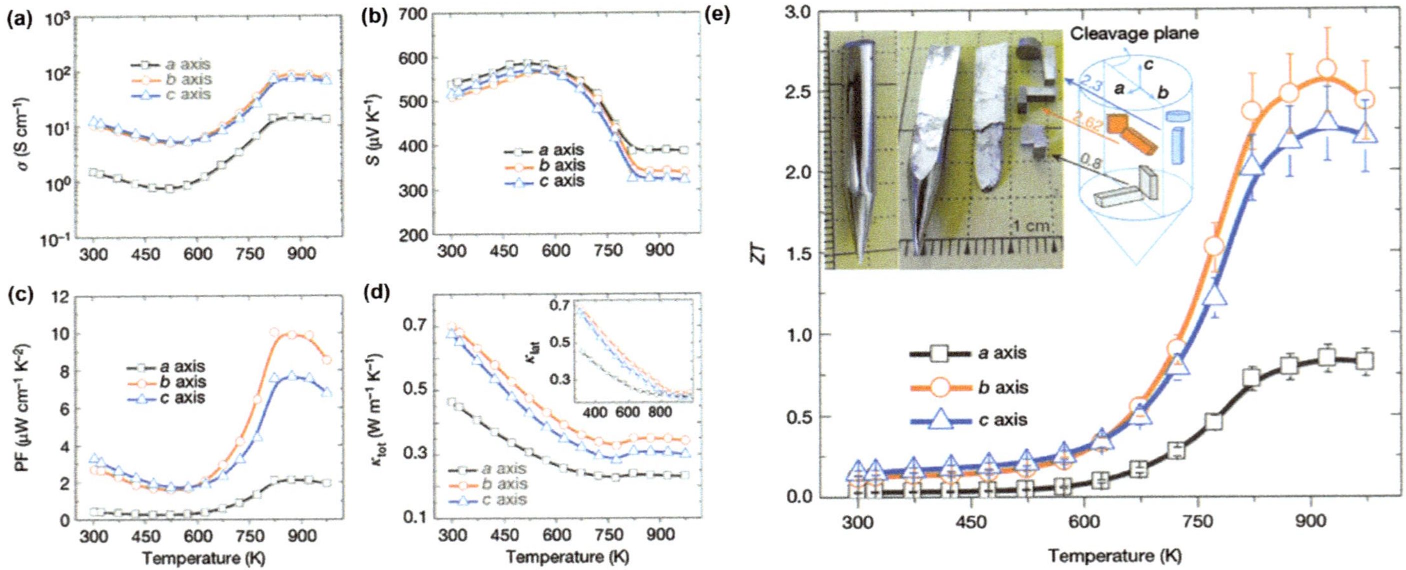

Figure 5.8 Thermoelectric properties of SnSe crystals. (a) Electrical conductivity. (b) Seebeck coefficient. (c) Power factor, PF. (d) Total thermal conductivity, κ_{tot}. Inset, lattice thermal conductivity, κ_{lat}. (e) zT values along different axial directions. Reproduced from ref. 48 with permission from Springer Nature, Copyright 2014.

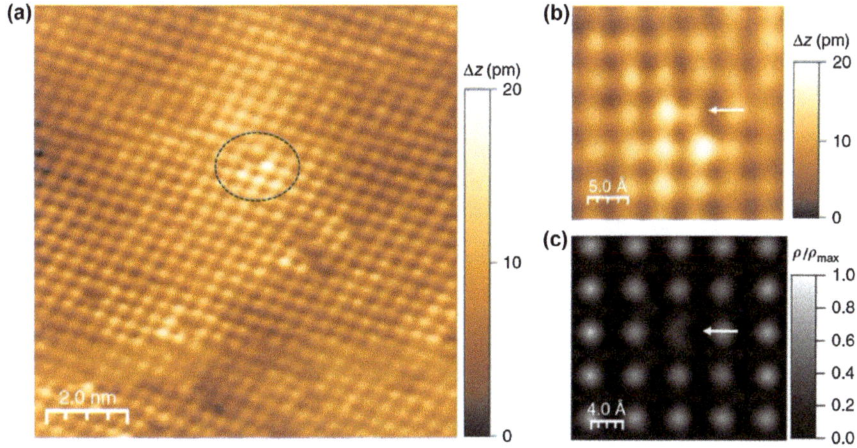

Figure 5.9 (a) STM topographic image of Bi-doped SnSe. The dotted ellipses indicate Bi dopants. (b) High-resolution image of a Bi dopant taken from one of the circled areas in (a). (c) Simulated STM image of a SnSe supercell with Bi dopant at the Sn site. The white arrows in (b) and (c) indicate the substitutional Bi atoms occupying the Sn site.

most suitable doping element. The Na-doped SnSe had an enhanced hole concentration of $4.4\times10^{19}\,cm^{-3}$ and a zT of 0.8 at 800 K.[52] Wu *et al.* first synthesized n-type $SnSe_{1-x}I_x$ polycrystals by melting and hot pressing.[56] When the I content x was more than 0.01, the $SnSe_{1-x}I_x$ polycrystals showed a stable n-type conductivity in the whole temperature range from 300 to 773 K. When $x=0.04$, the n-type $SnSe_{1-x}I_x$ sample had a carrier density of $2.0\times10^{17}\,cm^{-3}$ and a maximum zT of 0.8 at 773 K.

The most appealing property of SnSe is its intrinsic low thermal conductivity. According to the study by Zhao *et al.*, the thermal conductivity values of SnSe along the *a*, *b*, and *c* axes can be as low as 0.46, 0.70, and 0.68 $W m^{-1} K^{-1}$ at room temperature, respectively. Moreover, the κ decreases monotonically as the temperature increases and reaches a value of 0.23–0.34 $W m^{-1} K^{-1}$ at 973 K.[48] The ultra-low thermal conductivity of SnSe is due to its strong phonon anharmonicity. The mechanism of the phonon anharmonicity was investigated by inelastic neutron scattering (INS) measurements and first-principles calculations.[57] Figure 5.10 shows measured and calculated momentum-dependent phonon dispersions along $\Gamma-X$ and $\Gamma-Y$ directions in the first Brillouin zone. The transverse acoustic (TA) and transverse optical (TO) branches along $\Gamma-X$ are much softer than those along $\Gamma-Y$, which is consistent with the fact that the lattice thermal conductivity along the *a* axis is much lower than that within the *b*–*c* plane. When the temperature rises, a dramatic softening of phonon dispersions along $\Gamma-X$ and a modest softening of phonon dispersions along $\Gamma-Y$ are

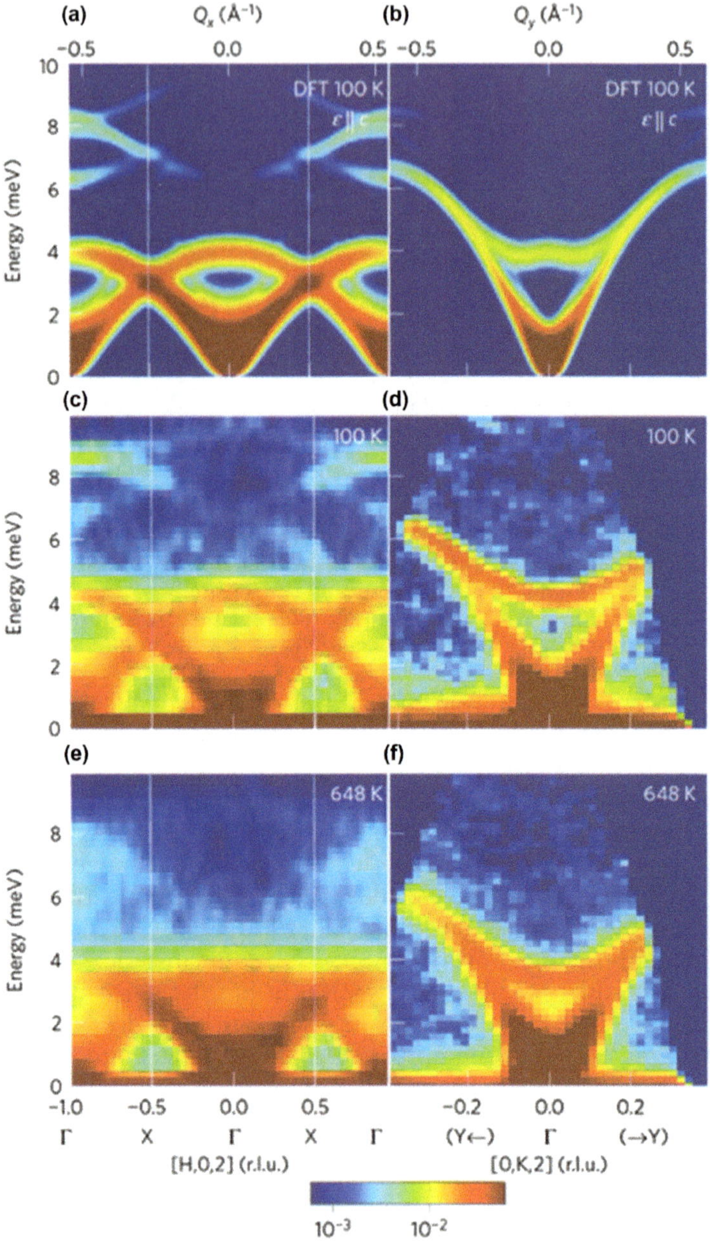

Figure 5.10 INS measurements and first-principles simulations reveal the strong dispersion anisotropy and softening with temperature of the low-energy optic phonons. (a), (b), $S(Q,E)$ calculated from first principles for the same directions, populated at $T = 100$ K, showing good agreement with INS data. (c), (d), Phonon dynamical structure factor, $S(Q,E)$ measured with INS for momenta along [H,0,2] and [0,K,2] (reciprocal lattice units—r.l.u.) at 100 K. (e), (f), Same as (c), (d) but at 648 K. Reproduced from ref. 57 with permission from Springer Nature, Copyright 2015.

observed, suggesting a strong anisotropy of anharmonicity phonons. In addition, all phonon branches are broadening as the temperature increases, also indicating phonon anharmonicity. The calculation further reveals that strong phonon anharmonicity is attributed to the lattice instability affected by the activity of a Sn lone pair.[57]

Defects also contribute to the ultra-low thermal conductivity in off-stoichiometry SnSe.[58] High-density Sn vacancies and Se interstitials were observed by advanced aberration corrected scanning transmission electron microscopy (STEM). For example, Figure 5.11(a) shows the annular bright field (ABF)-STEM images of off-stoichiometry SnSe synthesized by the Bridgman method. The interstitials are indicated by yellow arrows. For comparison, a stoichiometric SnSe single crystal was prepared by a horizontal gradient freezing method and no interstitial was observed in this sample by ABF-STEM (Figure 5.11(b)), which proves that the interstitials exist in the off-stoichiometric sample.[58] In addition, theoretical calculations revealed that the vacancies and interstitials can effectively increase phonon scattering thereby reducing thermal conductivity. The lattice thermal conductivity of stoichiometric SnSe was much higher than that of the off-stoichiometric one, which was coincident with the simulation results (Figure 5.11(c)).[58]

The S of doped SnSe is as high as 160 $\mu V K^{-1}$ at 300 K, which is superior to those of other thermoelectric materials such as PbSe and SnTe with similar carrier density, as shown in Figure 5.12(a). The high S value originates from the unique band structure of SnSe with multiple local band extrema. Figure 5.12(b) and (c) show the calculated S as a function of n_H at 300 K and the calculated band structure of SnSe, respectively. In the case of heavy

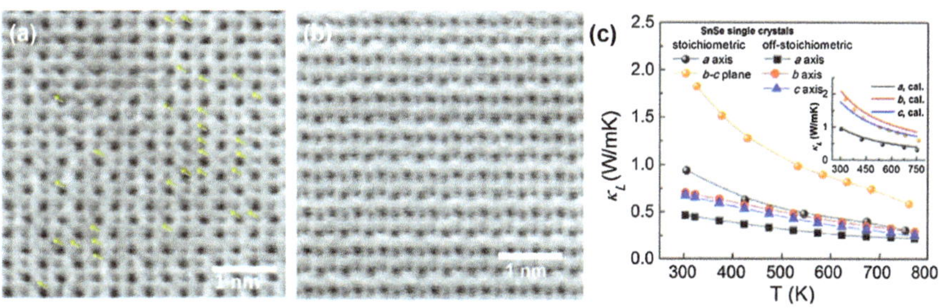

Figure 5.11 (a) ABF-STEM image of single crystalline $Sn_{0.985}Na_{0.015}Se$ viewed along the c axis, where yellow arrows indicate the interstitial atoms. (b) ABF-STEM image of a stoichiometric SnSe single crystal along the c axis exhibits a neat Sn–Se lattice; almost no interstitials can be seen. (c) A stoichiometric SnSe single crystal exhibits much higher lattice thermal conductivities than the off-stoichiometric SnSe sample; meanwhile, the lattice thermal conductivity values of a stoichiometric single crystal sample match the first-principles calculations based on a perfect Sn–Se lattice very well (inset).
Reproduced from ref. 58 with permission from Elsevier, Copyright 2017.

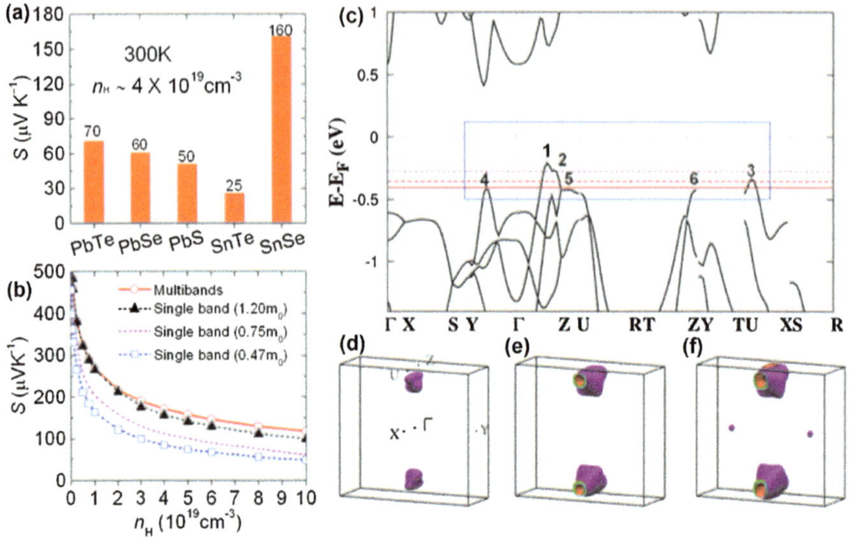

Figure 5.12 (a) Room temperature Seebeck coefficients comparisons for lead and tin chalcogenides. (b) Calculated Seebeck coefficients as a function of carrier density at 300 K. (c) Electronic band structure of hole-doped SnSe indicates non-parabolic, complex multi-band valence states. The red dotted lines from top to bottom represent the Fermi levels with the carrier density of 5×10^{17}, 5×10^{19}, 2×10^{20}, and 5×10^{20} cm^{-3}, respectively, indicating that heavy doping pushes the Fermi level deep into the multi-valence band structure. (d, e, and f) are the Fermi surfaces of SnSe at 5×10^{19}, 2×10^{20}, and 5×10^{20} cm^{-3}, respectively. The Fermi surface also illustrates the multiple types of pockets (or valleys) coming from the numerous valence bands all within a small energy window. Reproduced from ref. 50 with permission from AAAS.

doping, the Fermi level of SnSe is pushed deeply into multi-valence band structure because of the relatively small energy difference between valence band maximums. As a result, heavily doped SnSe has a large *S* due to band converagence.[50]

The electrical conductivity of SnSe is relatively low, although it has low κ and high *S*. Doping has been tried to increase the σ of SnSe, however, it is more challenging for SnSe in comparison with Pb-based chalcogenides due to the local distortion of Sn–Se bonds.[50] For example, Chen *et al.* synthesized Ag-doped p-type polycrystalline SnSe.[53] The $AgSnSe_2$ secondary phase appeared even when the doping concentration of Ag was as little as 1%, resulting in a reduction in electrical conductivity and Hall coefficient.[53]

The thermoelectric properties of 2D SnSe are also very interesting besides SnSe bulk materials. According to the first-principles calculations by Wang *et al.*, doped n-type monolayer SnSe had a high *zT* of 3.27 along the *b* axis under 700 K, which was enhanced seven times in comparison with its bulk counterpart under 700 K. The enhancement of *zT* arises from a large *S* due to the quantum confinement effect.[59] Recently, SnSe nanosheets were synthesized by a chemical exfoliation method and SnSe/poly(3,4-ethylene-dioxythiohene):poly-(styrenesulfonate) thermoelectric composites were fabricated. The composites showed a significantly enhanced power factor and a maximum zT of 0.32, as shown in Figure 5.13.[60] We produced SnSe nanosheets with a facile liquid exfoliation method.[61] The nanosheet thickness can be tailored through controlled centrifugation. The experimental data from a UV–Vis–NIR measurement showed that the bandgap of SnSe nanosheets increased when the nanosheet thickness decreased. This relationship between the bandgap and the thickness was verified by first-principles calculations.[61] Figure 5.14 shows the electronic structure of SnSe nanosheets and bulk. The calculated bandgap of SnSe bulk is 0.93 eV, while those of SnSe nanosheets with 1, 7, and 15 layers are 1.79, 1.19, and 1.13 eV, respectively. Compared to its bulk counterpart, 2D SnSe is expected to have lower κ, enhanced S, and tunable band structures. Therefore, SnSe 2D material provides a new opportunity to further improve the thermoelectric performance of SnSe.

SnSe is a high-performance medium-temperature thermoelectric material. Table 5.2 summarizes the current research status of SnSe crystals. The maximum zT of n- and p-type single crystals can be more than 2.0 and that of polycrystals is about 1.0. The major advantages of layered SnSe thermoelectric materials are listed here: (1) The intrinsic thermal conductivity of SnSe is low, thus the material's synthetic processes are simplified and it is not necessary to reduce κ by introducing a nanostructure or superlattice, which eliminates the shortcoming that nanostructured materials are usually not stable at high temperatures. (2) The band convergence occurs in heavily-doped SnSe due to its multi-valence band structure, which is helpful for the enhancement of S. (3) Both Sn and Se are environmentally friendly elements and their abundance of elements is higher than that of Bi and Te, which are beneficial for thermoelectric applications. The challenges remain for SnSe, including realizing heavy doping in SnSe, precisely tuning interstitials and vacancies in SnSe, achieving a large average ZT value in a broad temperature range, and improving the thermoelectric properties of polycrystalline SnSe.

5.4 BiCuSeO

Oxide-based thermoelectric materials such as ZnO,[66] $Ca_3Co_4O_9$,[67] $LaMnO_3$,[68] $SrTiO_3$,[69] and BiCuSeO[70] have been intensively studied in recent years. Due to the advantages of low toxicity, low cost, and high chemical and thermal stability, oxide-based thermoelectric materials are regarded as promising materials used at high temperatures. Among them, p-type BiCuSeO-based materials have shown superior thermoelectric performance and high zT over 1.[70] The zT of n-type BiCuSeO is very low, therefore it is rarely studied. In this chapter, we focus on the research progress of p-type BiCuSeO.

BiCuSeO belongs to the tetragonal crystal system with a space group of *P*4/*mm* (PDF 45-0296). In a BiCuSeO crystal, the Bi_2O_2 and Cu_2Se_2 layers are alternatively stacked along the c axis with a sequence of –Bi–O–Bi–Se–Cu–Se–,

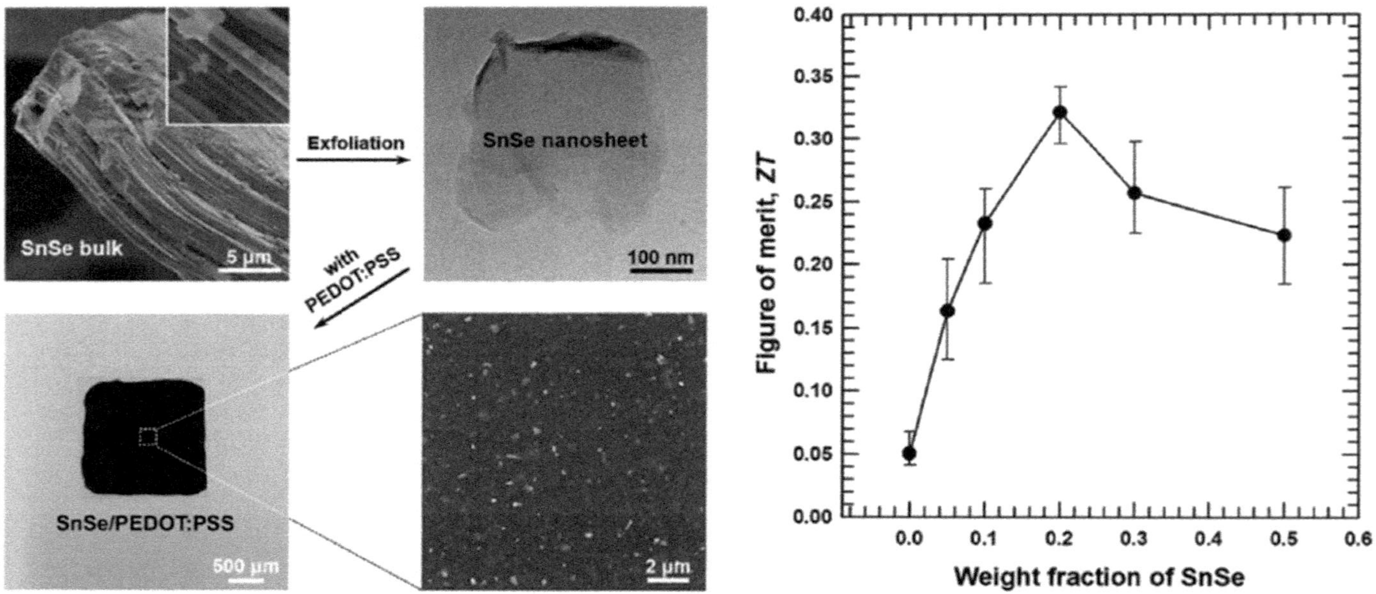

Figure 5.13 SnSe nanosheets synthesized by chemical exfoliation and SnSe/poly(3,4-ethylenedioxythiohene):poly-(styrenesulfonate) composites. The composites show a maximum zT of 0.32.
Reproduced from ref. 60 with permission from American Chemical Society, Copyright 2016.

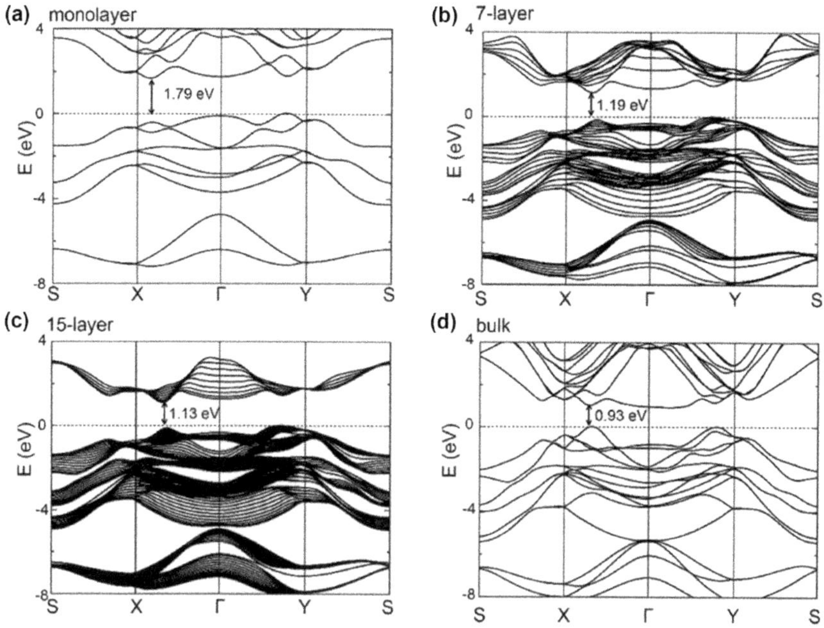

Figure 5.14 Calculated band structures for the SnSe monolayer (a), 7-layer (b), 15-layer (c), and bulk (d). The Fermi energies are fixed on the top of the valence bands.
Reproduced from ref. 61 with permission from American Chemical Society, Copyright 2017.

Table 5.2 The *zT* values of SnSe.

Reference	Year	Material	Type	Maximum *zT*	Temperature
49	2016	p-Type Na-doped SnSe	Single crystal	2.0	800 K
50	2015	p-Type Na-doped SnSe	Single crystal	2.0	773 K
48	2014	p-Type SnSe	Single crystal	2.6	923 K
9	2018	n-Type Br-doped SnSe	Single crystal	2.8	773 K
51	2016	n-Type Bi-doped SnSe	Single crystal	2.2	773 K
62	2017	Ag-doped $(SnSe)_{1-x}(SnS)_x$	Polycrystals	1.67	823 K
63	2017	p-Type Zn-doped SnSe	Polycrystals	0.96	873 K
52	2016	p-Type Na-doped SnSe	Polycrystals	0.8	800 K
54	2016	p-Type Ge-doped SnSe	Polycrystals	0.77	800 K
64	2016	p-Type textured SnSe	Polycrystals	0.92	873 K
65	2016	p-Type K-doped SnSe	Polycrystals	1.1	773 K
55	2016	n-Type Br-doped SnSe	Polycrystals	1.2	773 K
56	2015	n-Type I-doped $SnSe_{1-x}S_x$	Polycrystals	1.0	773 K
53	2014	p-Type Ag-doped SnSe	Polycrystals	0.6	750 K

as shown in Figure 5.15.[70] In Bi_2O_2 layers, a Bi atom is connected with four adjacent O atoms to form a Bi_4O tetrahedra. Similarly, a Cu atom is connected with four adjacent Se atoms to form a Cu_4Se tetrahedra in Cu_2Se_2

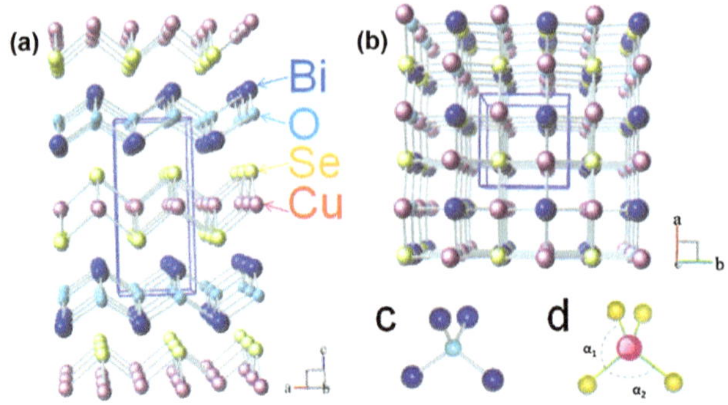

Figure 5.15 (a) Crystal structure of BiCuSeO along the *b* axis. (b) Structure along the *c* axis. (c) Bi_4O tetrahedra. (d) $CuSe_4$ tetrahedra.
Reproduced from ref. 70 with permission from the Royal Society of Chemistry.

layers. The Bi_2O_2 layers are insulating, while the Cu_2Se_2 layers are conductive and serve as carrier transport pathway.

BiCuSeO was first used as a thermoelectric material in 2010 and has attracted increasing attention since then.[71] The zT of BiCuSeO has been raised rapidly from 0.5 to 1.5 in the past few years.[71,72] Typically, bulk BiCuSeO-based thermoelectric materials are produced by a two-step process. First, BiCuSeO powders are prepared by solution phase synthesis,[73] mechanical alloying[74] or nonequilibrium self-propagating high-temperature synthesis (SHS).[75] Second, bulk materials are made of these powders by hot pressing,[76] SPS,[74] *etc.* The resulting bulk materials are mainly polycrystalline.[70]

In general, BiCuSeO-based materials have a high Seebeck coefficient, low electrical conductivity, and extremely low thermal conductivity.[70] Therefore, much research effort has been paid to improve the electrical conductivity. Pristine BiCuSeO bulks produced by the two-step process show p-type conduction due to the Cu vacancies that are introduced in the synthetic process. The electrical conductivity is about 1.12 S cm^{-1} at room temperature, which is much lower than those of conventional thermoelectric alloys such as Bi_2Te_3. The electrical conductivity is mainly contributed by holes. It can be estimated by $\sigma = n\mu e$, where μ is the carrier mobility, n is the carrier concentration, and e is the carrier charge.[77] Therefore, σ can be enhanced by increasing n or μ.

Doping is widely used to increase n. To understand the doping behavior, BiCuSeO can be simplified as an ionic crystal (strictly speaking, BiCuSeO is not an ionic crystal because of the hybridization of Cu and Se orbitals).[78] As a result, it can be considered that Bi is 3^+ valence, Cu is 1^+ valence, and both Se and O are 2^- valence in BiCuSeO. According to eqn (5.5), one can expect that more holes can be generated by substitution of Bi^{3+} with M^{2+} (M = Mg,[79] Ca,[80] Sr,[71] Ba,[74] *etc.*) dopants and the n can be increased.[70]

$$2Bi_{1-x}M_xCuSeO = (Bi_{2(1-x)}M_{2x}O_2)^{2(1-x)+} + (Cu_2Se_2)^{2-} + 2xh^+ \qquad (5.5)$$

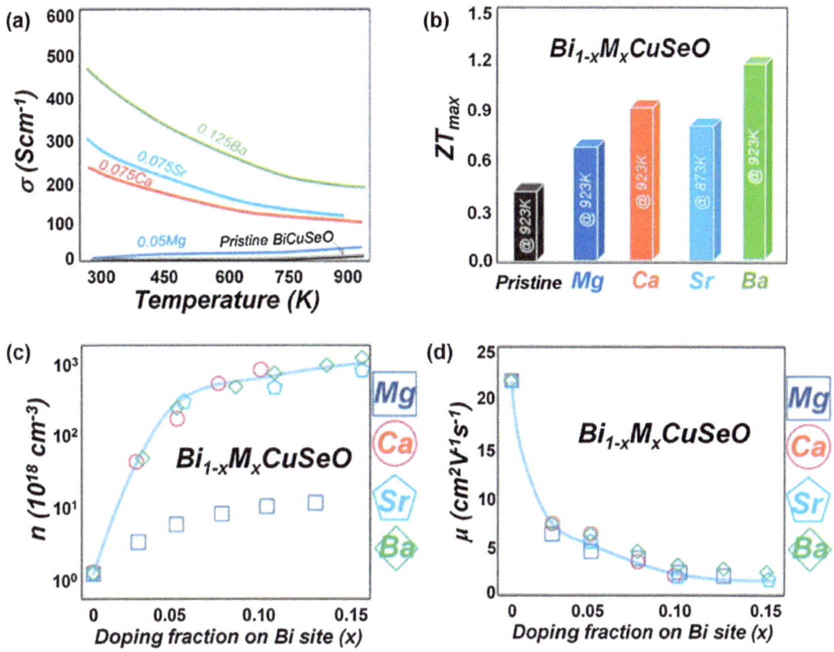

Figure 5.16 Thermoelectric properties of M^{2+} (M is Mg, Ca, Sr, or Ba) doped BiCuSeO. (a) Electrical conductivity. (b) Dimensionless zT for the $Bi_{1-x}M_xCuSeO$ (M is Mg, Ca, Sr, or Ba) systems. (c) Carrier concentration of BiCuSeO as a function of M^{2+} doping fractions. (d) Carrier mobility of BiCuSeO as a function of M^{2+} doping fractions.
Reproduced from ref. 70 with permission from the Royal Society of Chemistry.

Experiments have confirmed that M^{2+} doping can effectively enhance n, thus improving the electrical conductivity and zT, as shown in Figure 5.16(a) and (b). Particularly, with Ba doping, n is significantly promoted from 1.1×10^{18} cm^{-3} in undoped BiCuSeO to 1.1×10^{21} cm^{-3} in $Bi_{0.875}Ba_{0.125}$CuSeO, leading to an electrical conductivity of ~200 S cm^{-1} and a zT of 1.1 under 923 K.[74] On the other hand, M^{2+} doping increases scattering thereby reducing the μ. In $Bi_{1-x}M_xCuSeO$, when the M^{2+} doping fraction x increases from 0 to 0.15, the n increases rapidly, but the μ decreases gradually from ~22 to 2 cm^2 V^{-1} s^{-1}, as shown in Figure 5.16(c) and (d).

To alleviate the deterioration of μ, a method of 3D modulation doping was used in $Bi_{1-x}Ba_xCuSeO$.[81] $Bi_{0.875}Ba_{0.125}CuSeO$ with modulation doping was obtained by mixing pristine BiCuSeO and $Bi_{0.75}Ba_{0.25}CuSeO$ powders with a molar ratio of 1 : 1, following by SPS sintering. TEM analysis showed that the modulation doped sample was composed of Ba-rich areas marked by 1 in Figure 5.17(b) and Ba-poor areas marked by 2 in Figure 5.17(b).[81] Ba-rich areas provided a large number of charge carriers and Ba-poor areas served as charge carrier transport channels. A carrier mobility (4.1 cm^2 V^{-1} s^{-1}) was

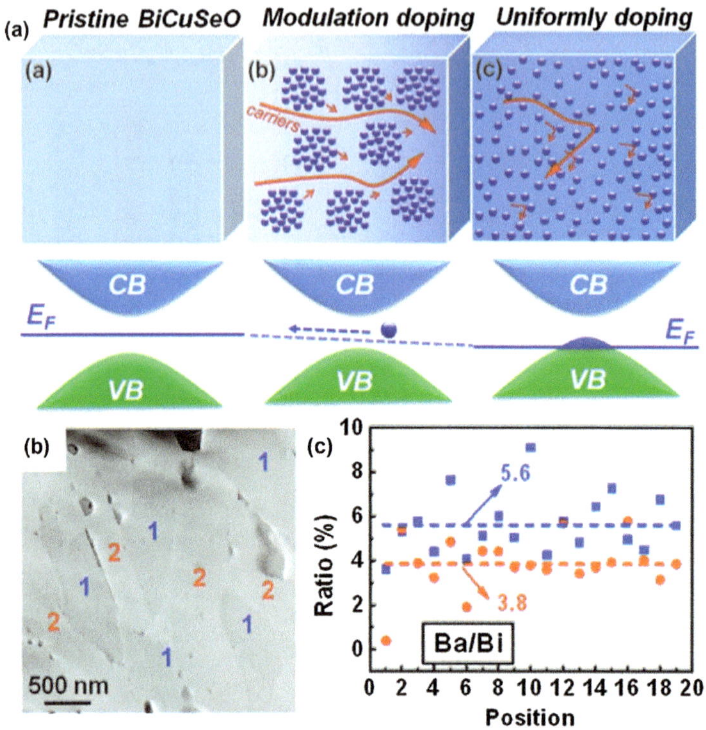

Figure 5.17 (a) Three-dimensional schematic showing the band structures and Fermi energy levels for the pristine BiCuSeO, modulation doped $Bi_{0.875}Ba_{0.125}CuSeO$, and uniformly doped $Bi_{0.875}Ba_{0.125}CuSeO$. For the modulation doped sample, the carriers transport preferentially in the low carrier concentration area. The modulation approach shows similar carrier concentration but higher carrier mobility compared to the uniform one. (b) Low-magnification high-angle annular dark-field image. (c) Ba/Bi content ratio in area 1 and area 2 in (b).
Reproduced from ref. 81 with permission from American Chemical Society, Copyright 2014.

obtained for the modulation doped sample, which was higher than that of the uniformly doped samples ($\sim$2.1 $cm^2 V^{-1} s^{-1}$), and a carrier concentration of 1.0×10^{21} cm^{-3} was still maintained. As a result, a high zT of 1.4 was achieved in $Bi_{0.875}Ba_{0.125}CuSeO$ with modulation doping.

Introducing extra vacancies in Cu sites is also beneficial to increase n, which can be explained by the following defect equation:[70]

$$2BiCu_{1-x}SeO = (Bi_2O_2)^{2+} + (Cu_{2(1-x)}Se_2)^{2(1+x)-} + 2xV_{Cu}^{-} + 2xh^{+}. \quad (5.6)$$

Liu *et al.* prepared $BiCu_{1-x}SeO$ ($x = 0$–0.1) samples with a two-step solid-state reaction method.[82] Figure 5.18 shows the electrical conductivity of samples as a function of temperature. When $x < 0.1$, the electrical conductivity of nonstoichiometric samples $BiCu_{1-x}SeO$ is much higher than that of

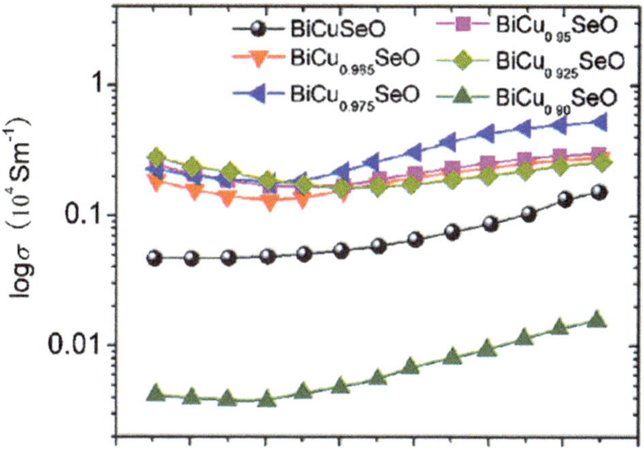

Figure 5.18 Temperature dependence of the electrical conductivity of $BiCu_{1-x}SeO$ ceramics.
Reproduced with permission from ref. 82 American Chemical Society, Copyright 2011.

pristine BiCuSeO because of the increase of n. However, when $x > 0.1$, the electrical conductivity drops significantly, which is caused by the emergence of the Bi_2O_3 phase.

Li *et al.* reported that the thermoelectric performance of BiCuSeO can be further enhanced by introducing Bi and Cu dual-vacancies.[83] In their experiments, positron annihilation spectrometry was used to investigate carrier transport behaviors between Bi/Cu dual vacancies. Figure 5.19 shows the positron density distribution in pristine BiCuSeO, $Bi_{0.975}CuSeO$, $BiCu_{0.975}SeO$, and $Bi_{0.975}Cu_{0.975}SeO$.[83] Since the injected positrons are annihilated by negative charges in the samples, the positrons tend to concentrate around negatively charged centers of Bi and Cu vacancies. It is interesting to note that positrons around Bi vacancies in $Bi_{0.975}Cu_{0.975}SeO$ are more concentrated than those in $Bi_{0.975}CuSeO$, while positrons around Cu vacancies in $Bi_{0.975}Cu_{0.975}SeO$ are less concentrated than those in $BiCu_{0.975}SeO$. This finding indicates that charges are transferred from Bi vacancies to Cu vacancies in $Bi_{0.975}Cu_{0.975}SeO$, which is helpful for the increase of n, leading to high electrical conductivity.

Another way to raise the electrical conductivity is to prepare single crystalline BiCuSeO. Theoretically, μ along the a–b plane of BiCuSeO is higher than that along the c axis. Therefore, it is expected that a high zT can be achieved along the a–b plane of single crystal BiCuSeO.[84] However, it is very challenging to obtain BiCuSeO single crystals with a sufficient size. Single crystal BiCuSeO with a size of $2.5 \times 2 \times 0.05$ mm^3 was prepared *via* a flux method.[85] However, this size is still far away from being suitable for thermoelectric applications. As an alternative, textured BiCuSeO samples were prepared by repeated hot forging.[76] Since BiCuSeO grains showed a preferential growth direction along the a–b plane during the hot forging

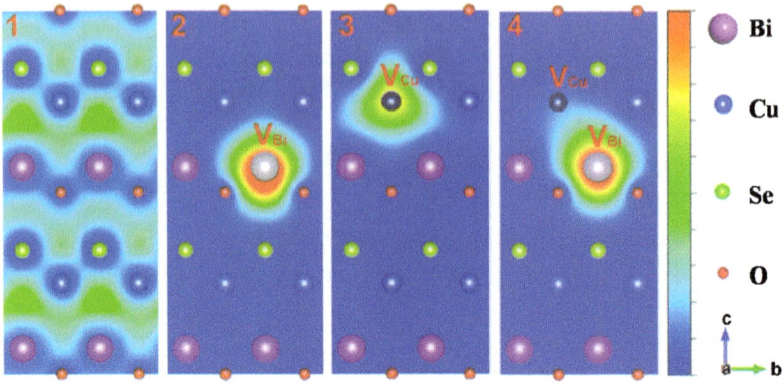

Figure 5.19 Schematic representation of trapped positrons for $Bi_{1-x}Cu_{1-y}SeO$ samples in (100) plane. 1: pristine BiCuSeO; 2: $Bi_{0.975}CuSeO$; 3: $BiCu_{0.975}SeO$; 4: $Bi_{0.975}Cu_{0.975}SeO$.
Reproduced from ref. 83 with permission from American Chemical Society, Copyright 2015.

process, they tended to orient the *c* axis to the pressing direction after three-times hot forging. Due to the high carrier mobility along the *a–b* plane, the carrier mobility of the resulting textured $Bi_{0.875}Ba_{0.125}CuSeO$ was twice as high as those of the $Bi_{0.875}Ba_{0.125}CuSeO$ samples synthesized in a previous study. Finally, the maximum zT was promoted to 1.4 in the textured $Bi_{0.875}Ba_{0.125}CuSeO$ produced by three-times hot forging.

The Seebeck coefficient of pristine BiCuSeO is ~350 $\mu V K^{-1}$ at 300 K. As mentioned previously, conductive Cu_2Se_2 layers are alternatively staked with insulating Bi_2O_2 layers, forming a natural superlattice structure. The high Seebeck coefficient may originate from this superlattice structure.[71] It is known that the Seebeck coefficient can be even higher in hole-doped SnSe in comparison with pristine SnSe due to band convergence. However, the Seebeck coefficient of BiCuSeO cannot be enhanced by hole doping. Figure 5.20 shows the band structure of pristine BiCuSeO.[78] The electronic band structure shows convergence of several heavy-hole and light-hole valence bands. Therefore, there is little space to further facilitate band convergence by hole doping.[86] Recently, Wen *et al.* suggested that extra spin entropy can be introduced through the incorporation of magnetic ions such as Ni and the Seebeck coefficient can be promoted.[86] They reported that the Seebeck coefficient of $Bi_{0.875}Ba_{0.125}Cu_{0.85}Ni_{0.15}SeO$ was raised to 403 $\mu V K^{-1}$ at 923 K by Ni doping, which was about two times larger than that of $Bi_{0.875}Ba_{0.125}CuSeO$ with modulation doping reported in previous study.

BiCuSeO-based materials have a low thermal conductivity. Theoretical calculations have suggested that the intrinsic low lattice thermal conductivity of BiCuSeO arises from anharmonic vibrations, which is caused by large displacement of Bi atoms.[87] In addition, defects, dislocations, and interfaces introduced in the synthetic process can also contribute to the low thermal

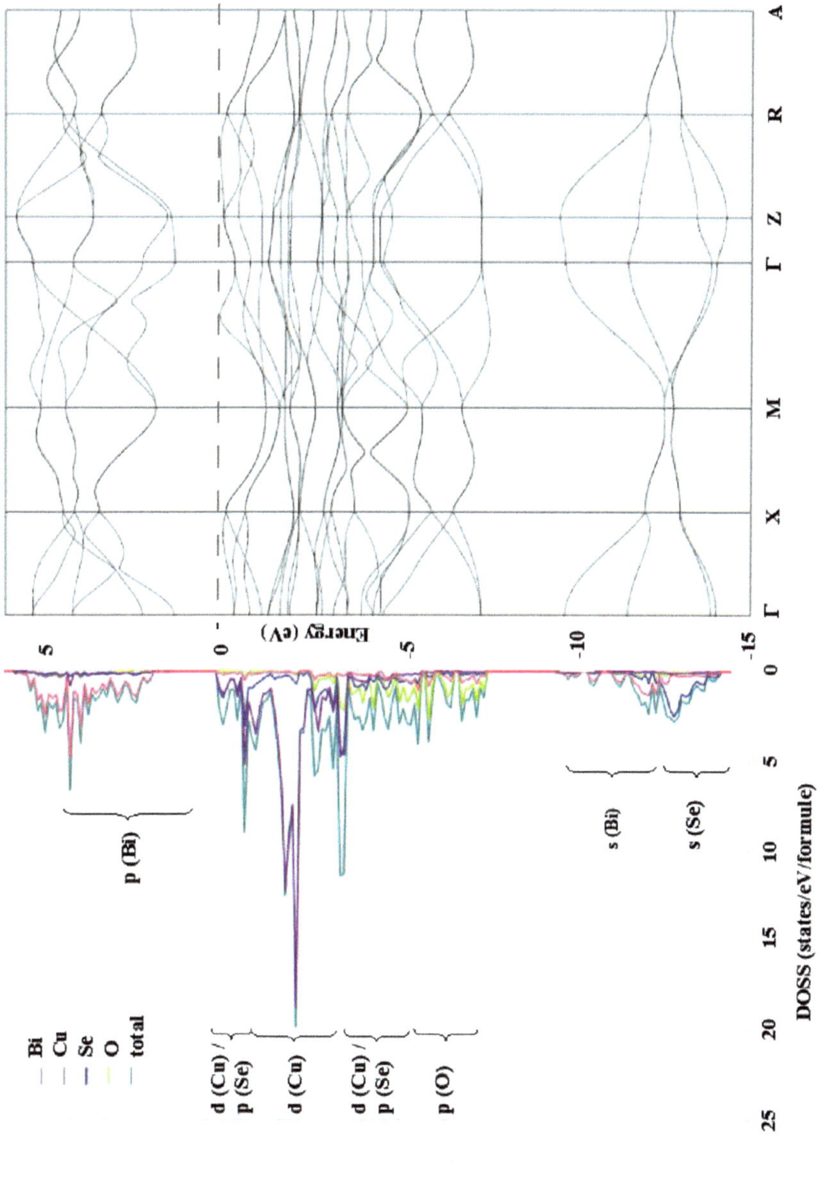

Figure 5.20 Electronic band structure (right) and projected DOS (left) for the undoped BiCuOSe. Reproduced from ref. 78 with permission from American Chemical Society, Copyright 2012.

Table 5.3 The zT values of BiCuSeO-based thermoelectric materials.

Reference	Year	Material	Type	Maximum zT	Temperature (K)
75	2017	SHS	p-Type	1.2	873
86	2017	Ba and Ni co-doped	p-Type	0.97	923
72	2016	Dual doping	p-Type	1.5	873
83	2015	Dual vacancies	p-Type	0.84	750
81	2014	3D modulation doping	p-Type	1.4	923
89	2013	Pb-doped	p-Type	1.14	823
80	2013	Ca-doped	p-Type	0.9	923
76	2013	Texture	p-Type	1.4	923
74	2012	Ba-doped	p-Type	1.1	923
82	2011	Cu deficiencies	p-Type	0.81	923
71	2010	Sr-doped	p-Type	0.76	873
88	2017	$Bi_{1.04}Cu_{1.05}Se_{0.99}Br_{0.01}O$ + 15% Ag	n-Type	0.05	475

conductivity of BiCuSeO-based materials. Recently, Ren *et al.* prepared $Bi_{0.96}Pb_{0.04}CuSe_{0.95}Te_{0.05}O$ samples by SHS.[75] The resulting samples contained multiple structural features such as point defects, nano-inclusions, and grain boundaries, which increased phonon scattering and reduced the lattice thermal conductivity approximately to the amorphous limit.

In summary, p-type BiCuSeO is a high-efficient thermoelectric material at median temperatures with low toxicity, low cost, and good reliability. Table 5.3 summarizes the research progress of p- and n-type BiCuSeO. The zT values of p-type BiCuSeO materials are more than 1.0 through doping, defect engineering, or texturing. N-type BiCuSeO is rarely investigated and its zT is limited. For example, Zhang *et al.* prepared n-type BiCuSeO by adding extra Bi and Cu and achieved a zT of 0.05.[88] Therefore, more effort is needed to improve n-type BiCuSeO, and thus both n- and p-type BiCuSeO materials can be used in thermoelectric devices. Additionally, single crystal BiCuSeO is predicted to have better thermoelectric properties in comparison with polycrystalline BiCuSeO by theoretical calculations; therefore, a possible research direction is to synthesize single crystal BiCuSeO with a large size.

5.5 Intercalated Layered Materials

Intercalated layered materials such as $AgCrSe_2$ are emerging thermoelectric materials.[10,90] Wu *et al.* reported the thermoelectric performance of $AgCrSe_2$ in 2016.[10] They achieved a zT of 0.55 in a nominal $Ag_{0.96}CrSe_2$ sample by introducing Ag vacancies. $AgCrSe_2$ can be a promising thermoelectric material because of its ultralow thermal conductivity, which arises from liquid-like thermal conduction behavior.[10]

Figure 5.21(a) shows the crystal structure of $AgCrSe_2$, in which Ag layers are alternatively stacked with $CrSe_6$ layers.[10] $AgCrSe_2$ undergoes an order-to-disorder transition at a critical temperature T_c of ~450 K. At the ground

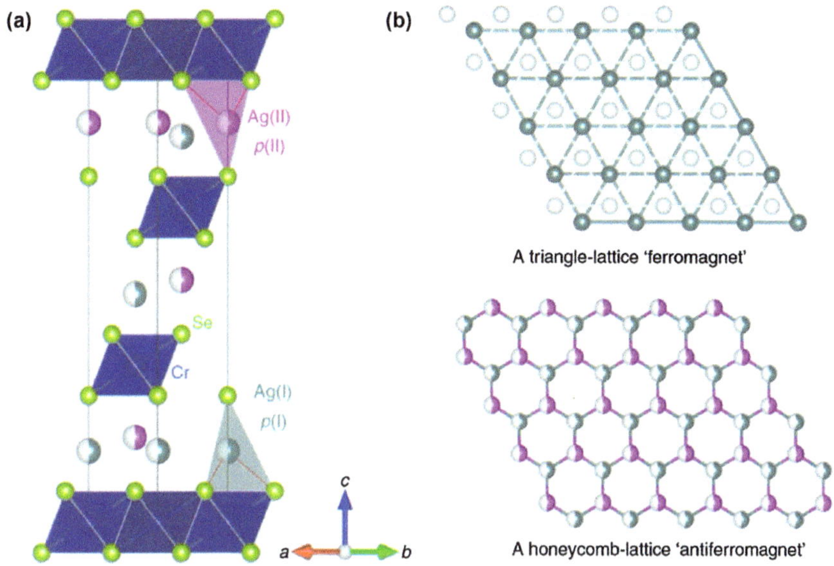

Figure 5.21 (a) The crystal structure of $AgCrSe_2$ with the $CrSe_6$ octahedra and $AgSe_4$ tetrahedra highlighted. (b) The occupational orders of tetrahedral interstitial sites. The uniform occupation leads to a "ferromagnetic" ground state (upper panel), such as $AgCrSe_2$, compared with an "anti-ferromagnetic" one at random occupation (lower panel).
Reproduced from ref. 10 with permission from Springer Nature, Copyright 2018.

state, Ag ions only occupy site I in $AgCrSe_2$ crystal, as shown in the upper panel of Figure 5.21(b).[10] When the temperature increases, more Ag ions immigrate from site I to site II. When the temperature is above T_c, about half of Ag ions occupy site II, as shown in the lower panel of Figure 5.21(b).[10] As a result, the space group transforms from $R3m$ at the ground state to $R\bar{3}m$ in the high-temperature phase. In the high-temperature phase, the fluctuation of Ag ions suppresses the transverse acoustic (TA) phonons, making the thermal conduction behavior of $AgCrSe_2$ similar to that of liquid materials.

INS measurements were performed to reveal the phonons' behavior in $AgCrSe_2$.[10] Figure 5.22(a) and (b) show the $S(Q,E)$ surface plots at 150 and 520 K, respectively.[10] At 150 K, the diffuse scattering exists very near to the elastic line, and a sharp peak of TA phonons is observed. By contrast, when the temperature rises to the 520 K, diffuse scattering is significantly broadened, and TA phonons are completely suppressed. Figure 5.22(c) shows the evolution of $S(Q,E)$ as a function of temperature. The TA branch is softened as the temperature increases and merges into diffuse scattering near T_c.[10] Figure 5.22(d) summarizes the temperature dependence of the half-width at the half-maximum of $\Gamma_{Diffuse}$ and Γ_{TA}.[10] As the temperature increases, the Γ_{TA} exhibits a linear temperature dependence in the whole temperature range, while the $\Gamma_{Diffuse}$ shows an obvious slope change near T_c.

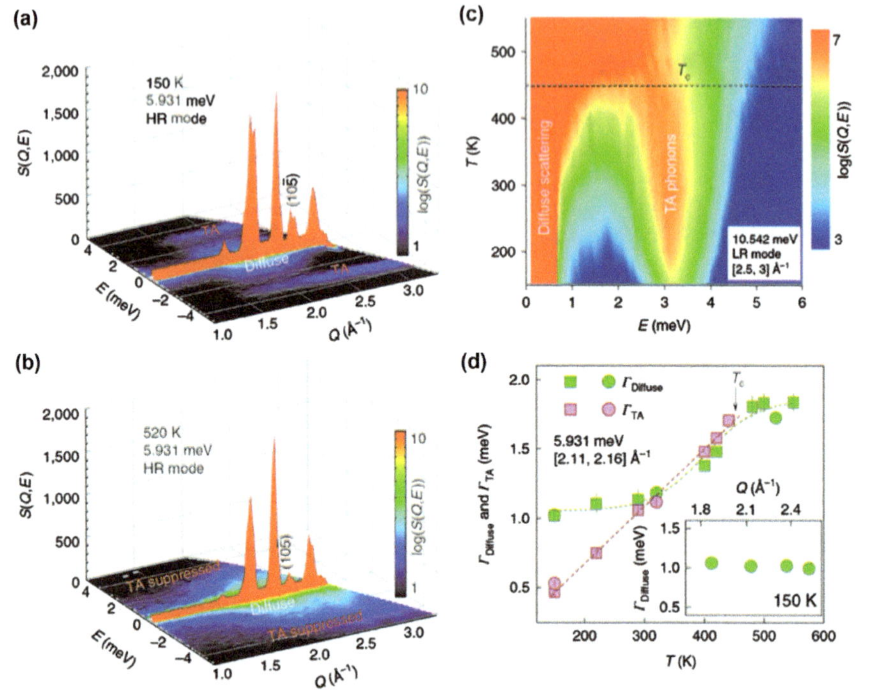

Figure 5.22 Suppression of TA phonons. $S(Q,E)$ surface plots at 150 (a) and 520 K (b) at $E_i = 5.931$ meV in the high-resolution mode. (c) Contour plot of $S(Q,E)$ as a function of temperature with $E_i = 10.542$ meV in the low-resolution mode at Q of [2.5, 3] $Å^{-1}$. (d) The half-width at half-maximum of the TA phonons and the diffuse scattering determined in spectral fitting. The Q-dependence of the latter is shown in the inset. The squares and circles are data points obtained in the low- and high-resolution modes, respectively. The dashed lines are guide for eyes. T_c is labelled in both (c) and (d).
Reproduced from ref. 10 with permission from Springer Nature, Copyright 2018.

When the temperature approaches T_c, the timescales of the TA phonons and the diffuse scattering are very similar; therefore, the TA phonons are completely suppressed due to the strong dynamic competition.[10]

It is suggested that the suppression of TA phonons can be observed in other layered materials intercalated with heavy atoms. Promising candidates may have a formula of $A^+TM^{3+}X^-$, where A is heavy atoms such as Ag, In, or Tl, TM is 3d transition metals except Zn, and X is chalcogen such as S or Se. Figure 5.23 shows the calculated probabilities of $A^+TM^{3+}X^{2-}$ crystallizing in the $R3m$ structure.[10] The $R3m$ structure is desirable because the heavy atoms occupy tetrahedral interstitials in this structure and liquid-like conduction is more likely to occur. Several compounds are predicted to have high probabilities to crystallize in this structure, which provides a promising route to the discovery of new thermoelectric materials with low thermal conductivity.

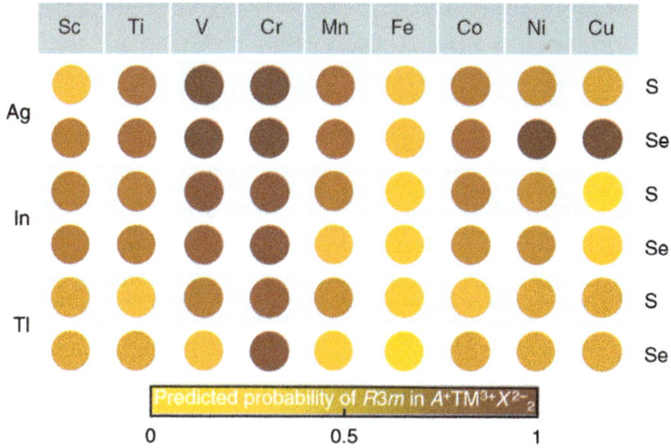

Figure 5.23 Predicted probabilities of the $R3m$ phase normalized to the probability of the most probable phase. Here, A = Ag, In and Tl, TM = 3d transition metals except Zn, and X = S and Se.
Reproduced from ref. 10 with permission from Springer Nature, Copyright 2018.

5.6 Summary

This chapter summarizes the research progress of four kinds of layered thermoelectric materials: Bi_2Te_3, SnSe, BiCuSeO, and $AgCrSe_2$. Bi_2Te_3-based thermoelectric devices are also briefly discussed. Figure 5.24 compares the Seebeck coefficient, electrical conductivity, thermal conductivity, and zT of these four materials. It can be seen that the Seebeck coefficient is high for all these layered materials because of the quantum confinement effect along the out-of-plane direction. Bi_2Te_3, a room-temperature thermoelectric material, has relatively high electrical conductivity and thermal conductivity; therefore, alloying and nanostructures are usually adopted to reduce the thermal conductivity in order to increase its zT. SnSe, BiCuSeO, and $AgCrSe_2$ show good thermoelectric properties at median temperatures. Their thermal conductivity and electrical conductivity are low; therefore, doping and vacancies are often manipulated to enhance the electrical conductivity in order to increase their zT values.

Much research progress has been achieved in layered thermoelectric materials. For example, Bi_2Te_3-based materials are widely commercialized; the zT value of n-type SnSe reaches 2.8. All of these achievements show that layered thermoelectric materials can play an important role in energy conversion and management in the near future. On the other hand, researchers gain insights in the design and synthesis of high-performance thermoelectric materials after investigating the transport properties of the layered materials. For instance, both high electrical conductivity and enhanced Seebeck coefficient can be obtained through the quantum confinement

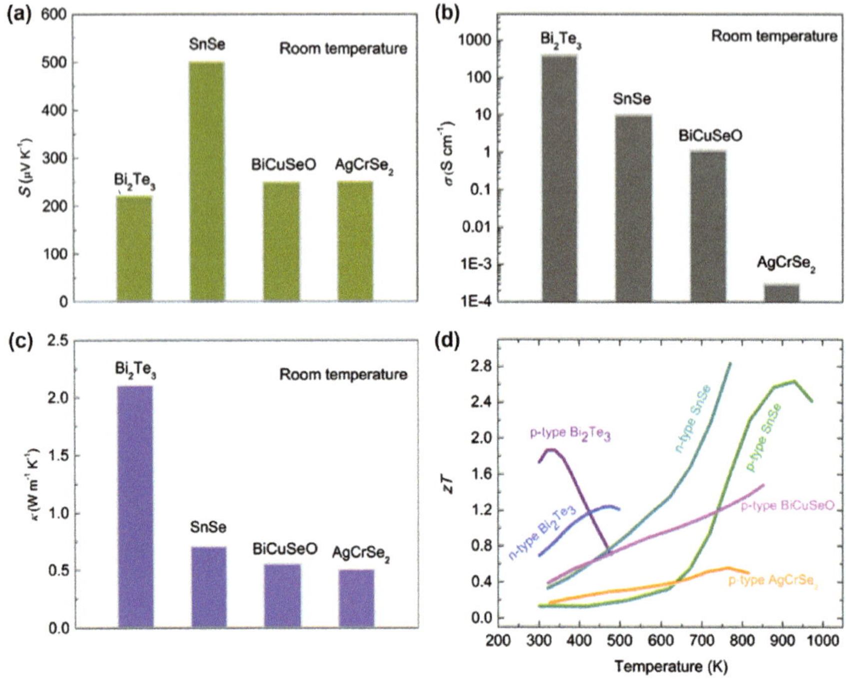

Figure 5.24 Thermoelectric properties of Bi_2Te_3, SnSe, BiCuSeO, and $AgCrSe_2$. (a–c). *S* (a), *σ* (b), and *κ* (c) of pristine Bi_2Te_3 single crystals,[8] SnSe single crystals,[48] BiCuSeO polycrystals,[70] and $AgCrSe_2$[90] polycrystals. *S*, *σ*, and *κ* are measured perpendicularly to the *c* axis of single crystals or the pressing direction of polycrystals. (d) the *zT* values of state-of-the-art Bi_2Te_3,[24,30] SnSe,[9,48] BiCuSeO,[88] and $AgCrSe_2$[90] as a function of temperature.

effect in 2D structures; thermal conductivity can be significantly reduced due to the suppression of TA phonons in the intercalated layered materials with heavy atoms.

Acknowledgements

This work was supported by the National Natural Science Foundation of China (Grant No. 51572149), National Key Research and Development Program of China (Grant No. 2016YFA0201003), National Basic Research Program of China (Grant No. 2013CB632504).

References

1. H. J. Goldsmid, *Introduction to thermoelectricity*, ed. R. Hull, R. M. Osgood, J. Parisi and H. Warlimont, Springer, Berlin, 2010.
2. G. Schmidt, L. Dudzinski and T. Sutliff, *6th International Energy Conversion Engineering Conference (IECEC)*, 2008, 5640.

3. A. Minnich, M. Dresselhaus, Z. Ren and G. Chen, *Energy Environ. Sci.*, 2009, **2**, 466.
4. Y. Yang, G. D. Xu and J. Liu, *J. Med. Devices*, 2014, **8**, 014507.
5. F. J. DiSalvo, *Science*, 1999, **285**, 703.
6. C. J. Vineis, A. Shakouri, A. Majumdar and M. G. Kanatzidis, *Adv. Mater.*, 2010, **22**, 3970.
7. M. Carle, P. Pierrat, C. Lahalle-Gravier, S. Scherrer and H. Scherrer, *J. Phys. Chem. Solids*, 1995, **56**, 201.
8. H. Goldsmid and R. Douglas, *Br. J. Appl. Phys.*, 1954, **5**, 386.
9. C. Chang, D. He, Y. Pei, C.-F. Wu, X. Wu, H. Yu, F. Zhu, K. Wang, Y. Chen, L. Huang, J.-F. Li, J. He and L.-D. Zhao, *Science*, 2018, **360**, 778.
10. B. Li, H. Wang, Y. Kawakita, Q. Zhang, M. Feygenson, H. L. Yu, D. Wu, K. Ohara, T. Kikuchi, K. Shibata, T. Yamada, X. K. Ning, Y. Chen, J. Q. He, D. Vaknin, R. Q. Wu, K. Nakajima and M. G. Kanatzidis, *Nat. Mater.*, 2018, **17**, 226.
11. S. Mishra, S. Satpathy and O. Jepsen, *J. Phys.: Condens. Matter*, 1997, **9**, 461.
12. B. Y. Yavorsky, N. Hinsche, I. Mertig and P. Zahn, *Phys. Rev. B*, 2011, **84**, 165208.
13. X. Li, H. Ren and Y. Luo, *Appl. Phys. Lett.*, 2011, **98**, 083113.
14. H. Goldsmid, *J. Electronics*, 1955, **1**, 218.
15. B. Poudel, Q. Hao, Y. Ma, Y. Lan, A. Minnich, B. Yu, X. Yan, D. Wang, A. Muto and D. Vashaee, *Science*, 2008, **320**, 634.
16. Y. Pan and J.-F. Li, *NPG Asia Mater.*, 2016, **8**, e275.
17. L. P. Hu, H. J. Wu, T. J. Zhu, C. G. Fu, J. Q. He, P. J. Ying and X. B. Zhao, *Adv. Energy Mater.*, 2015, 5.
18. J.-F. Li, W.-S. Liu, L.-D. Zhao and M. Zhou, *NPG Asia Mater.*, 2010, **2**, 152.
19. L. Hicks and M. S. Dresselhaus, *Phys. Rev. B*, 1993, **47**, 12727.
20. L.-D. Zhao, B.-P. Zhang, J.-F. Li, M. Zhou, W.-S. Liu and J. Liu, *J. Alloys Compd.*, 2008, **455**, 259.
21. N. A. Heinz, T. Ikeda and G. J. Snyder, *Acta Mater.*, 2012, **60**, 4461.
22. S. Chen, N. Logothetis, L. Ye and J. Liu, *Mater. Today: Proc.*, 2015, **2**, 610.
23. G. Zhang, B. Kirk, L. A. Jauregui, H. Yang, X. Xu, Y. P. Chen and Y. Wu, *Nano Letter.*, 2011, **12**, 56.
24. M. Hong, T. C. Chasapis, Z.-G. Chen, L. Yang, M. G. Kanatzidis, G. J. Snyder and J. Zou, *ACS Nano*, 2016, **10**, 4719.
25. R. Venkatasubramanian, E. Siivola, T. Colpitts and B. O'Quinn, *Nature*, 2001, **413**, 597.
26. B. Xu, T. Feng, M. T. Agne, L. Zhou, X. Ruan, G. J. Snyder and Y. Wu, *Angew. Chem., Int. Ed.*, 2017, **129**, 3600.
27. T. Xing, R. Liu, F. Hao, P. Qiu, D. Ren, X. Shi and L. Chen, *J. Mater. Chem. C*, 2017, **5**, 12619.
28. F. Hao, P. Qiu, Y. Tang, S. Bai, T. Xing, H.-S. Chu, Q. Zhang, P. Lu, T. Zhang, D. Ren, J. Chen, X. Shi and L. Chen, *Energy Environ. Sci.*, 2016, **9**, 3120.

29. G. Zheng, X. Su, X. Li, T. Liang, H. Xie, X. She, Y. Yan, C. Uher, M. G. Kanatzidis and X. Tang, *Adv. Energy Mater.*, 2016, 6.
30. S. I. Kim, K. H. Lee, H. A. Mun, H. S. Kim, S. W. Hwang, J. W. Roh, D. J. Yang, W. H. Shin, X. S. Li and Y. H. Lee, *Science*, 2015, **348**, 109.
31. L. Hu, T. Zhu, X. Liu and X. Zhao, *Adv. Funct. Mater.*, 2014, **24**, 5211.
32. Y. Min, J. W. Roh, H. Yang, M. Park, S. I. Kim, S. Hwang, S. M. Lee, K. H. Lee and U. Jeong, *Adv. Mater.*, 2013, **25**, 1425.
33. F. Li, X. Huang, Z. Sun, J. Ding, J. Jiang, W. Jiang and L. Chen, *J. Alloys Compd.*, 2011, **509**, 4769.
34. L. D. Zhao, B. P. Zhang, J. F. Li, H. L. Zhang and W. S. Liu, *Solid State Sci.*, 2008, **10**, 651.
35. J. Jiang, L. Chen, S. Bai and Q. Yao, *J. Alloys Compd.*, 2005, **390**, 208.
36. http://www.komatsu.com/CompanyInfo/press/2009012714011528411.html (last accessed May 2018).
37. X. Hu, P. Jood, M. Ohta, M. Kunii, K. Nagase, H. Nishiate, M. G. Kanatzidis and A. Yamamoto, *Energy Environ. Sci.*, 2016, **9**, 517.
38. Q. Zhang, J. Liao, Y. Tang, M. Gu, C. Ming, P. Qiu, S. Bai, X. Shi, C. Uher and L. Chen, *Energy Environ. Sci.*, 2017, **10**, 956.
39. S. Li, J. Pei, D. Liu, L. Bao, J.-F. Li, H. Wu and L. Li, *Energy*, 2016, **113**, 35.
40. R. Roth, R. Rostek, K. Cobry, C. Köhler, M. Groh and P. Woias, *J. Microelectromech. Syst.*, 2014, **23**, 961.
41. M. Tan, Y. Deng and Y. Hao, *Energy*, 2014, **70**, 143.
42. C. Navone, M. Soulier, J. Testard, J. Simon and T. Caroff, *J. Electron. Mater.*, 2011, **40**, 789.
43. J. Wang, H. Li, R. Liu, L. Li, Y.-H. Lin and C.-W. Nan, *Compos. Sci. Technol.*, 2018, **157**, 1.
44. F. Kim, B. Kwon, Y. Eom, J. E. Lee, S. Park, S. Jo, S. H. Park, B.-S. Kim, H. J. Im and M. H. Lee, *Nat. Energy*, 2018, 1.
45. G. J. Snyder, J. R. Lim, C.-K. Huang and J.-P. Fleurial, *Nat. Mater.*, 2003, **2**, 528.
46. W. Zhang, J. Yang and D. Xu, *J. Microelectromech. Syst.*, 2016, **25**, 744.
47. D. Champier, *Energy Convers. Manage.*, 2017, **140**, 167.
48. L. D. Zhao, S. H. Lo, Y. Zhang, H. Sun, G. Tan, C. Uher, C. Wolverton, V. P. Dravid and M. G. Kanatzidis, *Nature*, 2014, **508**, 373.
49. K. Peng, X. Lu, H. Zhan, S. Hui, X. Tang, G. Wang, J. Dai, C. Uher, G. Wang and X. Zhou, *Energy Environ. Sci.*, 2016, **9**, 454.
50. L.-D. Zhao, G. Tan, S. Hao, J. He, Y. Pei, H. Chi, H. Wang, S. Gong, H. Xu, V. P. Dravid, C. Uher, G. J. Snyder, C. Wolverton and M. G. Kanatzidis, *Science*, 2015.
51. A. T. Duong, V. Q. Nguyen, G. Duvjir, V. T. Duong, S. Kwon, J. Y. Song, J. K. Lee, J. E. Lee, S. Park, T. Min, J. Lee, J. Kim and S. Cho, *Nat. Commun.*, 2016, **7**, 13713.
52. T. R. Wei, G. Tan, X. Zhang, C. F. Wu, J. F. Li, V. P. Dravid, G. J. Snyder and M. G. Kanatzidis, *J. Am. Chem. Soc.*, 2016, **138**, 8875.
53. C.-L. Chen, H. Wang, Y.-Y. Chen, T. Day and G. J. Snyder, *J. Mater. Chem. A*, 2014, **2**, 11171.

54. T. A. Wubieneh, C.-L. Chen, P. C. Wei, S.-Y. Chen and Y.-Y. Chen, *RSC Adv.*, 2016, **6**, 114825.
55. C. Chang, Q. Tan, Y. Pei, Y. Xiao, X. Zhang, Y.-X. Chen, L. Zheng, S. Gong, J.-F. Li, J. He and L.-D. Zhao, *RSC Adv.*, 2016, **6**, 98216.
56. Q. Zhang, E. K. Chere, J. Sun, F. Cao, K. Dahal, S. Chen, G. Chen and Z. Ren, *Adv. Energy Mater.*, 2015, **5**, 1500360.
57. C. W. Li, J. Hong, A. F. May, D. Bansal, S. Chi, T. Hong, G. Ehlers and O. Delaire, *Nat. Phys.*, 2015, **11**, 1063.
58. D. Wu, L. Wu, D. He, L.-D. Zhao, W. Li, M. Wu, M. Jin, J. Xu, J. Jiang, L. Huang, Y. Zhu, M. G. Kanatzidis and J. He, *Nano Energy*, 2017, **35**, 321.
59. F. Q. Wang, S. Zhang, J. Yu and Q. Wang, *Nanoscale*, 2015, **7**, 15962.
60. H. Ju and J. Kim, *ACS Nano*, 2016, **10**, 5730.
61. Y. Huang, L. Li, Y.-H. Lin and C.-W. Nan, *J. Phys. Chem. C*, 2017, **121**, 17530.
62. C.-C. Lin, R. Lydia, J. H. Yun, H. S. Lee and J. S. Rhyee, *Chem. Mater.*, 2017, **29**, 5344.
63. J. C. Li, D. Li, X. Y. Qin and J. Zhang, *Scr. Mater.*, 2017, **126**, 6.
64. Y. Fu, J. Xu, G.-Q. Liu, J. Yang, X. Tan, Z. Liu, H. Qin, H. Shao, H. Jiang, B. Liang and J. Jiang, *J. Mater. Chem. C*, 2016, **4**, 1201.
65. Y.-X. Chen, Z.-H. Ge, M. Yin, D. Feng, X.-Q. Huang, W. Zhao and J. He, *Adv. Funct. Mater.*, 2016, **26**, 6836.
66. M. Ohtaki, K. Araki and K. Yamamoto, *J. Electron. Mater.*, 2009, **38**, 1234.
67. Y. Liu, Y. Lin, Z. Shi, C.-W. Nan and Z. Shen, *J. Am. Ceram. Soc.*, 2005, **88**, 1337.
68. W. Khan, A. H. Naqvi, M. Gupta, S. Husain and R. Kumar, *J. Chem. Phys.*, 2011, **135**, 054501.
69. N. Wang, H. Chen, H. He, W. Norimatsu, M. Kusunoki and K. Koumoto, *Sci. Rep.*, 2013, **3**, 3449.
70. L.-D. Zhao, J. He, D. Berardan, Y. Lin, J.-F. Li, C.-W. Nan and N. Dragoe, *Energy Environ. Sci.*, 2014, **7**, 2900.
71. L.-D. Zhao, D. Berardan, Y. L. Pei, C. Byl, L. Pinsard-Gaudart and N. Dragoe, *Appl. Phys. Lett.*, 2010, **97**, 092118.
72. Y. Liu, L.-D. Zhao, Y. Zhu, Y. Liu, F. Li, M. Yu, D.-B. Liu, W. Xu, Y.-H. Lin and C.-W. Nan, *Adv. Energy Mater.*, 2016, **6**, 1502423.
73. M. Samanta, S. N. Guin and K. Biswas, *Inorg. Chem. Front.*, 2017, **4**, 84.
74. J. Li, J. Sui, Y. Pei, C. Barreteau, D. Berardan, N. Dragoe, W. Cai, J. He and L.-D. Zhao, *Energy Environ. Sci.*, 2012, **5**, 8543.
75. G.-K. Ren, S.-Y. Wang, Y.-C. Zhu, K. J. Ventura, X. Tan, W. Xu, Y.-H. Lin, J. Yang and C.-W. Nan, *Energy Environ. Sci.*, 2017, **10**, 1590.
76. J. Sui, J. Li, J. He, Y.-L. Pei, D. Berardan, H. Wu, N. Dragoe, W. Cai and L.-D. Zhao, *Energy Environ. Sci.*, 2013, **6**, 2916.
77. H. Fröhlich and G. Sewell, *Proc. Phys. Soc.*, 1959, **74**, 643.
78. C. Barreteau, D. Bérardan, E. Amzallag, L. Zhao and N. Dragoe, *Chem. Mater.*, 2012, **24**, 3168.
79. J. Li, J. Sui, C. Barreteau, D. Berardan, N. Dragoe, W. Cai, Y. Pei and L.-D. Zhao, *J. Alloys Compd.*, 2013, **551**, 649.

80. Y.-L. Pei, J. He, J.-F. Li, F. Li, Q. Liu, W. Pan, C. Barreteau, D. Berardan, N. Dragoe and L.-D. Zhao, *NPG Asia Mater.*, 2013, **5**, e47.
81. Y. L. Pei, H. Wu, D. Wu, F. Zheng and J. He, *J. Am. Chem. Soc.*, 2014, **136**, 13902.
82. Y. Liu, L. D. Zhao, Y. Liu, J. Lan, W. Xu, F. Li, B. P. Zhang, D. Berardan, N. Dragoe, Y. H. Lin, C. W. Nan, J. F. Li and H. Zhu, *J. Am. Chem. Soc.*, 2011, **133**, 20112.
83. Z. Li, C. Xiao, S. Fan, Y. Deng, W. Zhang, B. Ye and Y. Xie, *J. Am. Chem. Soc.*, 2015, **137**, 6587.
84. X. Zhang, C. Chang, Y. Zhou and L. D. Zhao, *Materials*, 2017, 10.
85. S.-T. Dong, Y.-Y. Lv, B.-B. Zhang, F. Zhang, S. Yao, Y. B. Chen, J. Zhou, S.-T. Zhang, Z.-B. Gu and Y.-F. Chen, *CrystEngComm*, 2015, **17**, 6136.
86. Q. Wen, C. Chang, L. Pan, X. Li, T. Yang, H. Guo, Z. Wang, J. Zhang, F. Xu, Z. Zhang and G. Tang, *J. Mater. Chem. A*, 2017, **5**, 13392.
87. S. K. Saha, *Phys. Rev. B*, 2015, **92**, 041202.
88. X. Zhang, D. Feng, J. He and L.-D. Zhao, *J. Solid State Chem.*, 2018, **258**, 510.
89. J. L. Lan, Y. C. Liu, B. Zhan, Y. H. Lin, B. Zhang, X. Yuan, W. Zhang, W. Xu and C. W. Nan, *Adv. Mater.*, 2013, **25**, 5086.
90. D. Wu, S. Huang, D. Feng, B. Li, Y. Chen, J. Zhang and J. He, *Phys. Chem. Chem. Phys.*, 2016, **18**, 23872.

CHAPTER 6

Graphene–Carbon Nanotube Hybrid Films for High-performance Photovoltaic Devices

O. E. GLUKHOVA*[a,b] AND M. M. SLEPCHENKOV[a]

[a] Saratov State University, Department of Physics, Astrakhanskaya Street 83, Saratov, 410012, Russia; [b] I.M. Sechenov First Moscow State Medical University, Laboratory of Biomedical Nanotechnology, Trubetskaya street 8-2, Moscow, 119991, Russia
*Email: glukhovaoe@info.sgu.ru

6.1 Introduction

Due to its unique optical, electrical and mechanical properties, graphene is a new base material for applications of electronics and power engineering such as flexible transparent electrodes, optoelectronics, field effect transistors, energy storage and biosensors.[1–5] In recent years, the number of papers devoted to the study of hybrid architectures based on one-dimensional (1D) carbon nanotubes (CNTs) and 2D-dimensional graphene has increased substantially. Among these new hybrid materials, there are those distinguished as 3D composites with a vertical orientation of nanotubes, and 2D films with a horizontal orientation of tubes between the graphene layers.[6–13] Both varieties of hybrid graphene–CNT structure are of undoubted interest from the fundamental and applied points of view. In particular, 3D graphene–CNT material has found its application as an elemental base of supercapacitors.[14,15] In turn, 2D graphene–CNT films have broad prospects for use in photovoltaic devices.[16–18] Theoretical and

Smart Materials No. 34
Layered Materials for Energy Storage and Conversion
Edited by Dongsheng Geng, Yuan Cheng and Gang Zhang

Published by the Royal Society of Chemistry, www.rsc.org

experimental studies of the structural and physical properties of graphene–CNT material have been intensively carried out. A large number of works, published recently in leading scientific publications, indicate the potentially wide range of physical properties of graphene–CNT films with a horizontal orientation of nanotubes. In particular, their photoelectric, optical and electromechanical properties are noted.[13,17] The variety of properties of 2D graphene–CNT films are determined mainly by the features of their atomic structure. At present, there are several structural modifications of the films that differ in the way of binding the tube and graphene. Films, which are monolayers of graphene with individual single-walled CNTs (SWCNTs) located between them, are most often synthesized in practice.[11,19–21]

This chapter presents new interesting results of *in silico* studies of electrical conductivity and optical properties of graphene–CNT hybrid films, and also an assessment of the prospects for use of these films in photovoltaics. The adaptive intermolecular reactive empirical bond order (AIREBO) potential, the self-consistent charge density functional tight-binding (SCC DFTB) method, the nonequilibrium Green function (NEGF) method, the Kubo theory for the quantum description of irreversible processes and the linear response of the electronic population to an external electromagnetic field are used.[22–25] The programs Kvazar, Mizar and DFTB+ were used.[26,27]

6.2 Mathematical Methods for Finding the Equilibrium Configurations of the Super-cells of Graphene–CNT Hybrid Films

To create super-cells of graphene–CNT hybrid films, we used an effective method to construct unit cells of hybrid structures, the so-called "magnifying glass method".[28] The main features of this method can be formulated as follows: (1) at the initial stage, a model is constructed in the form of a large fragment of a graphene–CNT composite structure with a number of atoms of several tens of thousands, and the atomic network is optimized using the empirical AIREBO potential; (2) a fragment of a smaller size is cut from the optimized fragment (from the middle part to avoid the influence of the edges) and re-optimized using the SCC DFTB quantum method in a periodic box, and the box dimensions are also optimized to find the configuration corresponding to the minimum of the total energy; (3) at the final stage, the super-cell is extracted from the previous optimized fragment, which is again optimized in the periodic box by the SCC DFTB method; the optimization parameters are again the coordinates of the atoms and the dimensions of the box. We will describe the AIREBO and SCC DFTB methods very briefly.

6.2.1 AIREBO

The AIREBO force field is a mathematical model based on the Brenner potential, which was developed to take the van der Waals interaction and torsion interaction into account (including the dihedral angles) for those structures where these components make a significant contribution. The expression for the total energy of the structure has the following form:[22]

$$E = \frac{1}{2}\sum_{i}\sum_{j\neq i}\left(E_{ij}^{\text{REBO}} + E_{ij}^{LJ} + \sum_{k\neq i,j}\sum_{l\neq i,j,k} E_{ijkl}^{\text{tors}}\right). \tag{6.1}$$

Here, the first term is the energy of chemically interacting atoms, which is written in the form:

$$E_b = \frac{1}{2}\sum_{i=1}^{N_{\text{at}}}\left(\sum_{j\neq i}\left(V_R(r_{ij}) - b_{ij}V_A(r_{ij})\right)\right), \tag{6.2}$$

where $V_R(r_{ij})$ and $V_A(r_{ij})$ are the pair potentials of attraction and repulsion between the atoms of the first group with respect to the i-th atom, determined by the types of atoms and the distance between them r_{ij} (i, j are the numbers of atoms), N_{at}, the number of atoms; the index j takes the values of all the numbers of atoms of the first group that are located around the i-th atom. The multi-particle term b_{ij} is called the bond order; it changes the interaction energy of a given pair of atoms i–j, taking the interaction of σ- and π-electron clouds into account. The second term is the energy responsible for the van der Waals interaction. This energy is represented by the Lennard-Jones potential with several factors (weighting coefficients):

$$E_{ij}^{LJ} = S(t_r(r_{ij}))S(t_b(b_{ij}))C_{ij}V_{ij}^{LJ}(r_{ij}) + [1 - S(t_r(r_{ij}))]C_{ij}V_{ij}^{LJ}(r_{ij}). \tag{6.3}$$

The main task of the multipliers is to switch this potential on and off at different distances of atoms i and j from each other, because at a short distance the chemical interaction plays a greater role than the van der Waals one. The form of the potential itself is completely analogous to what was previously used in the Brenner force field.[28] The function $S(t)$ is called the switching function; its entire domain can be divided into three parts:

$$S(t) = \left\{\begin{array}{l} 1, t<0 \\ \Theta(-t) + \Theta(t)\Theta(1-t)(1-t^2(3-2t)), 0 \leq t \leq 1 \\ 0, t>1 \end{array}\right\}. \tag{6.4}$$

On the interval between 0 and 1, the function decreases to 0, and its form is determined by a smooth function, the cubic spline. In eqn (6.4), the function $\Theta(t)$ is the Heaviside function. The parameter b_{ij} is the hypothetical bond order in the case if $r_{ij} = r_{ij}^{\min}$. The distances between the atoms at the time of calculating this quantity remain unchanged. The function $t_r(r_{ij})$ is called the scaling function and has the following form:

$$t_r(r_{ij}) = \frac{r_{ij} - r_{ij}^{\min}}{r_{ij}^{\max} - r_{ij}^{\min}}. \tag{6.5}$$

The function $t_b(b_{ij})$ is needed to cut off the van der Waals interaction for a large value of the bond order. If the atoms i and j are not bonded through two or fewer atoms, then the following variable controls the function switching:

$$C_{ij}=1-\max\{\omega_{ij}(r_{ij}),\ \omega_{ik}(r_{ik})\omega_{kj}(r_{kj}),\ \forall k,\ \omega_{ik}(r_{ik})\omega_{kl}(r_{kl})\omega_{lj}(r_{lj}),\ \forall k,l\}, \quad (6.6)$$

where

$$\omega_{ij}(r_{ij})=S'\left(t_c(r_{ij})\right)=\Theta(-t)+\frac{1}{2}\Theta(t)\Theta(1-t)(1+\cos(\pi t)). \quad (6.7)$$

The third term in eqn (6.1) describes the energy of the torsion interaction, which is represented by a potential that depends on the dihedral angle ω_{kijl} for the interaction considered between atoms i–j:

$$V_{kijl}^{\text{tors}}=\frac{256}{405}\omega_{kijl}\cos^{10}\left(\frac{\omega_{kijl}}{2}\right)-\frac{1}{10}\omega_{kijl}. \quad (6.8)$$

The expression for the torsion energy is calculated through the potential as follows:

$$E_{kijl}^{\text{tors}}=\omega_{ki}(r_{ki})\omega_{ij}(r_{ij})\omega_{jl}(r_{jl})V^{\text{tors}}(\omega_{kijl}). \quad (6.9)$$

6.2.2 SCC DFTB

According to the SCC DFTB method, the total energy of a polyatomic system consists of three energy terms: E_{SCC}, E_{OCC}, E_{rep}.[23] These components correspond to the interaction of electrons with each other, to the energy of the occupied electronic states and to the repulsive energy of atomic nuclei, respectively. The energy of occupied states is determined by the expression:

$$E_{\text{OCC}}=2\sum_{\alpha}f(\varepsilon_\alpha,E_{\text{F}})\varepsilon_\alpha, \quad (6.10)$$

where f is the Fermi–Dirac function, E_{F} is the Fermi energy, and ε_α are the eigenvalues of the energy. Also this formula can be written in the following form:

$$E_{\text{OCC}}=2\sum_{\alpha}f(\varepsilon_\alpha,E_{\text{F}})\sum_{k,\nu}c_k^\alpha c_\nu^\alpha H_{k\nu}^0. \quad (6.11)$$

Here the complex coefficients c_k^α, c_ν^α are weighting coefficients in the representation of the wave function through a linear combination of atomic orbitals and are determined as a result of the solution of the eigenvalue problem for the energy operator:

$$E_{\text{SCC}}=\sum_{i,j}\gamma_{ij}(U_i,U_j,R_{ij})\Delta q_i\Delta q_j, \quad (6.12)$$

where U_i is the Hubbard parameter (half of the second derivative of the energy of the i-th particle by charge), $\Delta q_i \Delta q_j$ are the Mulliken charges, the parameter γ_{ij} is responsible for the Coulomb interaction. Different forms of this parameter are used in semi-empirical approaches. One of these forms is the Klopman–Ono form:[29]

$$\gamma_{ij}(U_i, U_j, R_{ij}) = \frac{1}{\sqrt{R_{ij} + 0.25(1/U_i + 1/U_j)^2}}. \tag{6.13}$$

Another expression for the parameter γ was introduced by Mataga–Nishimoto:[29]

$$\gamma_{ij}(U_i, U_j, R_{ij}) = \frac{1}{\sqrt{R_{ij} + 0.5(1/U_i + 1/U_j)^2}}. \tag{6.14}$$

However, the SCC DFTB method uses a more complex form for electronic interaction and the parameter is represented in the following form:

$$\gamma_{ij} = \frac{1}{R_{ij}} - S_{ij} f_{ij}. \tag{6.15}$$

Here, S_{ij} is the component responsible for the exponential decrease of the parameter at a close distance between atoms i and j, f_{ij} is an additional factor for light atoms such as hydrogen, which increases the repulsion between particles at a distance of 0.1–0.2 nm, allowing a more precise description of the interaction with them.

The last term of the total energy is the paired repulsive potential of the atomic nuclei of the multi-particle system. The E_{OCC} and E_{SCC} components are computed in a self-consistent procedure, since the Hamiltonian of the system depends on the magnitude of the Mulliken charges, which in turn are expressed in terms of the overlap matrix and the expansion coefficients of one-electron wave functions over atomic orbitals. The coefficients of the expansion can be obtained by finding the eigenvalues of the Hamiltonian matrix.

6.3 Results and Discussion

6.3.1 Atomistic Models and Band Structure of Mono- and Bilayer Graphene–CNT Hybrid Films

Atomistic models of a monolayer graphene–CNT hybrid film were constructed using the magnifying glass method.[28] This hybrid carbon structure is formed by two graphene layers with nanotubes between them. Tubes have sp^3-hybridized atoms and form covalent bonds with graphene. First of all, it was found that covalent bonds can be formed only when zigzag nanotubes contact the graphene. Therefore, we constructed a number of models with tubes (n, 0) and different distances between them. These inter-tube distances are measured in the number of hexagons of the graphene monolayer. Thus, models with the number $n=6$–20 (with step equal to 2) and inter-tube distance of 6–16 h (here and below h means hexagons; the increment of the number of hexagons is equal to 1) were constructed and investigated. σ shows the super-cells of some models of graphene–CNT hybrid films with tubes of different diameters and with different inter-tube distances. A pink color indicates atoms forming covalent bonds between the CNT and graphene. It can be seen from this figure that the tubes are deformed in the film, and graphene becomes corrugated. It is established that the hybrid film models with number $n=10$–20 and inter-tube distance of 8 h–16 h are energetically favorable. For these models, the heat of formation of H_f is less than -0.1 kcal mol^{-1} atom^{-1}. The bond length between the graphene atoms and CNT atoms is in the range of 1.61–1.62 Å in all cases. The compression ratio of the tubes is 1.57–1.69, which corresponds to the golden section. Its value depends on the inter-tube distance. More information about the translation vectors of the super-cell and other metric information are given in our previous paper.[28] A general view of the monolayer hybrid film is shown in Figure 6.1b for the case with CNTs (12,0) and an inter-tube distance of 7 h.

Similarly, a bilayer graphene–CNT hybrid film with the same tubes (n, 0) at $n=10$–20 and inter-tube distance of 8 h–16 h was constructed using the magnifying glass method. It was established that for a bilayer hybrid film there are few thermodynamically stable topological models. According to our

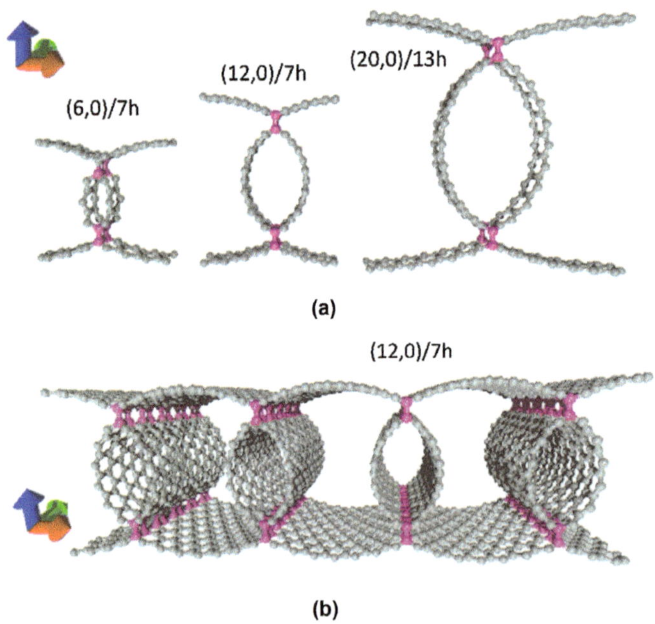

Figure 6.1 Atomic structure of a monolayer graphene–CNT hybrid film: (a) the super-cells of the hybrid film models (6,0)/7 h, (12,0)/7 h, and (20,0)/13 h; (b) the fragment of the hybrid film (12,0)/7 h (the atoms forming covalent bonds between the CNT and graphene are marked by a pink color).

numerical estimations, the formation of a bilayer hybrid film with CNTs (10,0) and (12,0) will be energetically most profitable. In this case, for models with CNTs (10,0) the inter-tube distance is 10 h, 12 h, 14 h, 16 h; for CNTs (12,0) the possible inter-tube distance is only 10 h and 12 h. For these cases, the formation is exothermic with $H_f < -0.1$ kcal mol^{-1} atom^{-1}. Figure 6.2a shows the super-cells for two models of bilayer hybrid films: a model with CNTs (10,0) and inter-tube distance of 16 h; and a model with CNTs (12,0) and inter-tube distance of 12 h. The general view of a bilayer hybrid film is shown in Figure 6.2b. At the same time, mono- and bilayer hybrid films are characterized by the same length of CNT–graphene covalent bond equal to ~0.16 nm. The compression ratio of all the tubes is approximately the same and corresponds to the golden section of 1.618. As for a monolayer film, the translation vector for a bilayer film along the *Y* direction (along the tube axis) is equal to L_y ~0.43 nm for all models, and the translation vector along the *X* direction (perpendicular to the tubes) L_x varies depending on the diameter of the tubes and the number of hexagons between them. It was found that for the hybrid film models with CNTs (10,0) and (12,0) at the inter-tube distance of 10 h, the translation vector $L_x = 2.40$ nm, and 2.46 nm, respectively. As the

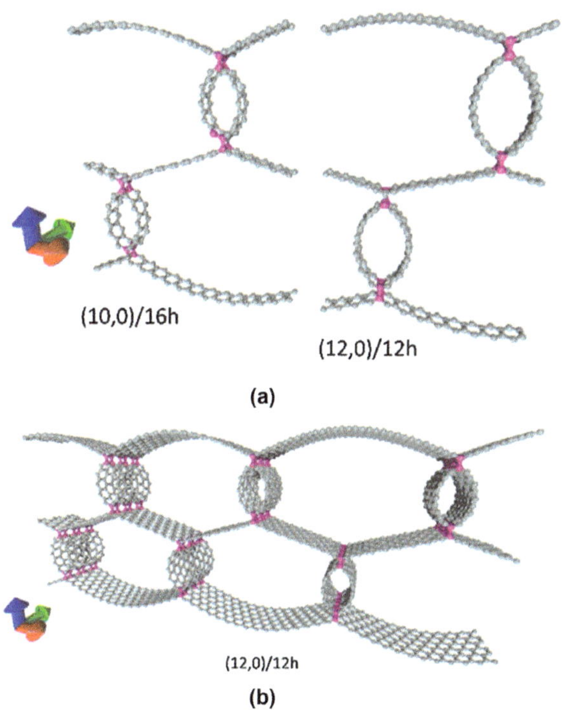

Figure 6.2 Atomic structure of a bilayer graphene–CNT hybrid film: (a) the super-cells of the hybrid film models (10,0)/16 h and (12,0)/12 h; (b) the fragment of a hybrid film model (12,0)/12 h (the atoms forming covalent bonds between the CNT and graphene are marked with a pink color).

inter-tube distance increases to 12 h, the value of the translation vector changes to $L_x = 2.89$ nm, and 2.92 nm, respectively. For other cases of hybrid film models with CNTs (10,0), the translation vector L_x amounts to 3.43 nm at an inter-tube distance of 14 h and 3.85 nm at 16 h.

The density of electronic states (DOS) distribution and the band structure of mono- and bilayer hybrid films were calculated by the SCC DFTB method. The calculations were performed in the basis of s and p orbitals. For monolayer hybrid films, the calculation results have shown that the samples with CNTs of metallic conductivity have a metallic type of conductivity like CNTs, and there is no energy gap between the valence band (VB) and conduction band (CB). This result is entirely expected, since graphene with a semi-metallic type of conductivity and CNTs with a metallic type of conductivity participate in the formation of hybrid films. Very interesting results were obtained for a planar hybrid film with CNTs of semiconductor conductivity type. The planar hybrid film is not uniquely a semiconductor. As our studies have shown, the type of conductivity is determined by two factors: (1) the diameter of CNT, and (2) the inter-tube distance. Figure 6.3 shows the calculated band diagrams as illustrative examples. The diagrams in Figure 6.3 show pictures of the distribution of sub-bands in VB and CB near the Fermi level E_F. The red color indicates the sub-bands of VB, the green color the sub-bands of CB. The dotted line denotes the Fermi level. The points M, Γ, J, and K are symmetry points of the Brillouin zone (BZ). The point Γ (0, 0) is the center of the BZ. The point M has coordinates (0, k_y), point J has coordinates (k_x, 0) and point K has coordinates (k_x, k_y).

The first three diagrams in Figure 6.3 are presented for CNTs (10,0), (14,0) and (16,0) of semiconductor conductivity type with an inter-tube distance of 11 h. By the example of these models, it can be seen that the value of the energy gap varies markedly with the change in the diameter of the nanotubes. As the diameter increases, the gap narrows. For a hybrid film sample with CNTs (10,0), the gap amounts to tenths of an electron volt, and for a sample with CNTs (16,0), the gap is only a few hundredths of an electron volt. Indeed, the distortion of the atomic framework is greatest in hybrid films with CNTs of subnanometric and nanometer diameter, so the properties of these samples differ markedly from others. Figure 6.3a–c shows that the hybrid film sample with the smallest CNTs (10,0) has the largest value of the energy gap between the VB and CB. Although at the same inter-tube distance in other hybrid film samples with CNTs (14,0) and (16,0), the gap sharply narrows. In addition to the diameter of CNTs, the inter-tube distance also has a big influence on the conductivity type of hybrid film. For example, Figure 6.3d–f shows the results of DOS calculations for the hybrid film samples with CNTs (14,0) at a small inter-tube distance and for a sample with CNTs (16,0) of large diameter and a large inter-tube distance of 13 h. It should be immediately noted that the gap is largest at a small inter-tube distance.

In order to make a final conclusion with respect to planar hybrid films with CNTs of semiconductor conductivity, we calculated the energy of all the samples with different topological models. The obtained result is shown in

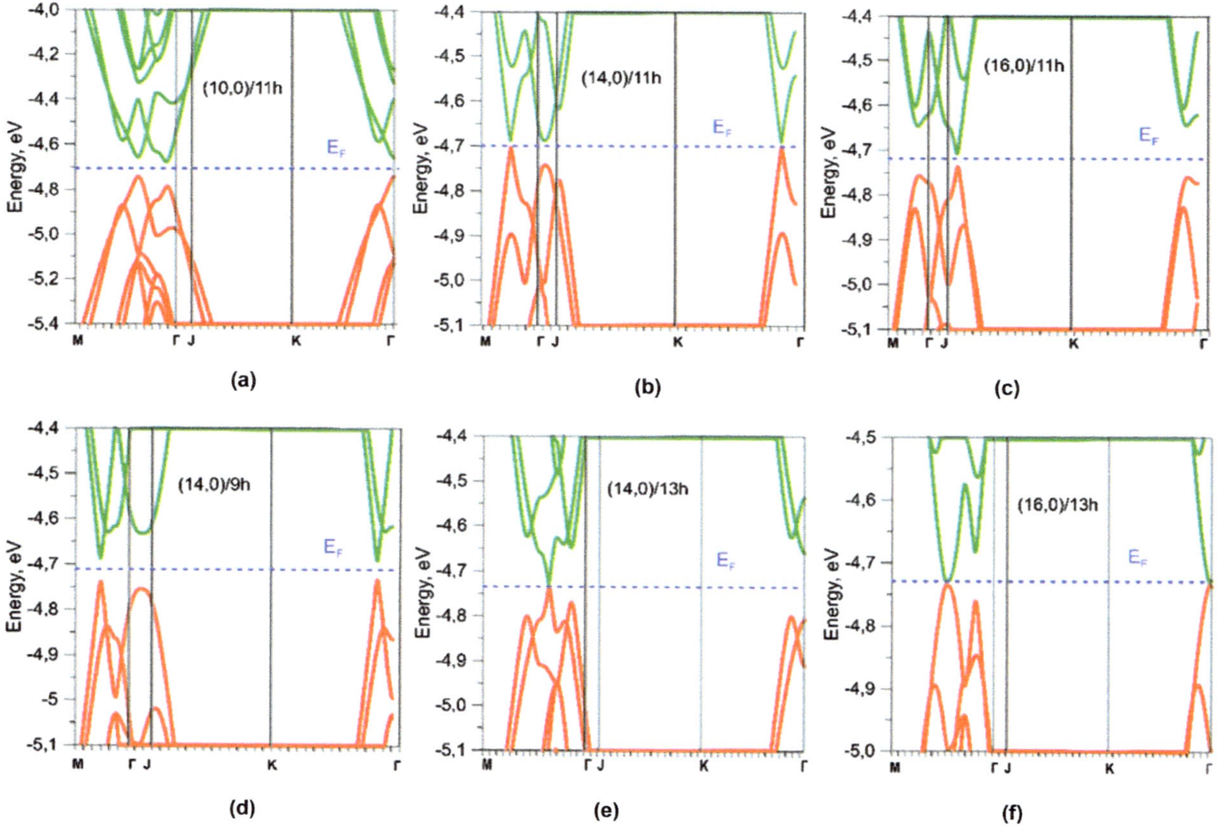

Figure 6.3 Diagrams of the band structure of a monolayer graphene–CNT hybrid film with CNTs of semiconductor conductivity: (a–c) for a model (10,0)/11 h; (b) for a model (14,0)/11 h; (c) for a model (16,0)/11 h; (d) for a model (14,0)/9 h; (e) for a model (14,0)/13 h; (f) for a model (16,0)/13 h.

Figure 6.4a. This figure shows the behaviour of the energy gap as a function of the inter-tube distance and the diameter of CNTs of semiconductor conductivity. As in Figure 6.3, it can be seen again that the planar monolayer hybrid film based on CNTs (10,0) with an inter-tube distance of 11 h has the largest energy gap value. In general, we can conclude that the planar hybrid films with

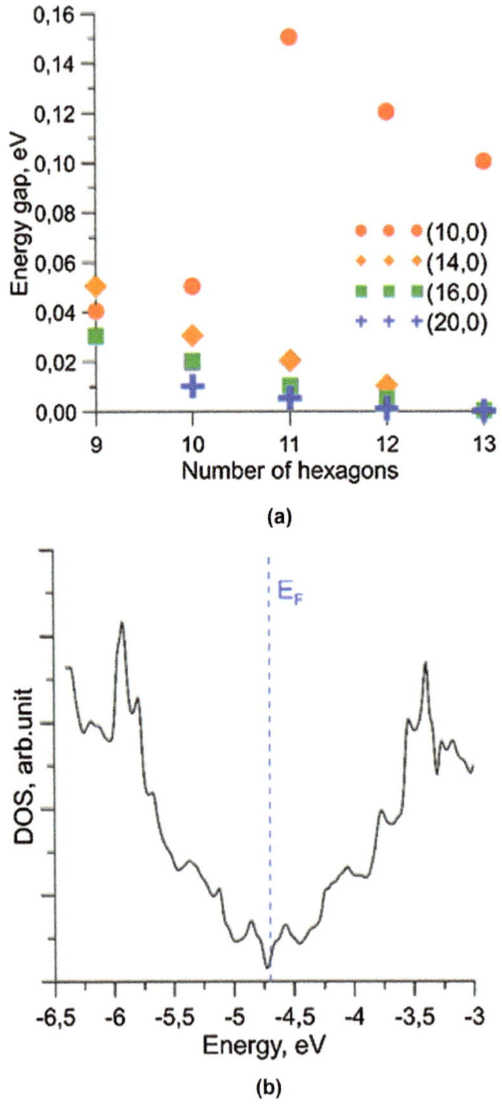

Figure 6.4 Energy characteristics of a monolayer graphene–CNT hybrid film: (a) the dependence of the energy gap on the inter-tube distance for hybrid film samples with CNTs of different diameters with semiconductor conductivity; (b) the DOS distribution for the hybrid film samples with CNTs (12,0) and inter-tube distance of 12 h.

CNTs of the smallest diameter will refer to semiconductor structures. The data in Figure 6.4a show that the samples with a tube of the smallest diameter always exhibit semiconductor properties regardless of the inter-tube distance, which is very important for possible applications of these materials in photovoltaics. At the same time, the hybrid film samples with CNTs (10,0) are characterized by the maximum value of the gap. The size of the gap depends on the distance between the CNTs. Its maximum value is observed at the inter-tube distance of 11 h and then decreases with inter-tube distance increasing. As noted above, the hybrid film samples with CNTs of metallic conductivity do not have an energy gap. An example of DOS for the hybrid film model with CNTs (12,0) and inter-tube of 12 h is shown in Figure 6.4b. Van Hove singularities and the nonzero DOS near the Fermi level are seen on the graph.

In the course of the investigation of bilayer films, we found that almost all the samples are characterized by a metallic type of conductivity, regardless of the chirality of the tube and the inter-tube distance. Figure 6.5 shows the

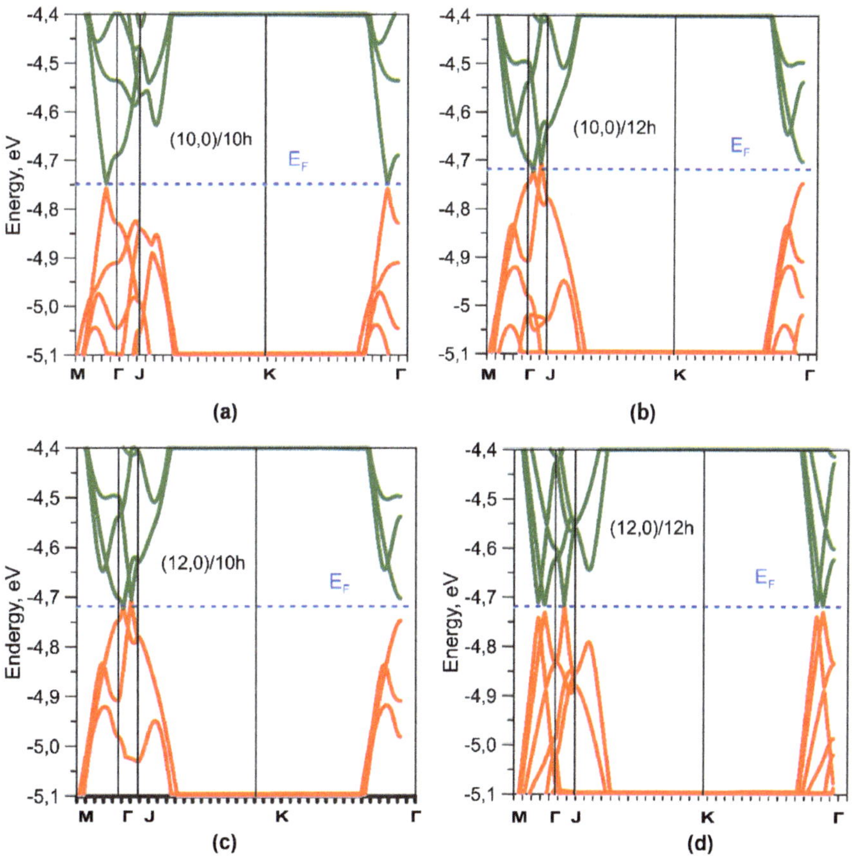

Figure 6.5 Diagrams of the band structure of a bilayer graphene–CNT hybrid film with semiconductor nanotubes: (a) for a model (10,0)/10 h; (b) for a model (10,0)/12 h; (c) for a model (12,0)/10 h; (d) for a model (12,0)/12 h.

band structures of several bilayer samples. Figure 6.5a shows the band structure of the bilayer film (10,0)/10 h. If the monolayer sample had a gap of ~0.06 eV, then the bilayer gap decreased to ~0.005 eV. Such an effect is observed for all the bilayer topological models. For all the other models presented in Figure 6.5b–d, there is no gap between the zones completely. The intersection of the VB and CB is observed in the gap between the points *T* and *J*. The Fermi level is in the interval −4.75 to −4.7 eV for all cases as in the monolayer models.

In addition to this information, we calculated the DOS distributions that are presented in Figure 6.6. For comparison, the images in Figure 6.6a and 6.6b show the DOS graphs for mono- and bilayer hybrid films of the same type. It is clearly seen that the electronic properties of a monolayer hybrid film are determined exclusively by the conductive properties of the

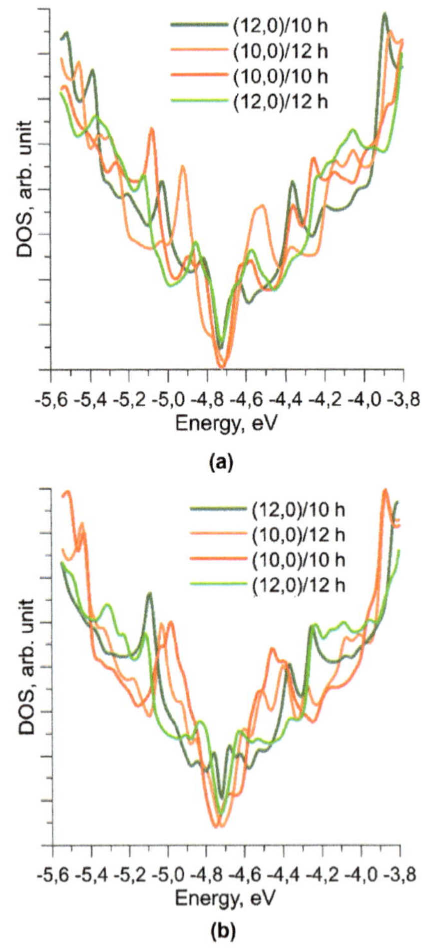

Figure 6.6 DOS of graphene–CNT hybrid films based on CNTs (10,0) and (12,0) with an inter-tube distance of 10 h and 12 h: (a) a monolayer film; (b) a bilayer film.

nanotubes themselves. All models of a monolayer hybrid film exhibit a gap between the VB and CB of several tenths of an electron volt. In this case, the bilayer hybrid film has no gap for all types of topological models.

One can draw a general conclusion that the bilayer graphene–CNT films are conductive, and the monolayer films with semiconductor nanotubes can act as semiconductor films.

6.3.2 Electrical Conductivity of Graphene–CNT Hybrid Film

To calculate the static electrical conductivity of a graphene–CNT hybrid film, we use the apparatus of the Keldysh–Green functions and the Landauer–Buttiker formalism.[30] Electrodes are considered to be massive compared to the conducting channel between them. They are in a state of equilibrium (the positions of the Fermi levels are equal) in the absence of the applied voltage. If a voltage is applied, the Fermi levels shift, and the electrons move from one contact to another through localized states of the conducting channel. The model of coherent electron transport was used. When the electron moves through the channel, only elastic collisions occur with the atoms of the crystal lattice. There is no energy loss during the process of electron transfer through the conducting channel. The current transfer is realized only by the states whose position is in the interval between the Fermi energy of the contacts. The electrons of the contact leaving through a channel are replenished due to the massive nature of the contacts themselves. The electrons that come to the other contact have energy higher than the Fermi energy for a given contact. They dissipate excess energy, moving on the lower energy levels. Due to the massiveness of the contacts, heat generation during such a process can be neglected. The energy and band structure calculations are performed using the SCC DFTB method within the Kvazar software package. Electrical conductivity is described by the expression:

$$G = \frac{I}{V} = \frac{e^2}{h} \int_{-\infty}^{\infty} T(E) F_T(E - E_F) \mathrm{d}E, \qquad (6.16)$$

where $T(E)$ is the transmission function characterizing the quantum mechanical transparency of the conducting channel (the region of the molecular structure enclosed between two contacts), depending on the energy of the electron moving along it, E_F is the Fermi energy of the contact material to which the object under investigation is connected, charge of electron, h is Planck's constant.

The thermal broadening function $F_T(E)$ is calculated by the formula:

$$F_T(E) = \frac{1}{4k_b T} \mathrm{sech}^2 \frac{E}{2k_b T}, \qquad (6.17)$$

where k_b is the Boltzmann constant, and T is the temperature. The integral for finding the static electrical conductivity is considered for the selected small energy interval due to the nature of the thermal broadening function. The largest contribution to the value of the obtained result is made by the transmission electrons near the Fermi energy and, as a rule, the interval $[E_F - 0.3, E_F + 0.3]$ is sufficient to calculate the static conductivity. A further increase in the energy interval under consideration does not lead to a change in the result of the calculation of the electrical conductivity.

The electron transmission function is given by:

$$T(E) = \frac{1}{N}\sum_{k=1}^{N} Tr\left(\Gamma_S(E)G_C^A(E)\Gamma_D(E)G_C^R(E)\right), \quad (6.18)$$

where $G_C^A(E)$, $G_C^R(E)$ are the advanced and retarded Green's matrices describing the contact with the electrodes, $\Gamma_S(E)$, $\Gamma_D(E)$ are the matrices of level broadening for the source and drain. N is the number of points in the inverse space.

The electron transmission function denotes the probability of a transition through the conducting channel of an electron with energy E, summed over all conduction channels. For crystal structures, the transmission function at one point of the inverse space has a step-like form with integer values on each step. The function $T(E)$ is found by a summation over all points of the first Brillouin zone (BZ), since materials possessing translational symmetry can be considered. The accuracy of the calculation of $T(E)$ is determined by the accuracy of decomposition of the inverse space for two-dimensional and three-dimensional structures possessing translational symmetry. The decomposition will be sufficient when the function $T(E)$ does not change with an increasing number of points of the inverse space k.

The calculated transmission functions $T(E)$ of mono- and bilayer hybrid films are shown in Figure 6.7 (units of measurement are $G_0 = e^2/h$). The π-electron approximation was used for calculations; the zero level was taken as the Fermi level. Since the graphs are symmetric with respect to the Fermi level, the figures represent only one part corresponding to the positive energy interval. A comparison of identical topological models of mono- and bilayer hybrid films shows that the $T(E)$ graphs coincide qualitatively. It is clearly seen that the conductivity at the Fermi level (0 eV) of the sample with CNTs (10,0) is practically zero. The periodicity of the CNT placement and the number of layers do not play a role. In the case of a hybrid film with CNTs (12,0), on the contrary, the connection between the value of the conductivity at the Fermi level and the inter-tube distance is clearly visible. A comparison of the same models of the hybrid films with a different number of layers shows that the bilayer hybrid film is characterized by conductivity almost 2 times higher. This result is quite expected, since the density of CNT packages in a bilayer hybrid film practically doubles in comparison with a monolayer one. We also found that a monolayer graphene–CNT hybrid film with CNTs of metallic conductivity exhibits the highest conductivity near the Fermi level ($T = 300$ K) at the inter-tube distance of 9 and 12 hexagons, that is, at a step multiple of three.

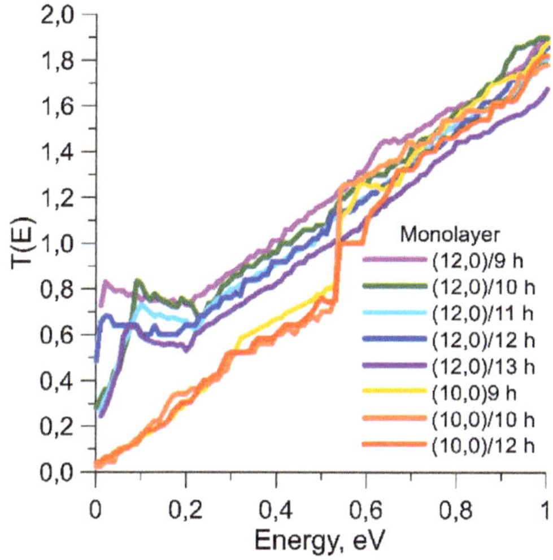

(a)

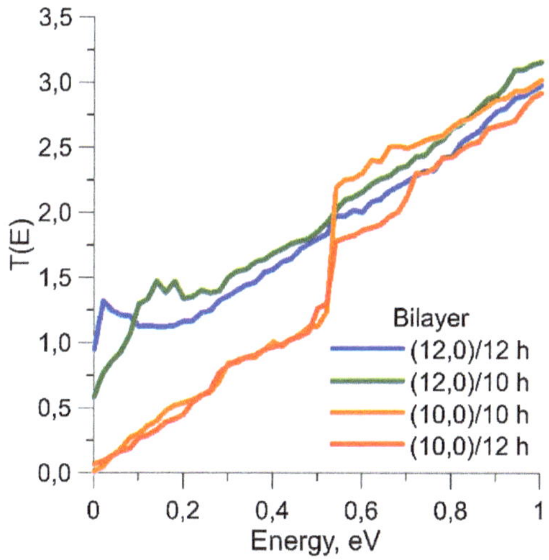

(b)

Figure 6.7 Graphs of the transmission function $T(E)$ for graphene–CNT hybrid films: (a) for models of monolayer films; (b) for models of bilayer films.

To clarify and verify the conclusions drawn above, we calculated the transmission functions of the hybrid films with different CNTs and different inter-tube distances. Also, we calculated the conductivity curves of ideal nanotubes $(n, 0)$ for $n = 10, 12, 14, 16, 18$. Figure 6.8a shows the transmission function $T(E)$ for the inter-tube distance of 13 h (the inset shows the $T(E)$ graphs of ideal nanotubes). This figure confirms the determining effect of nanotubes on the conductive properties of the composite. As is well known, the CNTs (10,0), (14,0) and (16,0) with a conductivity of semiconductor type have a gap of ~0.05 eV in the $T(E)$ spectrum near the Fermi level, while the graphene–CNT hybrid films with CNTs (12,0) and (18,0) do not have such a gap and their conductivity does not vanish. Taking the data of the graphs in Figure 6.7a and 6.8a into account, it becomes evident that all monolayer hybrid films with CNTs of semiconductor conductivity have a large active resistance regardless of their periodicity between the graphene sheets.

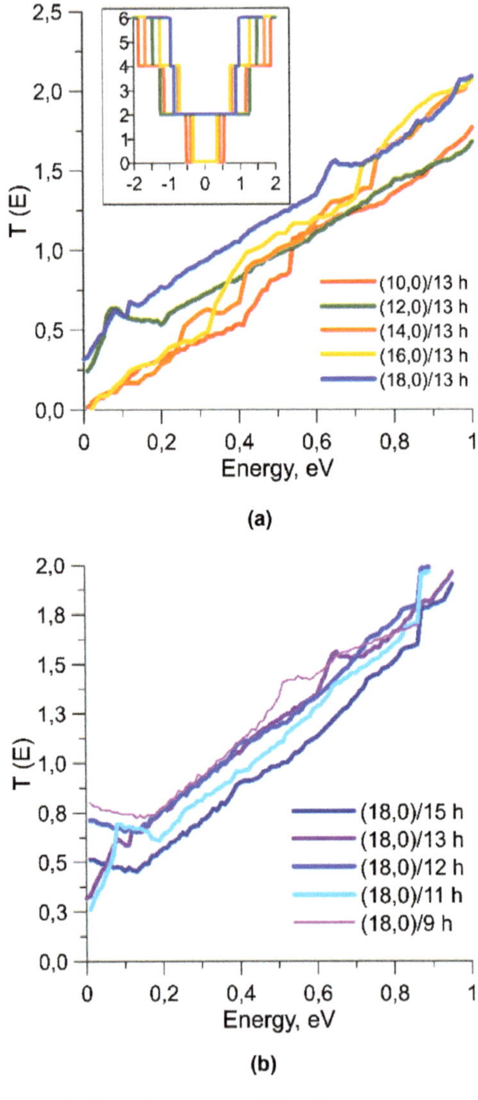

Figure 6.8 The transmission functions $T(E)$ of monolayer graphene–CNT hybrid films of various topological models: (a) for models with CNTs (10,0), (12,0), (14,0), (16,0), (18,0) and identical inter-tube distance of 13h; (b) for models with CNTs (18,0) and at inter-tube distances of 9, 11, 12, 13 and 15 h.

Figure 6.8b also clarifies the role of the distance between the CNTs of metallic conductivity. It can be seen again that a model of a graphene–CNT hybrid film with the smallest distance equal to 9 hexagons is characterized by the maximum conductivity at the Fermi level as for models of a hybrid film with CNTs (12,0). Figure 6.8b clearly demonstrates that the electrical conductivity reaches its highest value at an inter-tube distance of 9, 12 and 15 hexagons, that is, at a step multiple of three. It is important to note that at these distances the graphs $T(E)$ qualitatively repeat each other.

Next, we calculate the surface resistance of mono- and bilayer graphene–CNT hybrid films. As is well known, the electron transport decreases qualitatively with decreasing sizes of conductors. Indeed, for conductors of macroscopic dimensions, Ohm's law is satisfied. The conductive properties of the material are characterized by the specific conductivity σ. Accordingly, the electrical resistance R of the conductor from this material is determined by the conductivity σ and macroscopic parameters, such as the length of the conductor L and the cross-section S. The resistance R itself is defined as the reciprocal of the statistical electrical conductivity $R = 1/G$. The resistance is calculated on the surface area, so its value is valid for any film sizes.

For nanometer conductors, the electrical conductivity must be infinitely high. The limit of this quantity is determined by the mean free path λ of an electron in a given structure. If the length of the conductor is much larger than λ, then the electron transport can be considered as diffusion. If the length of the conductor is less than λ, then the electron transport is ballistic. In the case of a ballistic conductor, the resistance is independent of its length and is described by the expression $R = h/e^2M$, where the number of conduction channels M is the determining value. This number M can be determined from the number of intersections of the CB and VB at the Fermi level. In general, the generalized Ohm's law for active resistance can be written in the following form:[30]

$$R = \frac{h}{e^2 M}\left(1 + \frac{L}{\lambda}\right). \tag{6.19}$$

This formula takes the limiting case into account when $\lambda \gg L$, then the resistance is determined by the expression $R = h/e^2M$. In the opposite case, the resistance depends linearly on the length of the conductor. When considering the ballistic conductivity regime, it is possible to neglect inelastic scattering of electrons on the crystal lattice, which is accompanied by an exchange of energy between the electron and the atoms of the system under consideration. However, the elastic scattering on crystal lattice defects leads to quantum mechanical interference, which should be taken into account, for example, in the contact areas of different nanostructures. If the exchange of energy with the environment does not occur during the scattering of electrons, such an electronic transport is called coherent. In this case, the electrical resistance depends solely on the structure of the atomic grid and its features, and the best apparatus for calculating it is the nonequilibrium

Keldysh–Green functions (briefly presented above). Therefore, we calculate the resistance R on the basis of the calculated functions $T(E)$ from eqn (6.16), taking the temperature 300 K into account, described by the thermal broadening function in eqn (6.17).

Some of the most important and interesting results are shown in Figure 6.9. For example, Figure 6.9 shows the graphs of the change in the surface resistance of monolayer films for two variants of a topological structure. The red line shows the change in the resistance of the film based on CNTs (18,0), the distance between which varies from 9 to 15 hexagons. The film with the maximum packing of CNTs possesses the least resistance, more precisely, with a critically maximal one. In this case, the inter-tube distance is 9 h. With a smaller distance the structure will not be energetically stable. The small value of the resistance $R = 17$ kΩ is explained by the overlapping of π-clouds of neighboring CNTs, which increases the probability of passage of an electron through this structure. This is confirmed by the corresponding profile of the function $T(E)$ shown in Figure 6.8b. From this figure, it is seen that the $T(E)$ function assumes maximum values over the entire energy interval E in comparison with other variants of topological models. The case with an inter-tube distance of 12 h is characterized by a value of R close to this resistance value. In other cases, the resistance is higher by several tens of percent. This is due to the peculiarities of the graphene layer curvature, which changes abruptly with an increase in the inter-tube distance from 9 h to 13 h. At this interval of the change in the number of hexagons between the CNTs, the electronic structure of the hybrid film is rearranged, accompanied by the re-hybridization of the electron clouds of graphene. Therefore, the resistance varies abruptly. However, with a further increase in the inter-tube distance to 14 h, 15 h, "saturation"

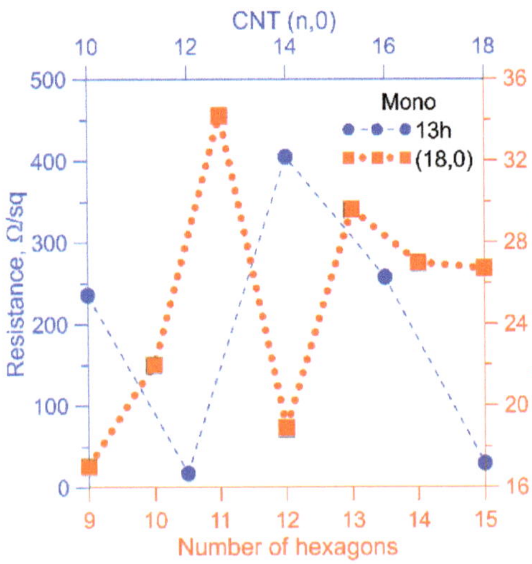

Figure 6.9 Resistance of a monolayer graphene–CNT hybrid film at different CNTs and their different location.

occurs when the resistance almost does not change any more. It can be said that at a distance of 14 h and above, $R \sim 26.5$ kΩ and more does not change. Starting from this distance, the electron π-clouds of neighboring CNTs no longer overlap, so the resistance of the sample does not change. The blue line in Figure 6.9 shows the change in the resistance of the film based on CNTs $(n, 0)$ as n varies from 10 to 18 in steps of "2" and the same inter-tube distance of 13 h. The course of the curve could be predicted in advance. The hybrid film samples with CNTs (12,0) and (18,0) of metallic conductivity type possess the minimum value of resistance. As was previously shown in Figure 6.8a, for these topological models, the profile of the function $T(E)$ does not have zero sections on the entire energy interval.

A similar picture of the resistance behaviour is also observed in a bilayer hybrid film. For hybrid film samples with CNTs of semiconductor conductivity, the resistance is also within the range of 100–450 kOhm, regardless of the position of the tubes. For hybrid film samples with CNTs of metallic conductivity, the resistance is much lower as compared to samples with CNTs of semiconductor conductivity. For them, the value of R lies in the range of 10–23 kOhm. It should be noted that in bilayer films with CNTs of metallic conductivity the resistance is slightly lower than for monolayer films. This is due to the increased packing density of conducting nanotubes in the case of the two layers and three layers of graphene.

In general, regarding the electrical conductivity of mono- and bilayer graphene–CNT hybrid films, these conclusions should be drawn: (1) hybrid films refer either to the metallic type of conductivity, or to narrow-band semiconductors; (2) the electrical resistance of mono-and bilayer films can differ from ideal CNTs if the hybrid film is based on CNTs of metallic conductivity with a short distance between them, close to the critically possible one.

6.3.3 Optical Conductivity and Optical Properties of Graphene–CNT Hybrid Film

To understand the prospects of the composite as elements of solar cells and photodetectors, it is necessary to investigate its optical properties, such as transmittance, reflection and absorption of electromagnetic waves. The parameters defining the listed properties are determined from Maxwell's classical theory of electromagnetism.

Let us consider the general case of an incidence of a wave of unit amplitude on a flat interface.[31] Let the wave be incident on the interface as shown in Figure 6.10. The electric field vector E lies in the XZ plane, respectively, the magnetic field vector is perpendicular to this plane. The angle of incidence is θ, Z_0 is the wave resistance, and η is the wave resistance.

Then for the incident wave, the field components E, H can be written as:

$$E_x = 1 \cdot e^{-jk_1 z},\ H_y = \varepsilon_1/\mu_1 \cdot e^{-jk_1 z} = 1/Z_1 \cdot e^{-jk_1 z}, \tag{6.20}$$

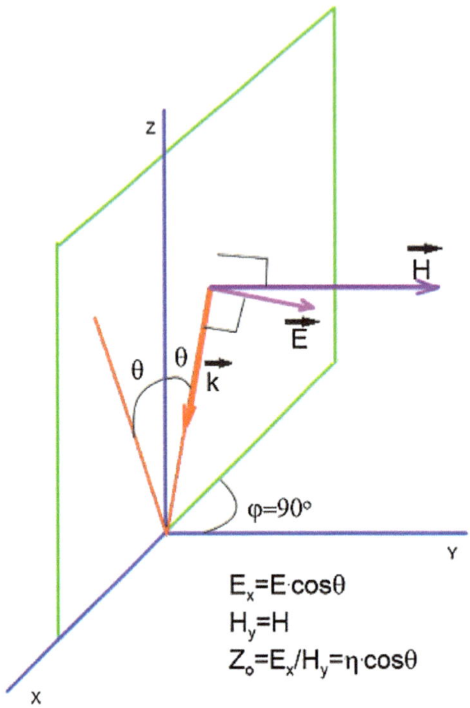

Figure 6.10 Schematic representation of the relative location of the wave vector *k* and the vectors of electric and magnetic field strengths in the case of p-polarization of the wave (*H*-wave).

where Z_1 is the wave impedance of the first medium. Herein after, we use the well-known relation in electrodynamics, which relates the electric and magnetic field strengths through the wave impedance:

$$\frac{E_x}{H_y} = \sqrt{\frac{\mu}{\varepsilon}} \text{ (for p-polarization or } H\text{-wave)},$$
$$\frac{E_x}{H_y} = -\sqrt{\frac{\mu}{\varepsilon}} \text{ (for s-polarization or } E\text{-wave)}. \tag{6.21}$$

When passing through the interface, the wave is partially reflected. For a reflected wave, one can write

$$E_x = R \cdot e^{jk_1 z} \quad H_y = -R\sqrt{\frac{\varepsilon_1}{\mu_1}} \cdot e^{jk_1 z} = -\frac{R}{Z_1} \cdot e^{jk_1 z}, \tag{6.22}$$

where R is the reflection coefficient. For a transmitted wave, one can write:

$$E_x = T \cdot e^{-jk_2 z}, \, H_y = T\sqrt{\frac{\varepsilon_2}{\mu_2}} \cdot e^{-jk_2 z} = \frac{T}{Z_2} \cdot e^{-jk_2 z}, \tag{6.23}$$

where T is the transmittance, and Z_2 is the wave impedance of the second medium. Next, we use the boundary conditions for the tangential components of the electric and magnetic fields (E_x and H_y). We match the interface with the plane $z = 0$. The terms responsible for the incident and reflected wave (E_x^+) will be on the left. The terms responsible for the transmitted wave (E_x^-) are on the right. Then, for the E_x field component we have:

$$1 \cdot e^{-jk_1 z} + R \cdot e^{jk_1 z} = T \cdot e^{-jk_2 z}. \tag{6.24}$$

Since $z = 0$ at the interface, we obtain

$$1 + R = T, \tag{6.25}$$

where $E_x^+ = 1 + R$, and $E_x^- = T$. Similarly, we write for H_y field component:

$$\frac{1}{Z_1} \cdot e^{-jk_1 z} - \frac{R}{Z_1} \cdot e^{jk_1 z} - \frac{T}{Z_2} \cdot e^{-jk_2 z} = J_x, \tag{6.26}$$

where J_x is the surface current density. It is known from Ohm's law in differential form that $J_x = \sigma_{xx} E_x$, where σ_{xx} is the x-component of the surface conductivity tensor. Taking that the medium is the same at the interface into account, we can set $Z_1 = Z_2 = Z_0 = \frac{\mu}{\varepsilon}$, $k_1 = k_2 = k$ for wave impedances. Then for $z = 0$ we have

$$\frac{1 - R - T}{Z_0} = \sigma_{xx} E_{xx}. \tag{6.27}$$

Taking that $E_x = E_x^+ = 1 + R$ into account, we have

$$\frac{1 - R - T}{Z_0} = \sigma_{xx}(1 + R). \tag{6.28}$$

Further, solving eqn (6.28), one can obtain expressions for the reflection coefficient R and transmittance T. To find transmittance T, we substitute the relation $R = T - 1$ into eqn (6.28):

$$R = \frac{-Z_0 \sigma_{xx}}{(2 + Z_0 \sigma_{xx})}, \tag{6.29}$$

$$T = \frac{2}{(2 + Z_0 \sigma_{xx})}. \tag{6.30}$$

By definition, the wave impedance for the case of p-polarization (H-wave) is $Z_0 = \eta \cos\theta$, where $\eta = 120\ \pi$ is the wave impedance of the vacuum.

Now we consider the case when a wave of unit amplitude with a vector E along the Y axis falls on the interface as shown in Figure 6.11.

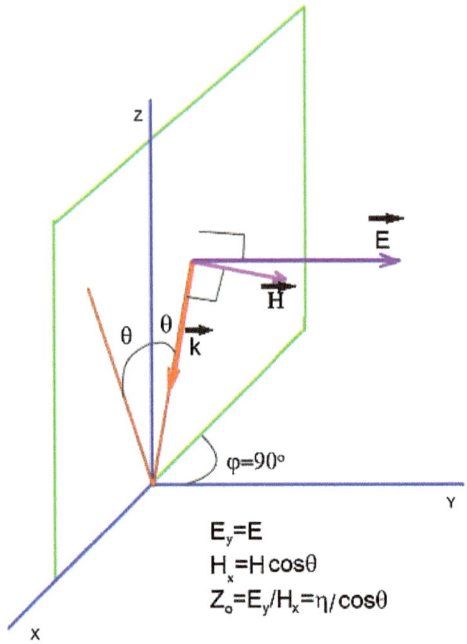

Figure 6.11 Schematic representation of the relative location of the wave vector k and the vectors of electric and magnetic field strengths in the case of s-polarization (E-wave).

In a similar way to the case of p-polarized plane waves, we can obtain the following equation

$$\frac{1-R-T}{Z_0}=\sigma_{yy}(1+R), \tag{6.31}$$

where the wave impedance for the case of s-polarization is $Z_0=-E_y/H_x=\eta/\cos\theta$. Accordingly, the parameters R, T are now determined by the y-component of the surface conductivity tensor.

Since a part of the electromagnetic wave can be absorbed or scattered in the medium (in the presence of inhomogeneities), it is necessary to introduce another parameter – the absorption coefficient A. From the law of conservation of energy,

$$A=1-|R|^2-|T|^2. \tag{6.32}$$

For the case of normal incidence of a wave, it should be written:

$$R=\frac{-\sigma_{\alpha\beta}Z_0}{2+\sigma_{\alpha\beta}Z_0}, \quad T=\frac{2}{2+\sigma_{\alpha\beta}Z_0}, \tag{6.33}$$

where Z_0 is the wave resistance in the free space $Z_0=120\ \pi$ Ohm; $\sigma_{\alpha\beta}$ is an element of the complex optical conductivity tensor.

As can be seen from the presented formulas, the main issue is the calculation of optical conductivity. Its calculation is possible only with the use of the quantum theory of irreversible processes. In this study, the elements of the complex optical conductivity tensor $\sigma_{\alpha\beta}(\Omega)$ were calculated using the Kubo–Greenwood formula, which determines the conductivity as a function of the photon energy Ω:[32]

$$\sigma_{\alpha\beta}(\Omega) = \frac{2e^2\hbar}{im_e^2 S_{\text{cell}}}\frac{1}{N_{\vec{k}}}\sum_{k\in \text{BZ}}\sum_{m,n}\frac{\hat{P}_{\alpha}^{nm}(\vec{\boldsymbol{k}})\cdot\hat{P}_{\beta}^{mn}(\vec{\boldsymbol{k}})}{E_n(\vec{\boldsymbol{k}}) - E_m(\vec{\boldsymbol{k}}) + \Omega + i\eta} \times \frac{f_\beta[E_n(\vec{\boldsymbol{k}}) - \mu] - f_\beta[E_m(\vec{\boldsymbol{k}}) - \mu]}{E_n(\vec{\boldsymbol{k}}) - E_m(\vec{\boldsymbol{k}})}, \quad (6.34)$$

where $f_\beta(x) = 1/(1 + \exp[\beta(x - \mu)])$ is the Fermi–Dirac function with the chemical potential μ, $\beta = 1/k_B T$ the inverse of thermal energy; S_{cell} is the area of the super-cell section; $N_{\vec{k}}$ – the number of k-points in the first BZ; $\hat{P}_{\alpha}^{nm}(\vec{\boldsymbol{k}})$, $\hat{P}_{\alpha}^{mn}(\vec{\boldsymbol{k}})$ are the matrix elements corresponding to the α- and β-components of the momentum operator vector, m_e and e is the mass and charge of the electron; $E_{n/m}(\vec{\boldsymbol{k}})$ are the sub-bands of the VB and CB. The number "2" takes the electron spin into account. The phenomenological parameter η takes the broadening of the spectral line into account. To calculate the elements of the momentum matrix $\hat{P}_{\alpha}^{mn}(\vec{\boldsymbol{k}})$ we used the well-known relation $\hat{P}(\vec{\boldsymbol{k}}) \to (m_e/\hbar)\nabla_{\vec{k}}\hat{H}(\vec{\boldsymbol{k}})$, where $\hat{H}(\vec{\boldsymbol{k}})$ is the Hamiltonian. The Hamiltonian was constructed within the framework of the SCC DFTB method.

According to the above theoretical considerations, we first calculated the elements of the optical conductivity tensor $\sigma(\Omega)$ for both types of hybrid films. Two diagonal elements of the tensor $\sigma(\Omega)$ are very important, since they characterize the properties of the optical conductivity of a given material: (1) in the direction of the axis of symmetry in a nanotube, and (2) in the perpendicular direction of the axis of symmetry in a nanotube. The off-diagonal elements of the optical symmetry tensor $\sigma(\Omega)$ were not considered in this chapter. Of great interest is the element of the conductivity tensor associated with the direction of the vector E along the CNT axis (s-polarization, or E-wave). In this case, the chirality of CNT has a large and decisive influence on the profile of the spectrum $\sigma_{yy}(\Omega)$. Therefore, we first of all paid great attention to the element of the tensor $\sigma_{yy}(\Omega)$. The real and imaginary parts of the element of the conductivity tensor $\sigma_{yy}(\Omega)$ for the UV-visible-IR wavelength range are shown in Figure 6.12a and b for the mono- and bilayer graphene–CNT hybrid film, respectively. The wavelength range is 60 nm–30 μm (a logarithmic scale is used in Figure 6.12). The calculated spectra for the real and imaginary parts of the tensor element $\sigma_{yy}(\Omega)$ for the various topological modifications of the composite and similar spectra for ideal plane graphene (black dashed curves) are presented in Figure 6.12. By the example of Figure 6.12a for a monolayer hybrid film, it can be seen that a series of additional peaks appeared in comparison with

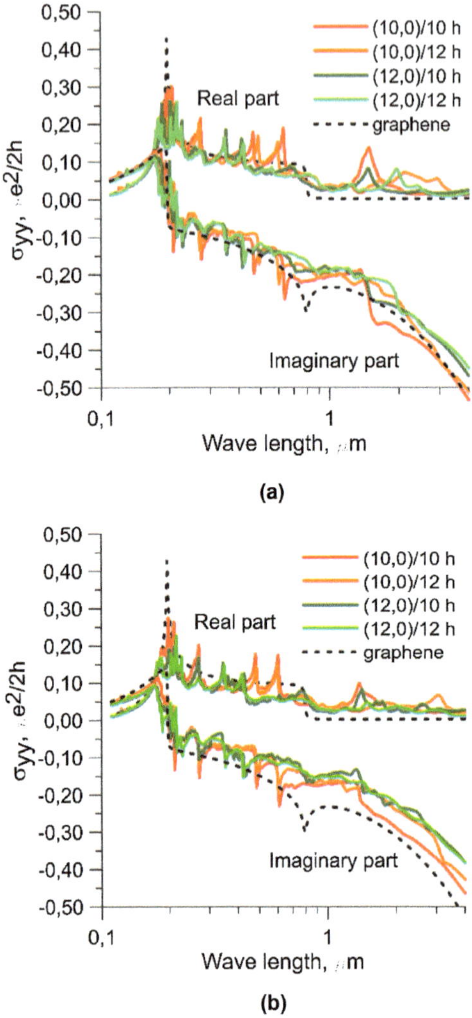

Figure 6.12 Element *YY* (along the axis of nanotubes) of the optical conductivity tensor: (a) for a monolayer graphene–CNT hybrid film; (b) for a bilayer graphene–CNT hybrid film.

ideal plane graphene in the visible and IR spectral regions. As it is known, the spectrum of ideal individual nanotubes exhibits similar bands but in the IR region. This allows us to conclude that the graphene–CNT hybrid film is very promising in modeling photodetective devices.

Similar investigations of bilayer hybrid films of various topological models have revealed the same nature in the dependence of optical conductivity. Figure 6.12b shows the spectra of the real and imaginary parts of $\sigma_{yy}(\Omega)$ for bilayer films of various topological models. It can be concluded that all the main conductivity peaks of the hybrid film samples with CNTs of semiconductor conductivity and metallic conductivity are repeated for bilayer films. Also, it should be noted that these peaks almost do not vary in intensity. This result is expected, since the bilayer film differs only in the packing density of nanotubes, and not in the tubes themselves. These facts make it possible to conclude that the mono- and bilayer hybrid films almost do not differ in optical conductivity for *E*-type waves (when the vector *E* is directed along the CNT axis). Such a result is explained by the large influence of the atomic structure of CNTs on the spectrum of optical conductivity of hybrid film. The spectra $\sigma_{xx}(\Omega)$ for mono- and bilayer films are not given here, since they do not represent any significant interest for analysis. In this case, the vector *E* is directed perpendicular to the CNT axis, so graphene plays a decisive role. As a result, the real part $\sigma_{xx}(\Omega)$ almost remains constant throughout the IR-visible-UV spectral regions, and the imaginary part decreases with increasing wavelength, as for $\sigma_{yy}(\Omega)$.

Based on the calculated elements of the optical conductivity tensor, the real and imaginary parts of the permittivity were calculated. Figure 6.13 shows the spectra of the permittivity tensor elements for the same direction (along the CNT axis). There is a sharp difference from similar characteristics of pure plane graphene. Several visible peaks are observed in the visible wavelength range. In general, the nature of the change in the permittivity repeats the properties of graphene. However, the increase in permeability with increasing wavelength is slower than in graphene.

The calculated elements of the optical conductivity tensor allow us to investigate the optical properties of hybrid films. Let us investigate the transmittance and absorption of electromagnetic waves over a wide wavelength range of 0.01–10 μm. To this end, we use eqn (6.29)–(6.31), and eqn (6.33). For comparison, we calculate the parameters *T* and *A* for graphene to understand how these parameters change for the graphene–CNT hybrid film. As is known, the transmittance of graphene varies nonlinearly with a change in wavelength. A clear peak is observed in the UV region at a wavelength of 200 nm, and the transmittance becomes 100% starting from ~1 μm, which can be seen in Figure 6.14a and b. This figure shows the *T* coefficient curves for mono- (Figure 6.14a) and bilayer films (Figure 6.14b). In all cases, the incident of electromagnetic waves is normal, the wave vector *k* is directed along the *Z* axis and s- and p- (or *E*- and *H*-wave) polarizations are considered. Therefore, in all the graphs in Figure 6.14, the solid curves denote cases with s-polarization, and dotted cases with p-polarization. The difference in the reaction of the films to these two types of polarization is very noticeable. We note that, in contrast to graphene, the graphene–CNT hybrid film exhibits transmittance of 99–100% for p-polarization waves over the entire wavelength range. But for s-polarized waves, the situation is very similar to the case with pure graphene. The difference is only in the presence of a set of individual peaks due to the CNTs. In general, the transmittance of all types of hybrid films is quite high. For a monolayer film, the transmittance exceeds 90%, for a bilayer film it is higher than 86%.

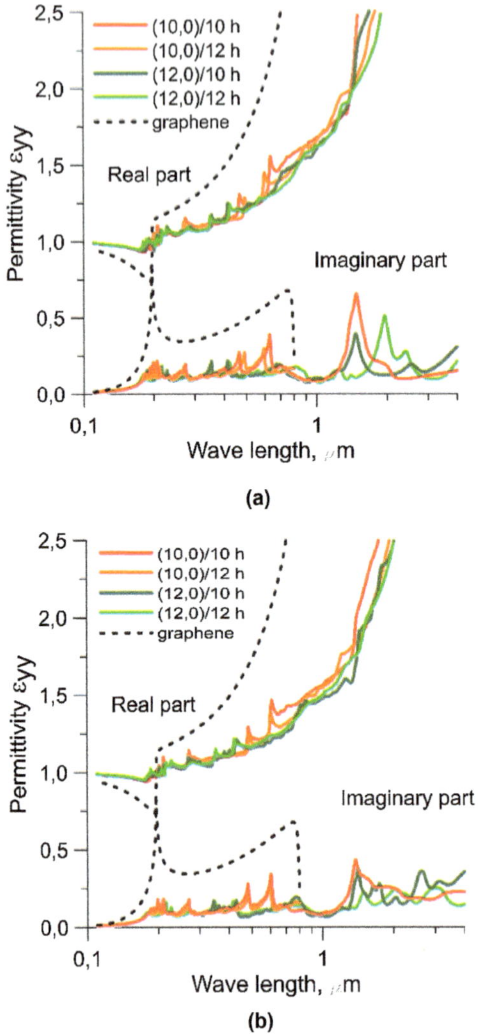

Figure 6.13 Graphs of complex permittivity: (a) for a monolayer graphene–CNT hybrid film; (b) for a bilayer graphene–CNT hybrid film.

Similar physical regularities were revealed for hybrid films in the absorption of electromagnetic waves. For s-polarized waves, there is a pronounced increase in absorption coefficient at wavelengths of 140 nm and 260 nm, as in pure graphene. However, these maxima are blurred in composite films because of the large number of peaks in this range. But clear peaks of the absorption coefficient, as well as transmittance coefficient are observed only for s-polarized waves. Absorption of p-polarized waves almost does not occur. The absorption coefficient remains within 1.5% over the entire wavelength range. In general, it can be concluded that the absorption

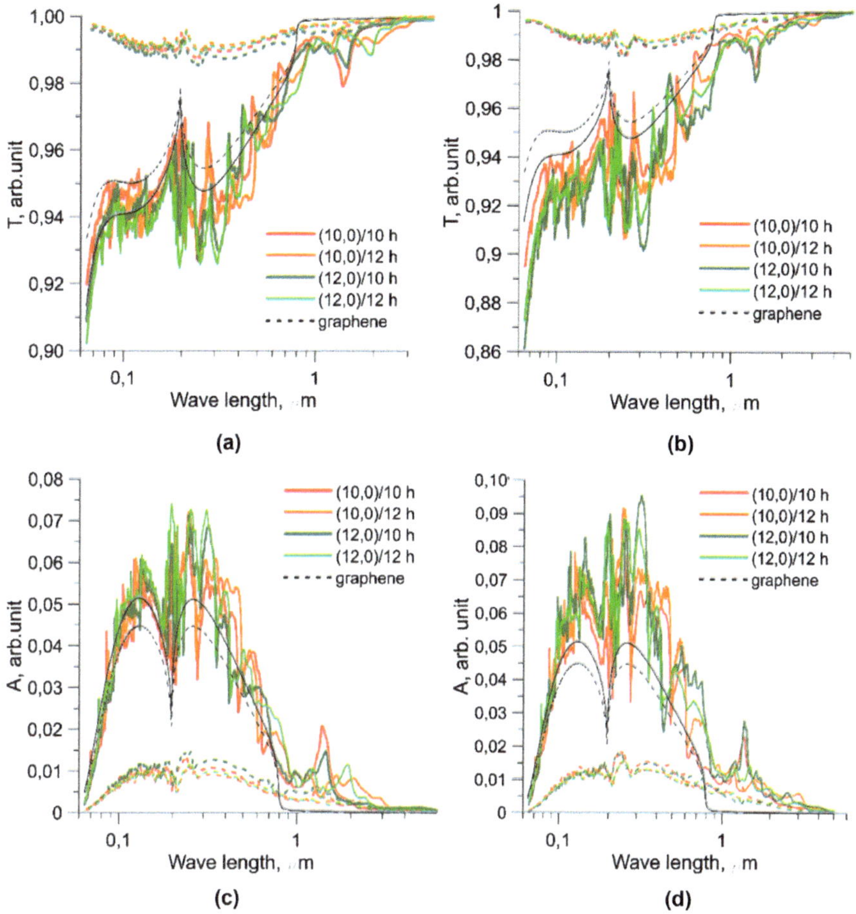

Figure 6.14 Optical characteristics of graphene–CNT hybrid film: (a) spectra of the transmittance coefficients of monolayer films; (b) spectra of the transmittance coefficients of bilayer films; (c) spectra of the absorption coefficients of monolayer films; (d) spectra of the absorption coefficients of bilayer films (logarithmic wavelength scale).

by the graphene–CNT hybrid films will not exceed 10%, regardless of the type of polarization of the waves.

Our results allow us to recommend a few-layer graphene-CNT hybrid film as a promising material for transparent electronics in the UV-visible-IR range. To understand how the transmittance of the films is related to their electrical conductivity, we calculated the transmittance for the middle part of the visible range (500 nm), red (700 nm) and violet (340 nm) boundaries, depending on the surface resistance of the hybrid film. Resistance changes due to the topology of the film, namely because of the chirality of the tube, and also because of the different number of layers of CNTs and graphene. That is, in this case, all possible configurations of mono- and

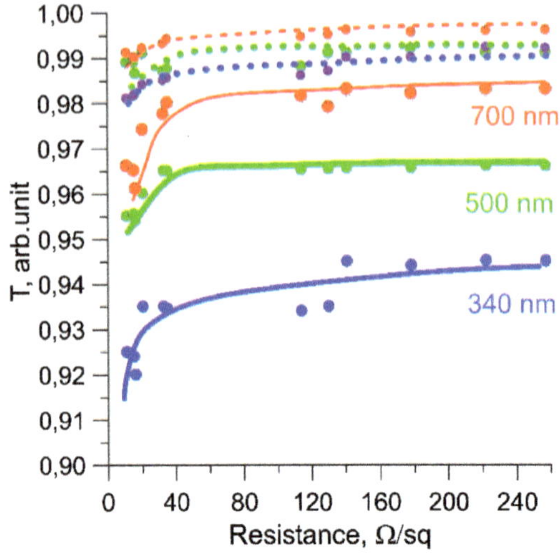

Figure 6.15 The relationship between the transmittance coefficient and the surface resistance of a hybrid graphene–CNT film (the dashed lines correspond to the s-polarized wave, solid lines to the p-wave).

bilayer films are taken into account. These dependencies are presented in Figure 6.15. As in Figure 6.14, the solid curves denote the dependences for p-polarized waves, the dotted curves for s-polarized waves. The overall trend for both types of waves is a rapid increase in the transmittance with an increase in resistance up to 30 kΩ and subsequent saturation. Saturation is achieved already at 110–120 kOhm.

6.3.4 Prospects for Using Graphene-CNT Hybrid Film in Photovoltaics

Traditional solar cells are able to absorb only part of the sunlight, and most of the light energy (especially the high-energy photons of the blue spectrum and the near-UV region) is lost as heat. The inability of conventional solar cells to absorb photons with any wavelength in the range from near-UV to near-IR determines the limits of battery performance and their low efficiency. In order to improve the efficiency of the photosensitive elements of solar cells, functionalized layered nanomaterials have been used in recent years. For example, the combination of lead sulfide (PbS) nanocrystals with pentacene provides effective absorption of electromagnetic waves in both the IR and visible ranges. The solar cells created on these laminates are called hybrid.[33] Polymer-functionalized CNTs can increase the absorptivity of solar cells and increase the photocurrent.[34]

In order to evaluate the prospects for the use of graphene–CNT hybrid films as a sensitive element of a photodetector or a solar cell, we calculated the absorbance spectra in the range from near-UV to near-IR. The results of

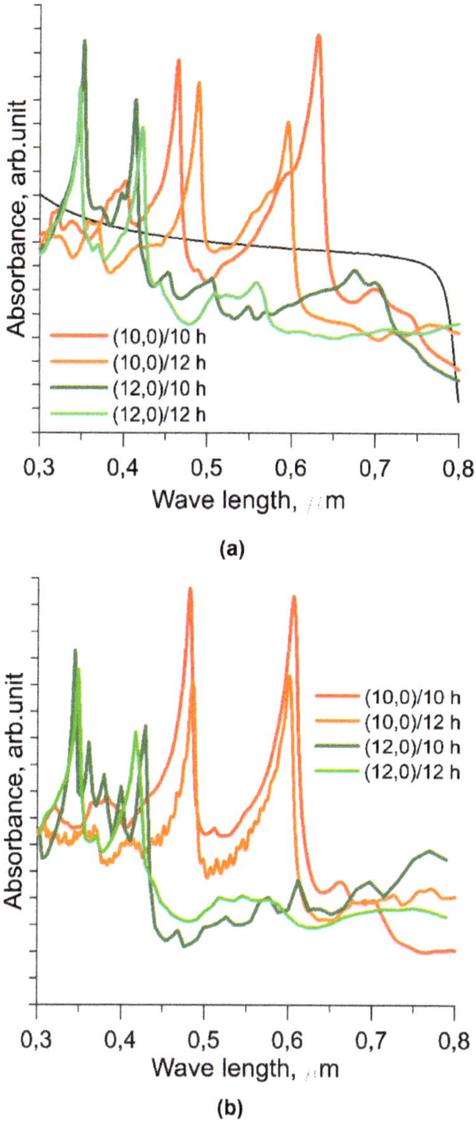

Figure 6.16 Absorbance spectra of graphene–CNT hybrid film: (a) a monolayer film; (b) a bilayer film.

the calculations are shown in Figure 6.16, where it can be seen that in the visible range there are two clearly expressed peaks, which are the contribution of mono- and bilayer films with CNTs of semiconductor conductivity. In the case of a bilayer film, the peaks coincide for CNTs (10,0) with intertube distances of 10 and 12 h. Hybrid film samples with CNTs of metallic conductivity have absorption peaks in the blue region of the spectrum and in the near-UV spectral regions. It can be concluded that by combining hybrid films, it is possible to obtain an effective laminate material, one layer of which will be very transparent, and the other will absorb visible light waves, providing a photocurrent.

6.4 Conclusion

Atomistic models of graphene–CNT hybrid films were constructed using the original "magnifying glass method". These hybrid films are layers of graphene with nanotubes between them, which have sp^3-hybridized atoms and form covalent bonds with graphene. Mono- and bilayer planar films were considered. It has been established that for monolayer hybrid films, the models with CNTs (n, 0) with the number $n = 10$–20 for the inter-tube distance of 8 h–16 h are energetically favorable. For bilayer films, the models with CNTs (10,0) and inter-tube distance of 10 h, 12 h, 14 h, 16 h, and also models with CNTs (12,0) and inter-tube distance of 10 h and 12 h turned out to be energetically favorable. For all the above-mentioned cases, the hybrid film formation is exothermic with $H_f < -0.1$ kcal mol^{-1} atom^{-1}.

The results of the calculations of DOS and the band structure for mono- and bilayer hybrid films showed that the electronic properties of monolayer films are determined exclusively by the conductive properties of the CNTs themselves. All models of a monolayer hybrid film exhibit a gap between the VB and CB of several tenths of an electron volt. The bilayer hybrid film has no gap for all types of topological models. The electrical resistance of mono- and bilayer films can differ a little from the resistance of ideal CNTs if the film is based on CNTs of metallic conductivity with a small inter-tube distance, close to the critically possible one. The calculated spectra of the real and imaginary parts of the element of the optical conductivity tensor for hybrid films of various topological modifications allowed us to establish that a series of additional peaks appeared in the visible and IR spectral regions due to the contribution of nanotubes in comparison with ideal planar graphene. The nature of the change in the permittivity of hybrid graphene–CNT films as a whole, repeats the properties of graphene.

The high sensitivity of graphene–CNT films to the type of polarization of electromagnetic waves is established. For p-polarization waves, graphene–CNT films exhibit 99–100% transmittance over the entire wavelength interval. And for s-polarized waves, the situation is very similar to the case with pure graphene, however, a number of individual peaks due to the CNTs are observed in the spectrum of the graphene–CNT film. In general, the transmittance of all types of films is quite high. For a monolayer film, it exceeds 90%, for a bilayer film it is higher than 86%.

The results of the calculation of the absorbance spectra of hybrid films in the spectral range from near-UV to near-IR showed that mono- and bilayer films with CNTs of semiconductor conductivity give two distinct peaks in the visible spectral region. Samples with CNTs of metallic conductivity have absorption peaks in the blue region of the spectrum and in the near-UV region. Therefore, by combining hybrid films it is possible to obtain an effective laminate material. One layer of this material will be very transparent, and the other layer will absorb visible light waves, providing a photocurrent.

Acknowledgements

Support from the Russian Presidential scholarship is acknowledged (project No СП-3135.2016.1).

References

1. B. H. Nguyen and V. H. Nguyen, *Adv. Nat. Sci: Nanosci. Nanotechnol.*, 2016, 7, 023002.
2. V. H. Nguyen, *Adv. Nat. Sci: Nanosci. Nanotechnol.*, 2016, 7, 023001.
3. D. Reddy, L. F. Register, G. D. Carpenter and S. K. Banerjee, *J. Phys. D: Appl. Phys.*, 2011, **44**, 313001.
4. R. Raccichini, A. Varzi, S. Passerini and B. Scrosati, *Nat. Mater.*, 2015, **14**(3), 271.
5. V. T. Dang, D. D. Nguyen, T. T. Cao, P. H. Le, D. L. Tran, N. M. Phan and V. C. Nguyen, *Adv. Nat. Sci.: Nanosci. Nanotechnol.*, 2016, 7, 033002.
6. G. K. Dimitrakakis, E. Tylianakis and G. E. Froudakis, *Nano Lett.*, 2008, **8**(10), 3166.
7. R. K. Paul, M. Ghazinejad, M. Penchev, J. Lin, M. Ozkan and C. S. Ozkan, *Small*, 2010, **6**(20), 2309.
8. X. Yang, D. Yu, B. Cao and A. C. To, *ACS Appl. Mater. Interfaces*, 2017, **9**(1), 29.
9. N. D. Kim, Y. Li, G. Wang, X. Fan, J. Jiang, L. Li, Y. Ji, G. Ruan, R. H. Hauge and J. M. Tour, *Nano Lett.*, 2016, **16**(2), 1287.
10. R. Lv, E. Cruz-Silva and M. Terrones, *ACS Nano*, 2014, **8**(5), 4061.
11. F. Tristán-López, A. Morelos-Gómez, S. M. Vega-Díaz, M. L. García-Betancourt, N. Perea-López, A. L. Elías, H. Muramatsu, R. Cruz-Silva, S. Tsuruoka, Y. A. Kim, T. Hayahsi, K. Kaneko, M. Endo and M. Terrones, *ACS Nano*, 2013, 7(12), 10788.
12. S. H. Kim, W. Song, M. W. Jung, M. A. Kang, K. Kim, S. J. Chang, S. S. Lee, J. Lim, J. Hwang, S. Myung and K. S. An, *Adv. Mater.*, 2014, **26**(25), 4247.
13. I. N. Kholmanov, C. W. Magnuson, R. Piner, J. Y. Kim, A. E. Aliev, C. Tan, T. Y. Kim, A. A. Zakhidov, G. Sberveglieri, R. H. Baughman Rodney and S. Ruoff, *Adv. Mater.*, 2015, **27**(19), 3053.
14. R. V. Salvatierra, D. Zakhidov, J. Sha, N. D. Kim, S. K. Lee, A. R. O. Raji, N. Zhao and J. M. Tour, *ACS Nano*, 2017, **11**(3), 2724.
15. W. Wang, M. Ozkanb and C. S. Ozkan, *J. Mater. Chem. A*, 2016, **4**, 3356.
16. Y. Liu, F. Wang, X. Wang, X. Wang, E. Flahaut, X. Liu, Y. Li, X. Wang, Y. Xu, Y. Shi and R. Zhang, *Nat. Commun.*, 2015, **6**, 8589.

17. A. A. Maarouf, A. Kasry, B. Chandra and G. J. Martyn, *Carbon*, 2016, **102**, 74.
18. X. Gan, R. Lv, J. Bai, Z. Zhang, J. Wei, Z. H. Huang, H. Zhu, F. Kang and M. Terrones, *2D Mater.*, 2015, **2**, 034003.
19. N. Jung, S. Kwon, D. Lee, D. M. Yoon, Y. M. Park, A. Benayad, J. Y. Choi and J. S. Park, *Adv. Mater.*, 2013, **25**(47), 6854.
20. Z. Zhang, L. Wang, J. Xiao, F. Xiao and S. Wang, *ACS Appl. Mater. Interfaces*, 2015, **7**(32), 17963.
21. S. Chen, W. Yeoh, Q. Liu and G. Wang, *Carbon*, 2012, **50**, 4557.
22. S. J. Stuart, A. B. Tutein and J. A. Harrison, *J. Chem. Phys.*, 2000, **112**, 6472.
23. M. Elstner, D. Porezag, G. Jungnickel, J. Elsner, M. Haugk, Th. Frauenheim, S. Suhai and G. Seifert, *Phys Rev B*, 1998, **58**, 7260.
24. A. Pecchia, G. Penazzi, L. Salvucci and A. Di Carlo, *New J. Phys.*, 2008, **10**, 065022.
25. M. P. Marder, *Condensed Matter Physics*, Wiley-VCH, Berlin, Germany, 2011.
26. http://nanokvazar.ru (last accessed March 2016).
27. http://www.dftb.org (last accessed May 2016).
28. V. V. Mitrofanov, M. M. Slepchenkov, G. Zhang and O. E. Glukhova, *Carbon*, 2017, **115**, 803.
29. M. Elstner and M. Elstner, *Theor. Chem. Acc.*, 2006, **116**, 316.
30. S. Datta, *Quantum Transport: Atom to Transistor*, Cambridge University Press, United Kingdom, 2005.
31. O. E. Glukhova, I. S. Nefedov, A. S. Shalin and M. M. Slepchenkov, *Beilstein J. Nanotechnol.*, 2018, **9**, 1321.
32. H. A. Le, S. T. Ho, N. D. Chien and V. N. Do, *J. Phys.: Condens. Matter*, 2014, **26**, 405304.
33. B. Ehrler, M. W. B. Wilson, A. Rao, R. H. Friend and N. C. Greenham, *Nano Lett.*, 2012, **12**(2), 1053.
34. M. Gong, T. A. Shastry, Y. Xie, M. Bernardi, D. Jasion, K. A. Luck, T. J. Marks, J. C. Grossma, S. Ren and M. C. Hersam, *Nano Lett.*, 2014, **14**(9), 5308.

CHAPTER 7

Metal–Organic Frameworks (MOFs) as Potential Hybrid Ferroelectric Materials

Y. SUN,[a] Y. CHENG*[b] AND K. Y. ZENG*[a]

[a] Department of Mechanical Engineering, National University of Singapore, 117576, Singapore; [b] Institute of High Performance Computing, Agency for Science Technology and Research, 138632, Singapore
*Email: mpezk@nus.edu.sg; chengy@ihpc.a-star.edu.sg

7.1 Introduction

Crystal structures can be categorized into 32 point groups, according to the number of rotational axes and reflection planes that leave the crystal structure unchanged. Twenty-one (21) of the 32 crystal point groups lack a centre of inversion symmetry, and 20 of them are found to be piezoelectric. Among the 20 piezoelectric crystal point groups, 10 groups are pyroelectric (polar). Theoretically, all the ferroelectric materials must belong to the 10 polar point groups, which are C_1, C_s, C_2, C_{2v}, C_3, C_{3v}, C_4, C_{4v}, C_6, and C_{6v} (Figure 7.1).[1] That is, ferroelectric materials are a special subgroup of piezoelectric and pyroelectric materials, featured by spontaneous electric polarization that can be reversed by the external electric field. Ferroelectric materials have a wide range of advanced physical properties, including ferroelectric hysteresis, high permittivity, high piezoelectric effect, high pyroelectric coefficient, strong electro-optic effect and anomalous temperature coefficients of resistivity, making them preferred candidates for non-volatile memories,[2] capacitors,[3] sensors,[4] actuators,[5] resonant wave devices,[6]

Smart Materials No. 34
Layered Materials for Energy Storage and Conversion
Edited by Dongsheng Geng, Yuan Cheng and Gang Zhang

Published by the Royal Society of Chemistry, www.rsc.org

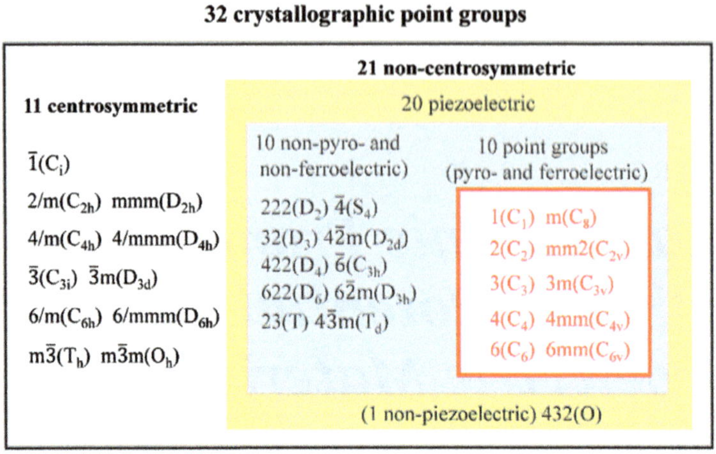

Figure 7.1 The 32 crystallographic point groups categorized by ferroelectric, pyroelectric, and piezoelectric properties, referring to W. Zhang, *etc.*[1] Reproduced from ref. 1 with permission from American Chemical Society, Copyright 2012.

infrared detectors,[7] optical switches,[8] electric-motor overload-protection circuits,[9] and many other applications.

The early history of ferroelectrics has been well described in a review by Busch.[10] Many researchers have been involved in the early exploitation of ferroelectricity. But the polarization of Rochelle salt ($KNaC_4H_4O_6{\cdot}4H_2O$), that could be switched by the application of an electric field observed by Valasek in 1920, is believed to be the pioneering discovery of ferroelectricity.[9] Since then, the ferroelectric field has been dominated by inorganic materials, such as barium titanate ($BaTiO_3$),[11] oxide compounds such as lead zirconate titanate ($Pb(Zr_xTi_{1-x})O_3$)[12,13] relaxor crystals and lithium niobite ($LiNbO_3$),[14] as well as some polymers[15] such as PVDF. The structural identification in inorganic ferroelectrics reveals two requirements for being ferroelectric, *i.e.* bistable electronic dipoles and a non-centrosymmetric lattice belonging to the 10 polar point groups. This informative structural instruction gives rise to the exploitation of ferroelectricity in organic crystals,[16] hybrid frameworks[17] and liquid crystals.[18–20]

Metal–Organic frameworks (MOFs) are newly emerging crystalline compounds comprised of metal ions/clusters directionally coordinated to organic ligands.[21–26] They are formed by self-assembly processes, leading to aggregation of metal ions into secondary-building units (SBUs), which are multinuclear metal clusters with more complex structural and electronic features than that of the single metal ion.[27] By selection of diverse SBUs and organic ligands, there are thousands of MOF crystals with different topology, pore size and shape, as well as chemical composition.[28] In fact, MOFs are one of the most versatile materials at present for many applications. For example, MOFs are featured by ultrahigh porosity (~90% free volume), large internal surface areas (>6000 $m^2 g^{-1}$) and optional choices for both organic ligands and inorganic ions/clusters,[29] leading to vast applications in gas adsorption and separation,[30,31] catalysis,[32,33] chemical sensing,[34] drug delivery,[35] and photochemical application,[36–38] and more. In addition, their applications in membranes,[39] thin film devices,[40] catalysis,[41] biomedical imaging[42] and supercapacitors are gaining more and more attention.[29] Despite all these advancements, few studies can be found on their ferroelectric properties. In fact, MOFs can be tuned practically for ferroelectric, magnetic and multiferroic applications[1,43,44] since they have versatile forms of metal-coordination systems as well as internal host–guest interactions. As a matter of fact, ferroelectric materials have been widely used as capacitors,[3] non-volatile memories,[2,45,46] piezoelectric devices for ultrasound imaging and actuators,[47] electro-optic materials for data storage applications,[48] thermistors,[49] switches known as transchargers or transpolarizers,[50] oscillators and filters,[51,52] light deflectors,[53] modulators and displays,[54,55] *etc.*

In 1969, Okada *et al.* proposed the first antiferroelectric MOF material $Cu(HCOO)_2{\cdot}4H_2O$, which consists of formic acid (HCOOH) as building blocks and protonated organic amines as templates.[56] It showed a first-order phase transition at about 235 K with a discontinuity in temperature dependent dielectric constants. Subsequently, the relative permittivity ε_r of three MOFs ($[Mn_3(HCOO)_6](H_2O)(CH_3OH)$, $[Mn_3(HCOO)_6](C_2H_5OH)$, and $[La_2Cu_3](H_2O)_n$) were found to show anomalies at 150 K ($[Mn_3(HCOO)_6]$-$(H_2O)(CH_3OH)$), 165 K ($[Mn_3(HCOO)_6](C_2H_5OH)$), 180 K and 255 K ($[La_2Cu_3\{NH(CH_2COO)_2\}_6](H_2O)_n$), respectively.[57–59] It was reported that the phase transition of the three MOFs were induced by the ordering and freezing of the internal polar guest molecules. Ferromagnetic transitions can occur because of the spin ordering of the M^{2+} (M = magnetic metals) incorporated in the host lattice based on the Goodenough–Kanamori rule.[60] Cheetham and his co-workers reported the first lead-free hybrid frameworks ($[(CH_3)_2NH_2]M(HCOO)_3$, M = Mn, Fe, Co, Ni) with a typical ABO_3 perovskites structure,[61] which presented antiferroelectric behaviours from 160 to 185 K because the organic molecules showed order–disorder electric ordering behaviours.[62] Zhang and Xiong reported an ABO_3 type perovskite MOF inclusion, $(Him)_2[KFe(CN)_6]$ (Him = imidazolim cation), which had an order–disorder structural transition due to the motions of the cationic guests (polar Him).[63] The dielectric constant showed two anomalies at 187 K and 158 K, respectively, revealing the possible switchable behaviour. Stroppa and his colleagues have conducted *ab initio* calculations on MOF $[C(NH_2)_3]M[(HCOO)_3]$, which had an ABX_3 structure and proposed that Cu and Cr-based MOF could exhibit multiferroic behaviours due to the Jahn–Teller effects and antiferro-distortion metals that could cause a switchable ferroelectric polarization by coupling to the $[C(NH_2)_3]^+$ group through hydrogen bonding.[64,65] Di Sante *et al.* also reported multiferroic behaviours of MOF $[(CH_3)_2NH_2]Mn(HCOO)_3$, of which the ferroelectric polarization up can be tuned to 6 $\mu C\,cm^{-2}$ by changing the organic cations by density functional theory (DFT) calculations.[66] Li *et al.* revealed that the indium

MOF, $[InC_{16}H_{11}N_2O_8]\cdot1.5H_2O$, possessed the coexistence of resistance switching and ferroelectricity.[67] A first principles molecular dynamics simulation revealed that the resistive switching effect resulted from the ferroelectric transition of the bridge structured dipoles (N···H–O···H–N) formed by the guest water molecules and the amino-tethered MOF nanochannel.[67] MOFs have increasingly drawn attention since they combine the advantages of both inorganic solids and organic linkers, including a satisfying mechanical strength, high structural flexibility, low density and solution processibility.[44] Despite these advancements, one of the current research hotspots is to synthesize MOFs that manifest promising ferroelectric properties (comparably or beyond inorganic ferroelectrics), while also being able to easily integrate to other devices.[68]

7.2 Experimental Characterization

Ferroelectricity is the ability of materials to have spontaneous polarization that can be reversed by the electric field. Experimentally, ferroelectricity can be captured by the polarization hysteresis loop (P–E loop, *i.e.* polarization *vs.* electric field). Ferroelectrics materials have a Curie temperature (T_c), a critical temperature beyond which materials lose their ferroelectric qualities. The T_c is denoted as the discontinuity in the material's temperature dependent dielectric constant curve.[68] Here, five commonly used experimental methods will be discussed. In most of the studies, combining different methods is usually recommended.

7.2.1 Phase Transition and Anomaly in Dielectric Constant

Phase transition refers to the transformation of a thermodynamic system that changes from one phase of matter to another.[1] In terms of ferroelectrics, the phase transition is from the paraelectric to ferroelectric and *vice versa* (Figure 7.2). Phase transition can be captured as the anomaly in the temperature dependent dielectric constant curve. In addition, this transition can be affected by many factors, such as pressure, electric field, shock wave, laser, *etc.*[69,70] and can lead to abrupt anomalies in elastic, thermal, and crystal structures of the materials.[12,71,72] The crystal lattice can transform to a non-centrosymmetric phase through displacive phase transition or order–disorder phase transition. In the former, the molecules displace relative to one another to form a polar, non-centrosymmetric lattice, thus showing a macroscopic ferroelectric polarization under an external electric field.[12] Many inorganic oxides such as $BaTiO_3$ exhibit displacive phase transition. In an order–disorder phase transition, the dipoles are randomly oriented but can spontaneously align in an ordered manner. In fact, many ferroelectrics often show both displacive and order–disorder phase transitions.[1]

Cheetham *et al.* found that the desolvated zeolitic imidazolate MOF ZIF-4(Zn) experienced a discontinuous porous-to-dense phase transition during cooling through 140 K. The phase transition of ZIF-4(Zr) involved a

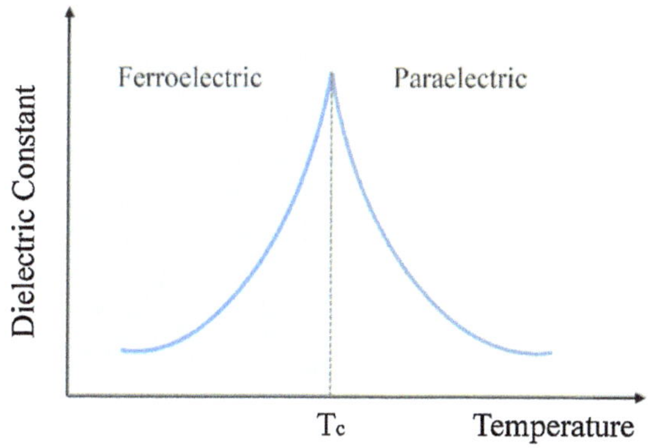

Figure 7.2 The Curie temperature (T_c) denoted as a discontinuity in the temperature dependent dielectric constant curve, beyond which the ferroelectric changes to paraelectric.

cooperative rotation of imidazolate linkers that caused the isotropic framework to become contracted and the pore space to be minimized.[73] Yadav *et al.* investigated the multiferroic behaviour of $(CH_3)_2NH_2Co(CHOO)_3$ MOF crystals, of which the organic cation $(CH_3)_2NH_2^+$ occupied the cavities in the formate framework through the N–H···O hydrogen bonds. The organic cation was ordered upon cooling, which resulted in a structural phase transition at 155 K. The phase transition was found to be a paraelectric–ferroelectric phase transition revealed by dielectric and pyroelectric measurements. Moreover, a polarization *vs.* electric field (*P–E*) hysteresis loop below 155 K was observed and hence confirmed the ferroelectric behaviour of the material.[74]

7.2.2 Scanning Second Harmonic Generation Microscopy

Second Harmonic Generation (SHG) measurement is based on a nonlinear ultrasonic method that has the ability to detect and monitor microstructural changes in materials. Scanning SHG microscopy can be used to obtain ferroelectric domain imaging.[75] In a SHG microscopy experiment, an incident light with frequency ω is focused on a sample through a microscope and scanned across the sample, as shown in Figure 7.3. A fundamental wave in the near-infrared range (wavelength 800–1000 nm) is adopted to scan the sample with a lateral resolution near the diffraction-limit (300–500 nm).[76] The outgoing SHG signal is subsequently collected either in transmission or reflection mode. In both SHG transmission and reflection modes (Figure 7.3a and b), the outgoing signal beam can be easily collected since it is separated from the incident beam in space. However, in reflection mode, only the very surface of the sample or internal domain boundary with approximately a coherence depth of a sample interface can be obtained.[76] Any of the two scanning modes can provide a two-dimensional (2D) map of the SHG response, revealing the domain structure of the sample. Performing

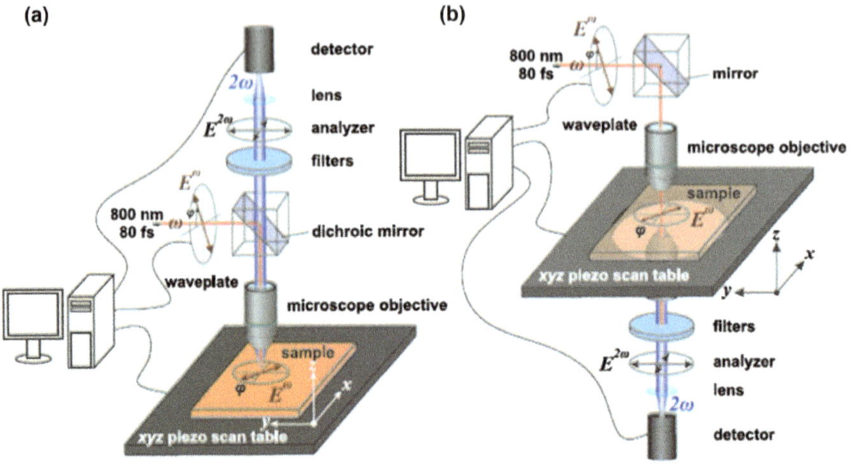

Figure 7.3 The two SHG microscopy experimental setups of (a) reflection-mode and (b) transmission mode.
Reproduced from ref. 76 with permission from John Wiley and Sons, © 2011 The American Ceramic Society.

multiple confocal 2D scans with the focal plane at different depths can realize the optical depth sectioning (SHG tomography) of the samples.[77] Detailed information about scanning SHG microscopy can be found in a number of papers[78–80] and reviews.[81–83]

The second-order contribution to the electric polarization of a material is proportional to the square of the electric field of the incident light wave with frequency ω. This will lead to a resulted nonlinear response with frequency 2ω.[84] According to the electric-dipole approximation, nonlinear optical effects exist only in non-centrosymmetric materials. Consequently, in order to make the SHG process happen, the inversion symmetry of the magnetically or electrically ordered phase of a material must be broken. Symmetry breaking plays a key role in studying the multiferroic phase in many different crystals and heterostructures and maintaining the extreme high sensitivity of SHG.[85–89]

Jain *et al.* used SHG to investigate the ferroelectric behaviour of dimethylammonium-Mn-formate (DMAMnF). The SHG signal was almost lost above the phase transition temperature, but was obviously detectable below the phase transition temperature. They also observed a hysteretic behaviour driven by the temperature characterized by two critical temperatures approximately at 160 K and 172 K. They successfully visualized the possible inhomogeneities in DMAMnF in the lateral spatial distribution of the SHG signal with a microscope collecting the output SHG signal. The obtained bright regions against the dark regions indicate the formation of ferroelectric domains in DMAMnF, as shown in Figure 7.4.[85]

Markey *et al.* used SHG microscopy to study the effect of fluoride that was incorporated in the MOF MIL-53(Fe) structure (MIL-53(Fe) = Fe(OH/F)(BDC)).

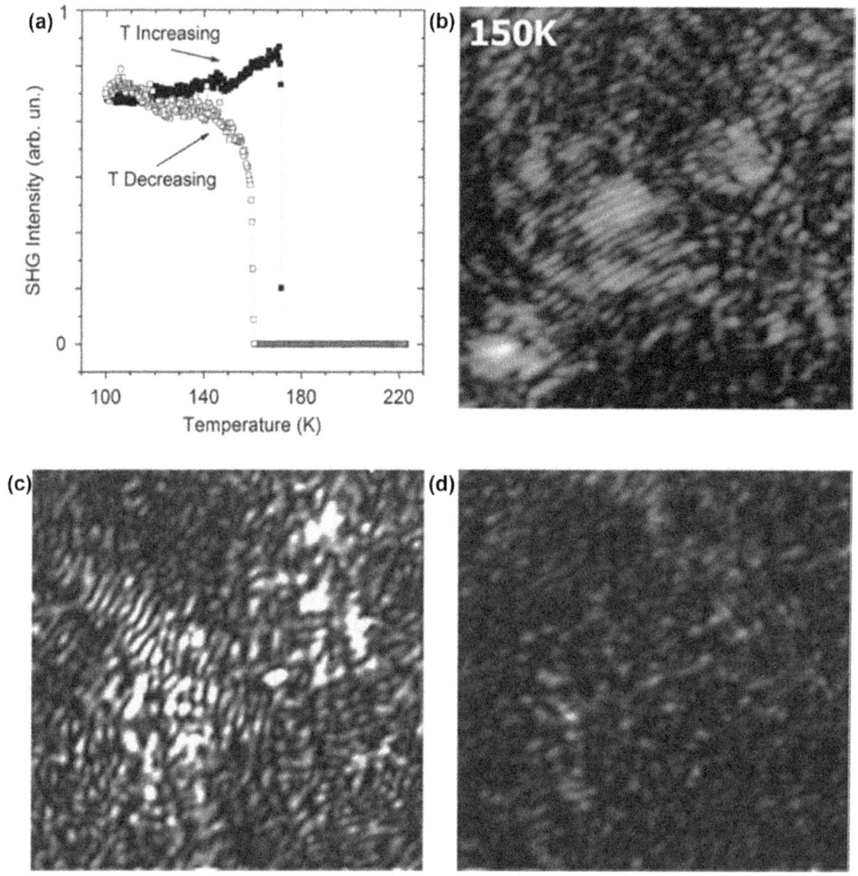

Figure 7.4 (a) The temperature dependent second harmonic generation (SHG). (b–d) SHG images of the same sample area (with no field applied) at 150 K (below both the critical temperatures, 160 and 172 K). Note: For each SHG measurement, the temperature is elevated above the critical temperature and then cooled to 150 K. The domain pattern changes. The greyscale is consistent for the three panels.
Reproduced from ref. 85, https://doi.org/10.1038/npjquantmats.2016.12, under the terms of the CC BY https://creativecommons.org/licenses/by/4.0/.

They found that the guest fluorine atoms significantly affected the non-centrosymmetric ordering.[90] Polarization distribution is obtained with a SHG polarization map. Furthermore, Zhou *et al.* successfully employed SHG to obtain the response of an acentric 3D Cd(II) MOF crystal, revealing the potential ferroelectric behaviours and photoluminescence of the materials.[91]

7.2.3 Sawyer–Tower Circuit

A traditional and commonly adopted macroscopic method to measure the *P–E* loop is the well-known Sawyer–Tower circuit.[92] It is comprised of an oscilloscope, a signal generator E, a reference capacitor C_0, and the unknown capacitor C_x of the ferroelectric sample to be determined. An illustration of the Sawyer–Tower circuit is shown in Figure 7.5. An alternating current (AC) signal is applied in the circuit through the signal generator E. The reference capacitor C_0 and sample capacitor C_x carry the same amount of charge Q. For the reference capacitor C_0, the amount of charge equals $Q = C_0 \times U$, in which U is the voltage on C_0. For the unknown capacitor C_x of the ferroelectric sample, the polarization P is proportional to charge Q ($Q = P \times A$) and A. The area of the electrode on the sample is denoted by A. Based on the above relations, it can be seen that the polarization P of the sample is proportional to the voltage U applied on the reference capacitor C_0. In most cases, the value of the reference capacitance C_0 is much larger than the value of the sample capacitance C_x. Thus, the voltage on the sample capacitance (U_x) is much larger than the voltage on the reference capacitance (U). In fact, U_x can be treated as being identical to the voltage produced by the signal generator E. Inputting the sample capacitance U_x (approximate to electric field E) as the x-channel and the reference capacitor U (approximate to polarization of the sample) as the y-channel of the oscilloscope, respectively, the ferroelectric hysteresis loop can be accordingly displayed on the oscilloscope.[93]

Ye *et al.* synthesized a new 3D ferroelectric MOF by using cycloaddition chiral ligand under hydrothermal reaction conditions. They obtained an electric hysteresis loop with a remnant polarization (P_r) and coercive field (E_c) of *ca.* 2.10 kV cm^{-1} and *ca.* 0.38 μC cm^{-2} respectively. The saturated polarization (P_s) was *ca.* 0.50 μC cm^{-2}, which was much larger than that of some typical ferroelectric compounds such as potassium sodium tartrate ($NaKC_4H_4O_6 \cdot 4H_2O$), and Rochelle salt.[94] Gupta *et al.* reported the ferroelectricity of a homochiral Cu(II) framework, $\{[Cu_2(L)(H_2O)_2] \cdot (4DMF)(4H_2O)\}_n$

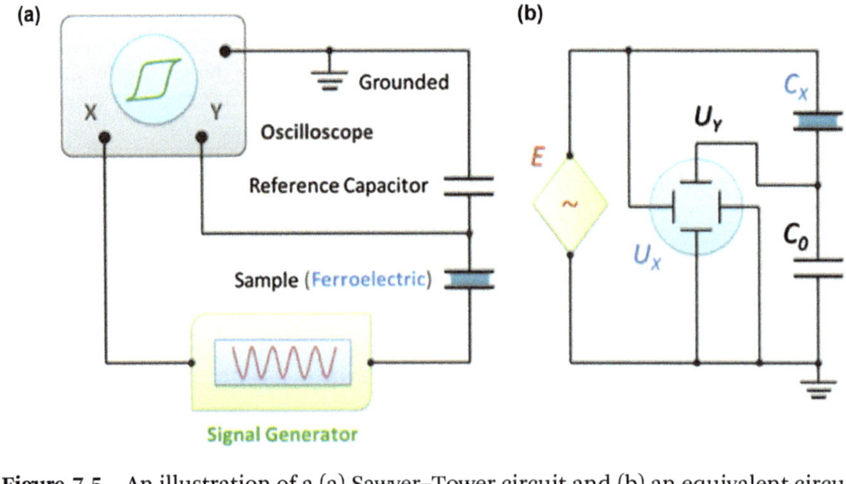

Figure 7.5 An illustration of a (a) Sawyer–Tower circuit and (b) an equivalent circuit. Reproduced from ref. 93 with permission from Springer Nature, Copyright 2014.

(LCu), which possessed a non-interpenetrated structure comprised of $[Cu_2(COO)_4]$ paddlewheel SBUs with NbO topology. The framework LCu was found to exhibit excellent ferroelectric properties with a remnant polarization (P_r) of approximately 3.5 μC cm^{-2} and a coercive field (E_c) of approximately 12 kV cm^{-1}, featured by a characteristic *P–E* loop.[95] Fu *et al.* obtained *P–E* loops (Figure 7.6) at different temperatures that demonstrated that the PD-DMACoF was a ferroelectric relaxer with a saturation polarization P_s of *ca.* 1.04 μC cm^{-2} (higher than Rochelle salt: 0.2 μC cm^{-2}, but smaller than TGS: 3.5 μC cm^{-2}), comparable to the P_s of $[NH_4][Zn(HCOO)_3]$ (1.0 μC cm^{-2}).[96]

7.2.4 Structural Analysis by X-ray Diffraction

X-ray diffraction (XRD) is one of the most widely used non-destructive techniques to analyse the structures of all kinds of materials, including fluids, powders and crystals based on the Debye–Scherrer method.[97] This technique uses X-ray, neutron or electron diffraction on powder or micro-crystalline samples, where ideally every possible crystalline orientation is represented equally.[85,98] It is worth noting that the *P–E* loop measurement can be integrated to the XRD beamline for the determination of the crystallographic response of the test material to the application of dynamic electric fields. In the integrated measurement system, *in situ* diffraction data are collected and the temporal evolution of the crystal structure can be tracked. Besides, the time-resolved diffraction data is also collected to

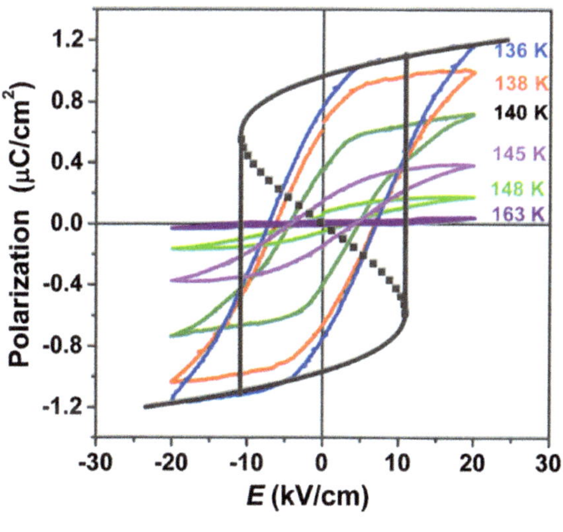

Figure 7.6 The dielectric hysteresis loop recorded at a series of temperatures below T_c (*ca.* 151 K). All the loops were measured at same frequency (50 Hz). The black solid line is the fitted hysteresis loop based on the Landau theory.
Reproduced from ref. 96 with permission from John Wiley and Sons, Copyright © 2011 WILEY-VCH Verlag GmbH & Co. KGaA, Weinheim.

measure the lattice strain in response to a square-wave electric field, which is applied to the sample to switch its polarization state. In such experiments, the effects of an applied 'on–off' field as a function of time can be studied.[99]

7.2.5 Nanoscale Spectroscopic Measurements

Binning and Rohrer[100] invented Scanning Tunnelling Microscopy (STM) based on quantum mechanics in 1981, which is the premier member of Scanning Probe Microscopy (SPM). But STM can only be applied to conductors and semiconductors, which confines its usefulness to non-conductive materials as well as materials with thick oxide layers. In order to overcome this limitation, Binning, Quate and Gerber[101] created the first Atomic Force Microscopy (AFM) in 1986. AFM can image insulators at the atomic scale with a horizontal resolution of 1 nm and vertical resolution of 0.01 nm. AFM has many advantages including ultra-high spatial resolution, nondestructive, simple sample handling and quantitative imaging. Besides, AFM can be performed in liquid or air and has a high signal-to-noise ratio.

Currently, Piezoresponse Force Microscopy (PFM) is the most widely used SPM method for imaging domain structures as well as bias-induced changes in the structure and properties of piezoelectric materials and ferroelectrics.[102] This technique is based on inverse piezoelectric effects, *i.e.* monitoring the surface displacement of piezoelectric materials induced by electric bias. The standard experimental setup is on the base of commercial SPM equipment with a four-quadrant photo detector, a conductive probe, a function generator and two lock-in amplifiers, as shown in Figure 7.7.

An ac voltage is applied between the conductive tip and the sample's surface by a functional generator. The bias induced deformation of the sample's surface is detected by a laser spotted on the beam and reflected on a four-quadrant photodiode. Two signals (a + c) − (b + d) and (a + b) − (c + d)

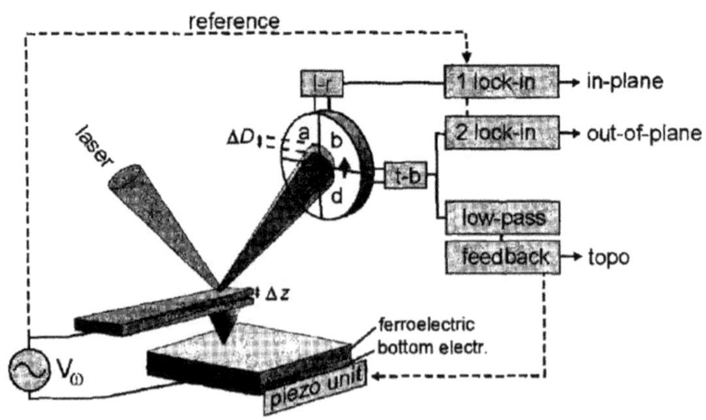

Figure 7.7 The experimental setup of PFM.[102]
Reproduced from ref. 102 with permission from Springer Nature, Copyright 2007.

are demodulated in two lock-in amplifiers that represent the in-plane and out-of-plane signals, respectively. The force on the cantilever stays unchanged by a feedback loop during the whole scanning process.

Dual AC Resonance Tracking PFM (DART-PFM) is a technique developed by an AFM manufacturer (Asylum Research, Oxford Instruments, CA, USA). It is capable of reducing the crosstalk with topography due to a shift in the resonant frequency. A feedback loop is used to adjust the two drive frequencies of the cantilever to match the resonance frequency. Figure 7.8 shows a schematic of the two frequencies, and the resulting amplitudes (A_1 and A_2) when the resonant frequency changes. For example, if the frequency shifts downward, A_1 moves to A_1' and A_2 moves to A_2'. The change in the A_2–A_1 signal causes the feedback loop to respond by changing the drive frequency until the A_2–A_1 signal is zero again.[103]

The obtained responses are directly from the cantilever rather than the sample and the cantilever motion is influenced by the oscillation of the sample according to the inverse piezoelectricity. The sample oscillation responses A_{drive} and φ_{drive} induced by the electric field can be extracted by fitting the responses to a damped (simple) harmonic oscillator (DHO or DSHO) model. The DHO (DSHO) is appropriate for modelling single eigenmode characterization if the tip–sample interaction doesn't significantly perturb the oscillation of the cantilever.[104,105] In this model, the responses amplitude A_{drive} and phase φ_{drive} of cantilever oscillation can be expressed by following formulas.[106]

$$A(f) = \frac{f_0^2 A_{\text{drive}}}{\sqrt{(f_0^2 - f^2)^2 + (f_0 f/Q)^2}} \tag{7.1}$$

and

$$\varphi(f) = \tan^{-1}\frac{f_0 f}{Q(f_0^2 - f^2)} + \varphi_{\text{drive}}. \tag{7.2}$$

By solving these two eqn (7.1) and (7.2), the sample responses A_{drive}, φ_{drive}, contact resonance frequency f and Q factor can be extracted from the four known parameters from the cantilever oscillation A_1, A_2, φ_1, φ_2. Therefore, the piezoresponse of the materials can be quantified by using the DART-PFM technique.

Sun and colleagues recently adopted the DART-PFM to probe the piezoresponses of NUS-6-(Hf/Zr) MOF nanocrystals. The NUS-6(M) (M = Zr or Hf, $C_{120}O_{291}S_{15}M_{21}$) were derived from UiO-66(M) (Figure 7.9a),[107,108] which can be best achieved by periodically replacing the 12 connected M_6 clusters (M = Zr or Hf) in the parental UiO-66-SO_3H framework (Figure 7.9b) with 9 connected M_6 clusters (Figure 7.9c–e).[108]

Figure 7.10 shows DART-PFM amplitude images and topography images as well as calculated d_{zz} images and 3D images of the piezoresponse amplitude overlaid on the topography. From Figure 7.10a and e, it can be seen that the

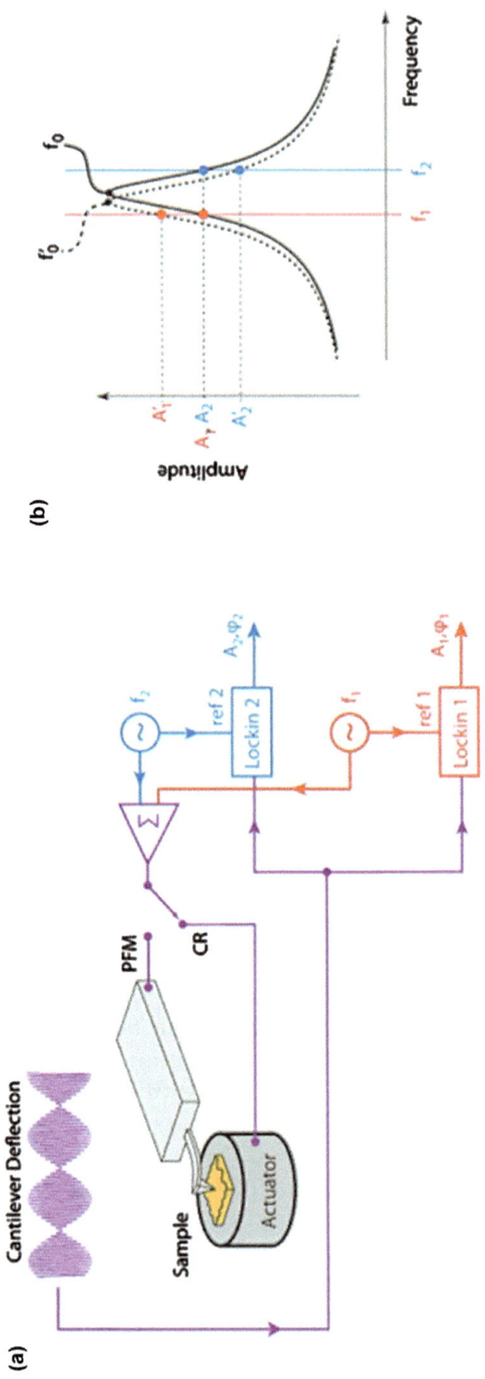

Figure 7.8 Principle of DART-PFM.[103] Reproduced from ref. 103 with permission from IOP Publishing.

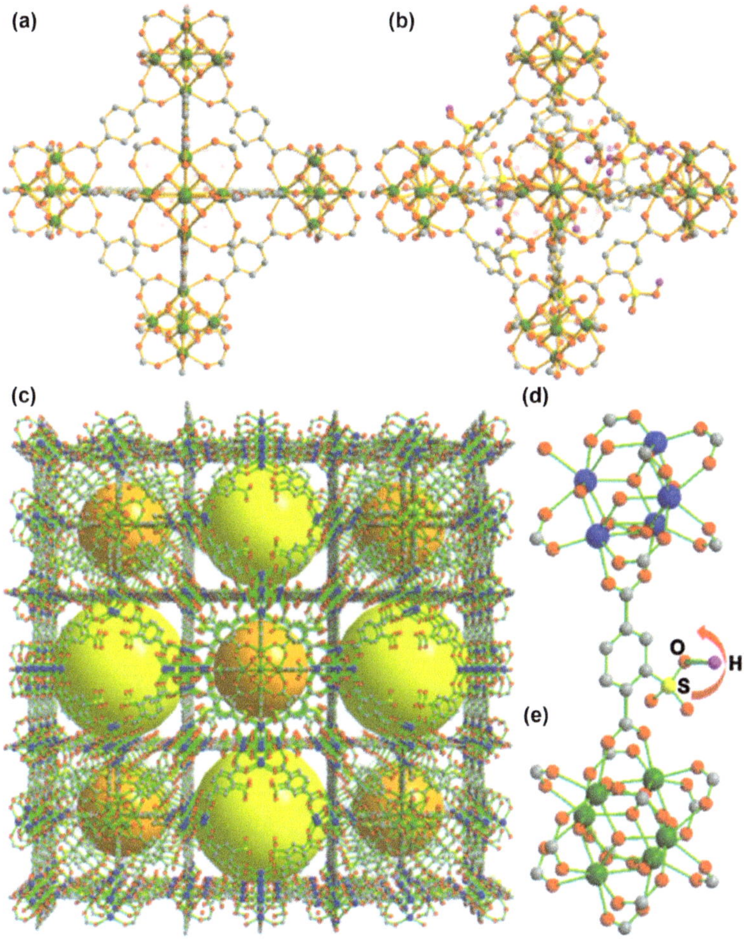

Figure 7.9 Crystal structure description: (a) UiO-66; (b) parental UiO-66-SO_3H; (c) NUS-6, which is featured by micropores (orange spheres) and mesopores (yellow spheres); (d) a 9-connected, and (e) 12-connected M_6 cluster (M = Zr or Hf).
Reproduced from ref. 44 with permission from the Royal Society of Chemistry.

particle sizes of both the NUS-6-(Hf) and NUS-6-(Zr) nanocrystals are approximately 150–200 nm with moderate variation. According to Figure 7.10b and f, either NUS-6-(Hf) or NUS-6-(Zr) nanocrystals exhibit distinct piezoresponses with several hundreds of picometer (100–400 pm) amplitudes under applied V_{ac} voltages ($V_{ac} = 5$ V). Besides, the NUS-6-(Hf) is found to exhibit a stronger piezoresponse, featured by a larger value of PFM amplitude (300–400 pm) than that of the NUS-6-(Zr) (200–300 nm) nanocrystals. Larger values of d_{zz} (2.0–3.5 pm V^{-1}) is derived for NUS-6-(Hf), which is much larger than that of 1.5–2.5 pm V^{-1} of NUS-6-(Zr) (Figure 7.10c and g). The calculated

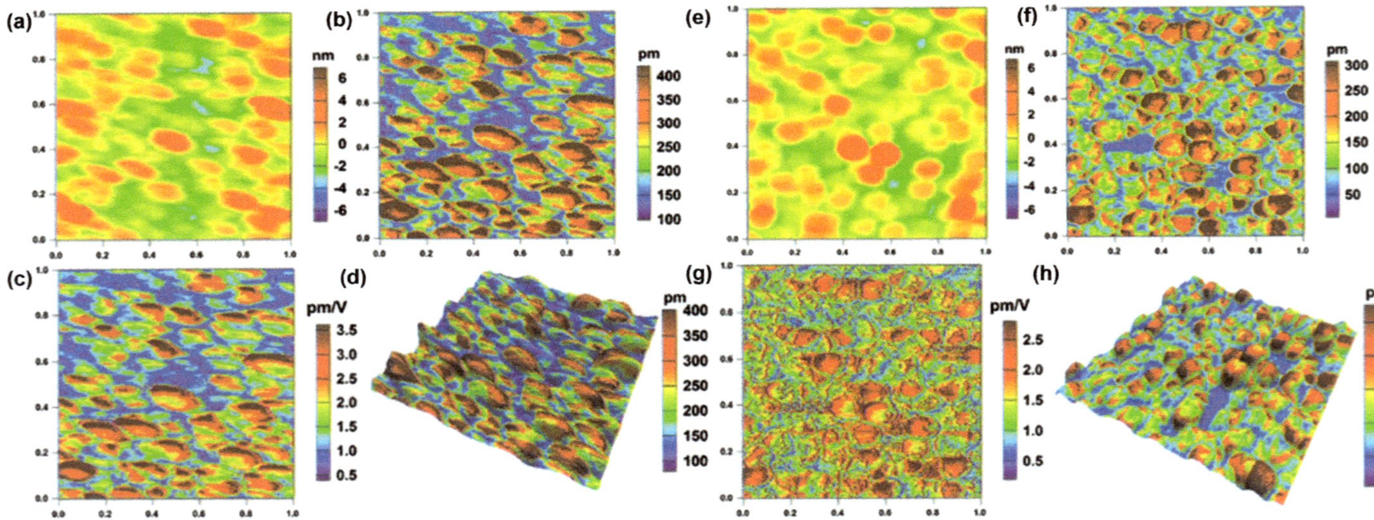

Figure 7.10 DART-PFM results (scan area: 1 μm×1 μm) of (a–d) NUS-6-(Hf) and (e–h) NUS-6-(Zr) nanocrystals. (a, e) Topography images; (b, f) piezoresponse amplitude images; (c, g) calculated d_{zz} according to DSHO, and (d, h) a 3D image that is formed by the piezoresponse amplitude overlaid on topography images.
Reproduced from ref. 44 with permission from the Royal Society of Chemistry.

d_{zz} values of NUS-6 series MOFs are found to be smaller than those of the typical piezoelectric materials such as lead zirconate titanate (PZT) film, which possesses a d_{zz} of ~18 pm V^{-1},[109] but larger than those of biomaterials such as bones that possesses a d_{zz} of ~0.7 pm V^{-1}.[110] Figure 7.10d and h show 3D images that are comprised of the PFM amplitude images overlaid on the topography images for NUS-6-(Hf/Zr) nanocrystals. In addition to the above results, it is worth noting that the piezoresponse distribution of each nanocrystal is non-uniform, which indicates that the NUS-6 series nanocrystals have possible structural anisotropy.[111]

Piezoresponse Force Spectroscopy (PFS) is a technique used to acquire local ferroelectric hysteresis loops from the surfaces of tested materials. During PFS measurements, the optimal signal-to-noise ratio is achieved at frequencies near the contact resonances of the cantilever. The AFM conductive tip approaches the sample's surface vertically with a deflection set point (trigger force) as feedback. A hysteresis loop is acquired by sweeping the bias once the set point is reached. The electric bias (Figure 7.11a) comprises a dc part coupled with an ac part applied to the AFM conductive tip. The $V_{dc}(t)$ consists of a sequence of pulses with amplitude V_i and length τ_1 (dc-on/high state) separated by zero biases with a duration of τ_2 (dc-off/low state).[44]

Usually the responses at dc-off pulses are used to calculate the polarization hysteresis loop (PR loop). The PR loop is calculated with the relation of $PR = A \times \cos(\varphi)$, where A and φ represent the butterfly-shaped amplitude loop and phase hysteresis loop, respectively. A typical PR *versus* bias is shown in Figure 7.11b. From the PR loop, many critical parameters describing the switching processes of the ferroelectric materials can be extracted and analysed, such as forward and reverse coercive biases (V_0^+ and V_0^-), remanent piezoresponses (R_0^+ and R_0^-), saturation piezoresponse (R_s^+ and R_s^-), nucleation bias (V_c^+ and V_c^-), and the work of switching (defined as the area enclosed by the hysteresis loop). The PR loop is produced by measuring the current through a ferroelectric capacitor when a triangular waveform voltage is applied. The dipole moment can be oriented along the positive electric field direction, reaching a saturated polarization state (position 4 in Figure 7.11b). When the positive voltage decreases to zero, the ferroelectric material retains a remnant polarization R_0^+. For standard dielectrics, R_0^+ cannot be sustained without a voltage. When the polarization decreases to zero, the voltage is called the coercive bias (V_0^+), where an equal number of dipoles orient in opposite directions, thus cancelling out the total polarization (position 6). A further increased negative bias can lead to the dipoles orientating in the opposite direction (position 0). As the electric field is swept towards the positive direction again, a hysteresis loop is formed. The PR loop itself denotes the energy consumed by switching the dipoles in the structure.[68]

7.3 Simulation Methods

Despite those experimental methods, theoretical studies or computational modelling including the Ginzburg–Landau,[112] atomistic modelling methods,[113]

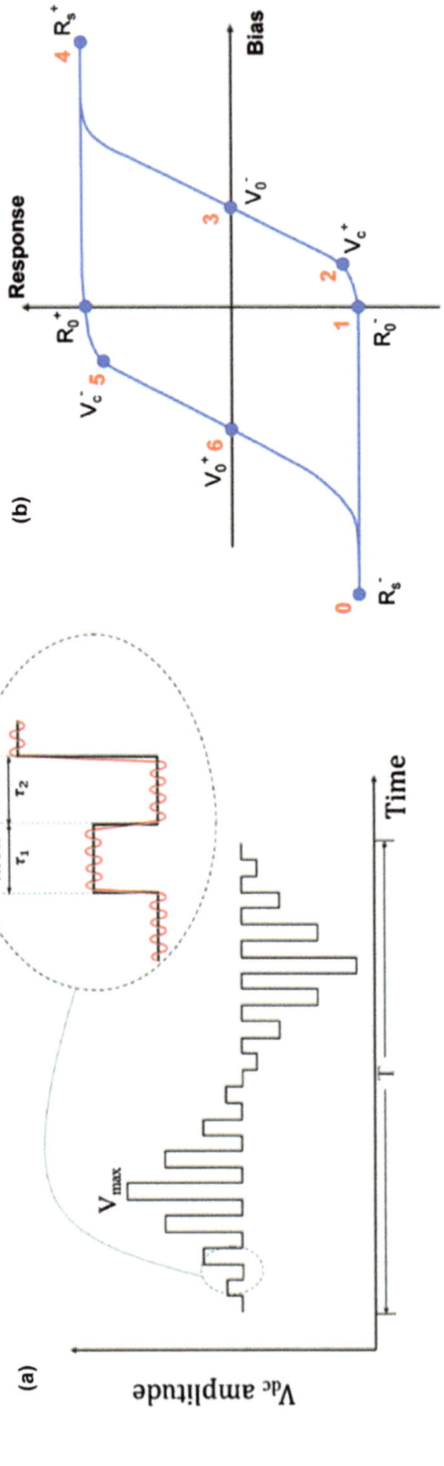

Figure 7.11 (a) The switching/driving electric bias used in PFS; (b) a typical PR loop.

effective potentials or Hamiltonians fit to the first-principles[114,115] and Monte Carlo approach[116] have been commonly adopted for investigating the ferroelectric properties of a vast number of materials.

From the microscopic point, ferroelectricity originates mainly from three aspects. Firstly, ferroelectricity is related to the short-range repulsions of electron clouds of adjacent ions favouring the paraelectric state and the long-range Coulomb forces favouring the ferroelectric state. Secondly, it depends on the balance between the long-range and short-range forces that govern the crystal structure. Thirdly, the energy balance can incline to a ferroelectric state due to short-range off-centring interactions such as chemical bonding, stereochemical activity and geometric effects.[117]

7.3.1 Monte Carlo Method with the Ginzburg–Landau Model

The formation and evolution of domain structures in ferroelectric materials can be performed by using a simulation model based on the time-dependent Ginsburg–Landau equations.[118–121] In this approach, it usually has pre-set assumptions on the domain topographies and their evolutional way. Simultaneously, the nonlocal elastic interactions, electric dipole–dipole interactions, and local interactions at domain walls should be considered. The core problem that the Landau approach deals with is a free energy function from which the location and nature of the phase transition and the evolution of the system from a nearby initial condition to the equilibrium state can be evaluated.[118] The total free energy function of a material with domain structures is the sum of the Landau bulk free energy F_L, the gradient energy F_G, the electro-strictive energy F_{es}, the elastic energy F_{ela}, the energy derived from the long-range electric dipole–dipole interaction F_{dip}.[118]

The Monte Carlo simulation has been widely used to simulate dipole alignment inside domains as well as on the domain walls.[112] The move attempted in Monte Carlo involves the random selection of a site in the lattice followed by random reorientation of the dipole at this site. The move is judged and accepted using the Metropolis importance sampling algorithm according to the change of free energy associated with the dipole reorientation and the simulation temperature. By judging the relative effect of the different energy terms in the free energy function and simulation conditions on the evolution of the dipole configuration, the model is further allowed to acquire ferroelectric hysteresis loops. In the process of acquiring ferroelectric hysteresis loops, the applied electric field is changed by an increment. The simulation allows the evolution from the configuration state developed at the previous field value.[122]

Ghoufi *et al.* used anisotropic isobaric/isothermal molecular dynamics (MD)[123,124] together with grand canonical Monte Carlo techniques to derive an osmotic Monte Carlo approach that is able to deal a guest-assisted structural transition of the MIL-53(Cr) MOF.[125] They found that the behaviour of the MIL-53(Cr) solid upon CO_2 adsorption could be divided into two "microscopic" transitions, *i.e.* consecutive structural switches between a

Lp and a Np form, both of which transitions were accompanied by a phase mixture (Np, Lp) domain, of which the composition varied with the pressure.

7.3.2 *Ab Initio* Supported Model Simulation

The Density Functional Theory (DFT)[126–128] calculation is a powerful tool to explore the dielectric and ferroelectric properties of materials. DFT is a computational quantum mechanical modelling method used in physics, chemistry and materials science to investigate the ground-state electronic structure of many-body systems, including atoms, molecules, and condensed phases. Using this theory with functionals, the properties of a system with many electrons can be determined. One of the parameter-free *ab-initio* descriptions of complex electronic systems has been achieved with the development of DFT by Kohn and Sham.[129] The core idea of the Kohn–Sham theory is to split the energy functional into several terms:

$$E[\rho]=T_{\mathrm{kin}}[\rho]+E_{\mathrm{H}}[\rho]+E_{\mathrm{xc}}[\rho]+E_{\mathrm{ext}}[\rho], \tag{7.3}$$

where T_{kin} is the kinetic energy of non-interacting electrons. E_{H} is the Hartree energy, *i.e.* the classical Coulomb energy of the electrons. E_{xc} is the exchange–correlation energy. E_{ext} is the interaction energy of the electrons with the ions, which can be described by the 1/r potential as in all-electron methods or by pseudo-potentials. The kinetic energy can be written as:

$$T_{\mathrm{kin}}[\rho]=-2\sum_{i=1}^{N}\int\psi_i^{*}(\mathrm{r})\frac{h^2}{2m}\nabla^2\psi_i(\mathrm{r})\mathrm{d}^3\mathrm{r}. \tag{7.4}$$

The energy functional can be minimized with respect to the wavefunctions ψ_i. Normalizing wavefunctions requires $\int|\psi_i(\mathrm{r})|^2\mathrm{d}^3\mathrm{r}=1$. This requirement is introduced by Lagrange parameters ε_i. Applying the variational principle, the Kohn–Sham equation can be written as:

$$\left\{\frac{h^2}{2m}\nabla^2+\hat{\nu}_{\mathrm{eff}}(\mathrm{r})\right\}\psi_i(\mathrm{r})=\varepsilon_i\psi_i(\mathrm{r}). \tag{7.5}$$

Where $\hat{\nu}_{\mathrm{eff}}=\hat{\nu}_{\mathrm{ext}}+\hat{\nu}_{\mathrm{H}}+\hat{\nu}_{\mathrm{xc}}$. In the real space representation, the individual terms have following forms:

$$\text{External potential:}\quad \hat{\nu}_{\mathrm{ext}}(\mathrm{R},\mathrm{r})=-\sum_{\mu=1}^{M}\frac{\mathrm{e}^2Z^2}{|\mathrm{r}_i-\mathrm{R}^{\mu}|}, \tag{7.6}$$

$$\text{Hartree potential:}\quad \nabla^2\hat{\nu}_{\mathrm{H}}(\mathrm{r})=4\pi\mathrm{e}^2\rho(\mathrm{r}), \tag{7.7}$$

$$\text{xc Potential (local density approximation):}\quad \hat{\nu}_{\mathrm{xc}}(\mathrm{r})=\frac{\delta}{\delta\rho(\mathrm{r})}\int\mathrm{d}\mathrm{r}\rho(\mathrm{r})\varepsilon_{\mathrm{xc}}(\rho(\mathrm{r})). \tag{7.8}$$

The charge density $\rho(r)$ can be solved iteratively until the input density (used to define the potential term in the Hamiltonian) equals the output density fulfilling the required calculation accuracy. R is the atom position which changes at certain steps to optimize the atomic structure or each time-step of a MD algorithm. The atom position R can be moved by a molecular static or molecular dynamics time-step once the self-consistency condition of the electron density has been fulfilled.[130]

The local density approximation (LDA) has been widely adopted to study ferroelectricity.[131–135] Based on the full-potential linear augmented plane-wave (FLAPW) calculations, Cohen and Krakauer proposed that ferro-electricity originated from the delicate balance between the long-range Coulomb force, which favoured the ferroelectric phase, and the short-range repulsion that favoured the paraelectric phase.[131–133,136] In terms of the long-range Coulomb interactions, the Born dynamic charges (Z^*) have been computed for a series of perovskites.[137,138]

Born effective charges are quantities that elucidate physical properties such as piezoelectric and ferroelectric properties since they represent the coupling between lattice displacement and the electric field. The Born effective charge tensor Z^* of an atom k is defined as,[139]

$$Z^{*}_{k,\alpha\beta} = V\frac{\delta P_{\alpha}}{\delta \tau_{k,\beta}} = \frac{\delta F_{k,\beta}}{\delta \xi_{\alpha}} = -\frac{\partial^2 E}{\partial \xi_{\alpha} \partial \tau_{k,\beta}}. \tag{7.9}$$

Herein, P_α represents polarization induced by the periodic displacement $\tau_{k,\beta}$, or by the force $F_{k,\beta}$ induced by an electric field ξ_α. E is the total energy of the unit cell.[139] In the calculation process, the parameters of the unit cell are along the right-handed Cartesian axes. Each ion is displaced by a small but finite distance along the three Cartesian axes at one time and the polarization is calculated. The polarization change with respect to the undistorted structure divided by the displacement gives the elements of Born charges along one particular axis of an ion.[139]

The polarization change ΔP is theoretically defined as[140]

$$\Delta P_{\alpha} = P_{\alpha}(\mathrm{t}) - P_{\alpha}(0) = \sum_{k} \frac{\mathrm{e}}{\Omega} Z^{*}_{k,\alpha\beta}\, \Delta\tau_{k,\beta}. \tag{7.10}$$

Herein, 0, t, and Ω correspond to the starting structure, end structure, and volume of unit cell, respectively. Polarization change is derived after the calculation of cell volume, Born effective charge and polarization distance according to the above equation.

Pan and co-workers explored the structural and functional properties of MOF RSMOF-1 with water molecules encapsulated in its lattice by using a first-principles Born–Oppenheimer Molecular Dynamics (BOMD) simulation based on DFT with the Perdew–Burke–Ernzerhof (PBE) functional and CP2K code.[118,141,142] In their work, the hydrogen bond interactions between water molecules and amino groups in RSMOF-1 was described by the Grimme dispersion correction.[143] Besides, the interaction between the valence electrons and atomic nucleus was studied by combining the Gaussian and planewave (GPW) basis set with the Goedecker–Teter–Hutter (GTH) pseudopotential.[144,145] From the MD trajectory, the movements of the dipole moments of the guest water molecules were visualized. They have found that the RSMOF-1 demonstrated both resistive switching (RS) behaviours and ferroelectricity at room temperature. Moreover, the controlled movements of hydrogen bond interactions between water molecules and the amine-tethered nanochannels under the external electric field resulted in tunable electrical properties of RSMOF-1.[67]

7.4 Ferroelectric MOFs

Ferroelectric MOFs can be divided into intrinsic ferroelectric and induced ferroelectric MOFs. The intrinsic ferroelectric MOFs should have non-centrosymmetric structures which belong to the 10 point groups as shown Figure 7.1. In terms of the induced ferroelectric MOFs, it is reported that the porous MOFs can encapsulate a myriad of polar guests that can self-organize within the lattice by supramolecular interaction with themselves or the host to produce a non-centrosymmetric network.[146,147]

7.4.1 Intrinsic Ferroelectric MOFs

For intrinsic ferroelectric MOFs, there exist ordered oriented intrinsic dipole moments inside a crystalline structure leading to spontaneous ferroelectric polarization. The bipolar state can cause a phase transition from a high symmetric paraelectric phase that occurs at high-temperature ($>T_c$) to a low symmetric ferroelectric phase at low-temperature ($<T_c$). The symmetry reduction upon temperature decrease is called symmetry breaking. Ferroelectric materials can usually be classified according to the order of phase transition. The first-order phase transition has a discontinuous change of order parameter P_s while a second-order phase transition has a continuous change of P_s, as shown in Figure 7.12.[1]

The ferroelectric phase transitions can also be classified into two categories according to the form of phase change occurring at the Curie Temperature T_c. One is the displacive-type phase transition and the other is the order–disorder-type phase transition, as shown in Figure 7.13.

In the displacive-type phase transition, the atoms displace relative to each other to form a polar, non-centrosymmetric lattice below T_c.[148] In the order–disorder-type phase transition, the randomly-oriented dipoles are spontaneously aligned in a correlated manner, below T_c.[68]

One typical example is the ABX_3 perovskite-type MOFs. A series of MOFs with perovskite-type structures that follow the ABX_3 pattern show ferroelectricity, where A is typically a quaternary ammonium cation, B is a divalent metal ion (M^{2+}), and X is (formate)$^-$. The formate$^-$ and M^{2+} transitional metal ions form a semicuboid anionic ReO_3-type structure.[149]

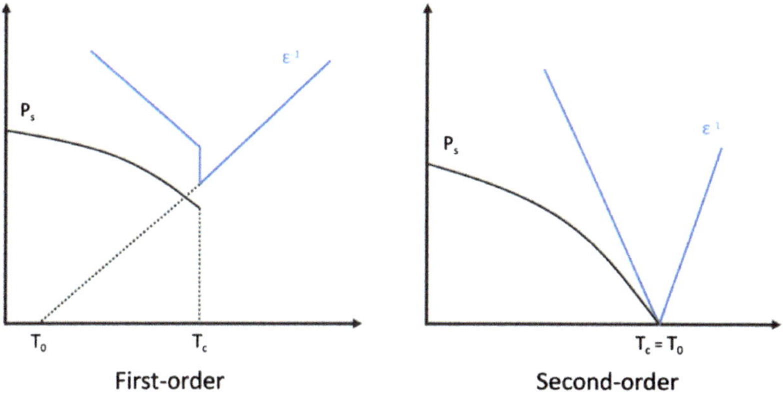

Figure 7.12 The temperature dependent dielectric constant and polarization of first-order and second-order phase transitions. The dielectric constant and temperature obey the following relationship: $\varepsilon = C/(T - T_0)$, herein, ε, C, T and T_0 denote the dielectric constant, the Curie constant, temperature, and the Curie–Weiss temperature, respectively. Reproduced from ref. 1 with permission from American Chemical Society, Copyright 2012.

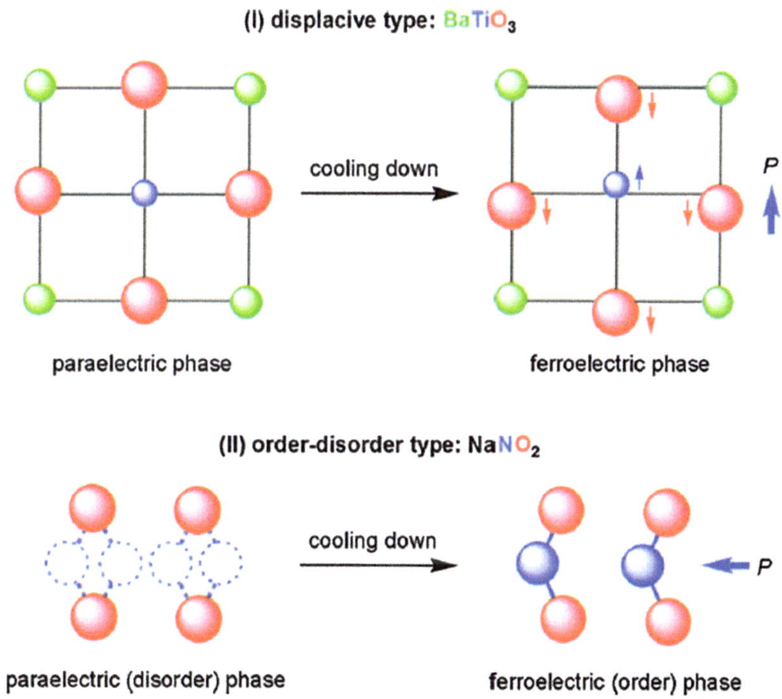

Figure 7.13 Schematics of displacive type phase transition and order–disorder type phase transition in ferroelectric materials. Reproduced from ref. 1 with permission from American Chemical Society, Copyright 2012.

In the perovskite-like MOFs, the ammonium cations are locked in the porous structure so that long-range diffusion is not possible.

Cheetham *et al.* have systematically investigated various properties of a series of MOFs with ABO_3 perovskite-type structures, $[(CH_3)_2NH_2][M(HOOC)_3]$ (DMMF), where A = H_2NMe_2, B = M (M is divalent metal ions, such as Zn^{2+}, Mn^{2+}, Fe^{2+}, Co^{2+}, Ni^{2+}), and O = HCOO.[62,150] Figure 7.14 shows the crystal structure of $[(CH_3)_2NH_2][Zn(HOOC)_3]$ perovskite MOF.

Temperature dependent dielectric constant measurements reveal that antiferroelectric behaviour occurs in the range of 160–185 K (DMFeF, 160 K, DMZnF, 160 K, DMCoF, 165 K, DmNiF, 180 K, DMMnF, 185 K), which possibly resulted from the order–disorder phase transition of the organic molecules. Senaris-Rodriguez *et al.* attributed the ferroelectricity of DMMnF to the dynamics of the DMA cations. Beyond the T_c, the DMA cations can rotate inside the cubo-octahedral cavity created by the $[Mn(HCOO)_3]^-$ framework. When the temperature is below T_c, the rotation of the DMA cations become frozen. The cooperative arrangement of the cations inside the cavities gave rise to the observed dielectric transition of DMMnF.[151] The M^{2+} transitional metal ions can be chosen to induce ferromagnetic behaviour in MOFs. For example, by adding a transition metal ion (Mn^{2+}, Fe^{2+}, Co^{2+}, and Ni^{2+}), the corresponding DMMF become magnetically ordered, leading to a series of exhilarating multifunctional materials.[62]

Wang *et al.* reported a new three-dimensional chiral metal formate framework, $[NH_4][Zn(HCOO)_3]$, which was the first MOF to possess a perfect ferroelectric *P–E* hysteresis loop. It has a ferroelectric phase transition at 191 K because of the disorder–order transition of NH_4^+ cations within the structure (Figure 7.15a).[152] Figure 7.15b confirmed the ferroelectricity of $[NH_4][Zn(HCOO)_3]$ by a series of *P–E* loops measured below T_c. The characteristic *P–E* loop clearly occurred at 189 K, below the Curie temperature 191 K. The spontaneous polarization increases with the decrease of temperature. The *P–E* loop saturates at 163 K with a remnant polarization (P_r) of 0.68 $\mu C\,cm^{-2}$, a coercive field (E_c) of 2.8 $kV\,cm^{-1}$, and a saturation spontaneous polarization (P_s) of 1.03 $\mu C\,cm^{-2}$.

Ferroelectricity in perovskite-type MOFs at room temperature is still an ongoing quest, since it requires T_c above 300 K.[149] The recently reported heterometallic perovskite-type MOF $[C_2H_5NH_3][Na_{0.5}Fe_{0.5}(HCOO)_3]$ shows a discontinuity in its temperature dependent dielectric constant at 360 K, which fulfils the Curie–Weiss law.[60] The observation of the order–disorder phase transition can be explained by the cation ordering at room temperature. This makes the $[C_2H_5NH_3][Na_{0.5}Fe_{0.5}(HCOO)_3]$ MOF crystal the first potential candidate for ferroelectric perovskite-type MOF at room temperature. However, polarization measurements have not been conducted for this MOF. In another heterometallic perovskite-type MOF, $[(C_2H_5)_2NH_2][Fe^{II}Fe^{III}(HCOO)_6]$, a dielectric anomaly in its dielectric constant has also been reported, with T_c of 240 K. However, it has not been confirmed that its phase below T_c is non-centrosymmetric.[153]

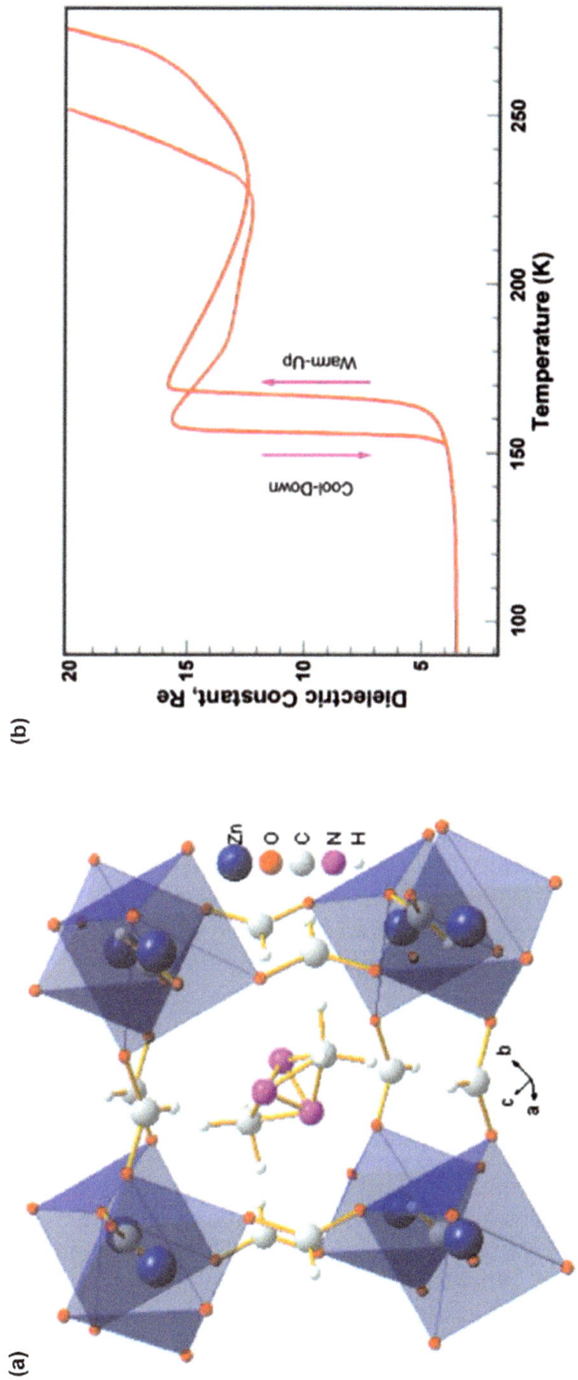

Figure 7.14 (a) The unit cell of $[(CH_3)_2NH_2][Zn(HOOC)_3]$ at room temperature. (b) The temperature dependent dielectric constant of the single crystal. A hysteresis loop exists at about 10 K.
Reproduced from ref. 150 with permission from American Chemical Society, Copyright 2008.

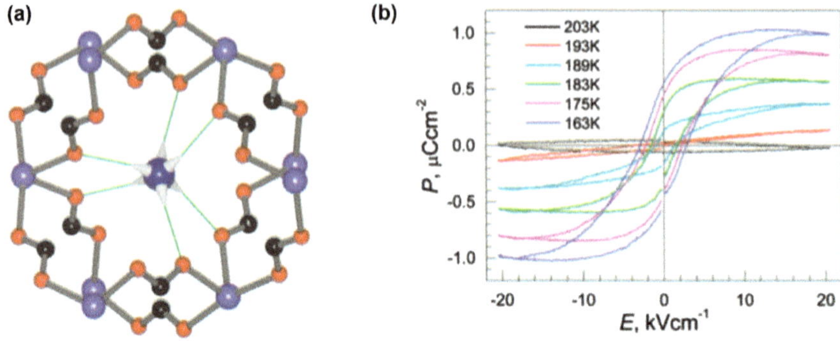

Figure 7.15 (a) One trigonally disordered NH_4^+ cation in the channel at 290 K. (b) Dielectric hysteresis loops at various temperatures.
Reproduced from ref. 152 with permission from American Chemical Society, Copyright 2010.

7.4.2 Induced Ferroelectric MOFs

Ferroelectricity of MOFs can also be induced by guest polar molecules such as water and ethanol molecules.[58,154] The host–guest interaction in the porous lattice of MOF crystals is a typical example. Because the ferroelectricity of MOF can be induced by guest polar molecules, more and more researchers have been attracted to this hotspot because the design of new functional inorganic–organic hybrid materials have highly promising applications.[155,156] The guest molecules must organize into non-centrosymmetric lattices that are dynamic under the influence of electric fields. For example, the guest molecules can organize into the host framework to form a polar structure through non-covalent interactions.

Marlow *et al.* found that the ordered arrangement of dipole molecules such as *p*-nitroaniline (pNA) in different molecular sieves (Silicalite-1, $AlPO_4$-5) within zeolite pores resulted in a lowering of the symmetry compared to the host, second harmonic generation, pyroelectric effects and special phase transition phenomena, *etc.*[147] Cui *et al.* reported that water and methanol were typical solvents with high polarizabilities. Figure 7.16 shows the dielectric constants ε_r of H_2O, and CH_3OH when cooling to 4.2 K.[157] The nonpolarizable C_6H_6 is also presented for comparison. It can be seen that the ε_r value of the liquid H_2O is as high as 10^2 just above the freezing point and then decreases very sharply. At low temperature, H_2O becomes a nonpolarizable material. Similar behaviour was also observed for CH_3OH.

Wang *et al.* subsequently measured the dielectric constants of $[Mn_3(HCOO)_6]$ and $[Mn_3(HCOO)_6](H_2O)(CH_3OH)$ in the temperature range 4.2–300 K. In fact, $[Mn_3(HCOO)_6]$ is a porous ferrimagnet with T_c at 5–10 K.[158] Without the guest polar molecules, the porous $[Mn_3(HCOO)_6]$ exhibited a small dielectric constant that was independent of temperature. In addition, there was not much difference between the maximum and minimum values (<2) of ε_r when the temperature changed from 4.2 to 300 K.

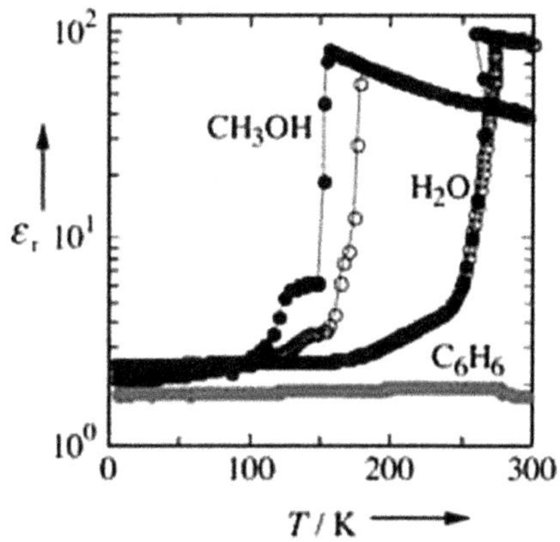

Figure 7.16 The temperature dependent dielectric constants ε_r of H_2O, CH_3OH and C_6H_6. The closed and open circles correspond to ε_r values during the cooling and heating processes, respectively.
Reproduced from ref. 57 with permission from John Wiley and Sons, Copyright © 2005 WILEY-VCH Verlag GmbH & Co. KGaA, Weinheim.

However, the value of ε_r of the $[Mn_3(HCOO)_6](H_2O)(CH_3OH)$ increased to 20 upon cooling to around 150 K and then the value of ε_r sharply decreased to 7 when the temperature further decreased. It was explained that the sharp decrease of ε_r resembled that of the liquid–solid phase transitions of H_2O and CH_3OH. Moreover, it was found that the guest molecules were somehow free at high temperature and almost freeze at low temperature. Adding polar guest molecules into the porous lattice could result in a bipolar state that is polarizable at high temperature and nonpolarizable at low temperature.

The encapsulation of polar guest molecules within the porous framework is crucial for realizing ferroelectric MOFs. But some small guest molecules, such as water and methanol, do not usually form a stable polar order in the lattice.[157] Therefore, complex phenomena can be observed. In some situations, the MOF can show multiferroic behaviour. For example, a charge-balanced MOF with dimethylammonium cations can hydrogen bond to the formate-based framework $[M^{II}(HCOO)][NR_2H_2]$ (M = Zn, Fe, Co, Ni).[62,150] Since the dimethylammonium cations can occupy three different positions within the lattice, the antiferroelectric behaviour occurs below 160–185 K according to the different metallic ions. Yadav *et al.* proposed a crystalline assembly that contained a cationic cage $\{[Zn_6(H_2O)_{12}][TPTA]_8\}(NO_3)_{12}\cdot 26H_2O$, which featured the tripodal ligand $[PS(NH^3Py)_3]$ (TPTA). The ferroelectric measurement was conducted on a single crystal of $\{[Zn_6(H_2O)_{12}][TPTA]_8\}(NO_3)_{12}\cdot 26H_2O$ that yielded a remnant (P_r) polarization of 1.2 $\mu C\ cm^{-2}$ at room temperature. They suggested that the ferroelectric response of $\{[Zn_6(H_2O)_{12}][TPTA]_8\}(NO_3)_{12}\cdot 26H_2O$ originated from the toggling of nitrate anions and solvate molecules found in the pockets between the cages.[159]

7.4.3 MOFs with Ferroelectric-like Behaviour

Some MOFs may show ferroelectric-like behaviours even though they do not belong to the 10 point group as shown in Figure 7.1. Ferroelectric-like behaviour is defined as the displacement *vs.* electric field (D–E) hysteresis loop exhibiting the same shape as that of typical ferroelectric materials.[160]

Recently, Sun *et al.* reported ferroelectric-like behaviours of NUS-6-(Hf) MOF nanocrystals with a grain size of 50–1000 nm.[44] By conducting PFS experiments, the ferroelectric-like behaviour of NUS-6 can be observed. The PR loops (Figure 7.17c and f) are transformed using the equation of $\mathrm{PR} = A \times \cos\varphi$, where A and φ denote butterfly-shaped amplitude (Figure 7.17a and d) and phase loops (Figure 7.17b and e), respectively. It can be seen that the PR loops probed from both NUS-6-(Hf) and NUS-6-(Zr) nanocrystals are non-symmetric. They showed the positive values of the remnant piezoresponses (PR when bias = 0 V). The internal bias must exist from the aligned dipoles since the PR loops were shifted along the horizontal voltage axis.[161]

As for the inorganic–organic hybrid MOF NUS-6-(Hf/Zr), the possible dipoles are the coordination bonds between the metal ion (Hf/Zr) and oxygen.[162] In fact, it is reported that the non-covalent bonds generally show polarity because of the asymmetry of electron clouds that occurs when two or more molecules interact.[68] According to the PR loop shown in Figure 7.17c, it is worth noting that NUS-6(Hf) has a slim and tilted shape that resembles some of the synthetic ferroelectric polymers for energy applications. Such ferroelectric polymers can be experimentally characterized by PR loops with large electric energy density and quite small remnant polarization.[163–166] It is believed that materials with high energy storage and release capabilities possess these kinds of PR loops.[167] Additionally, the coercive biases (biases when PR = 0) of NUS-6-(Hf) are less than those of NUS-6-(Zr). NUS-6-(Zr) shows almost no ferroelectric polarization switching because its PR value stays almost positive (PR > 0, Figure 7.17f). Based on these findings, it is suggested that NUS-6-(Hf) has much stronger ferroelectric-like characteristics than that of NUS-6-(Zr). Since both NUS-6-(Hf) and NUS-6-(Zr) share the same crystal structure and ligands (Figure 7.9),[108] the only difference is the metal ions (Hf/Zr). This is the only possible factor that may lead to the significant differences in the ferroelectric-like behaviour of NUS-6-(Hf) and NUS-6-(Zr). The Hf–O bond has larger polarity than the Zr–O bond, which is also demonstrated by the studies of NU-1000(Hf).[168] The HfO_2 is found to be more ionic than ZrO_2.[169] Therefore, the ferroelectric properties of MOFs can be possibly tuned by judicious selection of metal ions to form more polar coordination bonds in their asymmetric structures. However, the study could not exclude the effect of guest water molecules encapsulated in the MOF lattice to form polar ordering.[170] Moreover, molecular fragments of MOFs can exhibit free rotation of ligands around a static axis that can also be treated as bistable orientations,[68] which may also result in the ferroelectric-like behaviour of MOF-based materials.

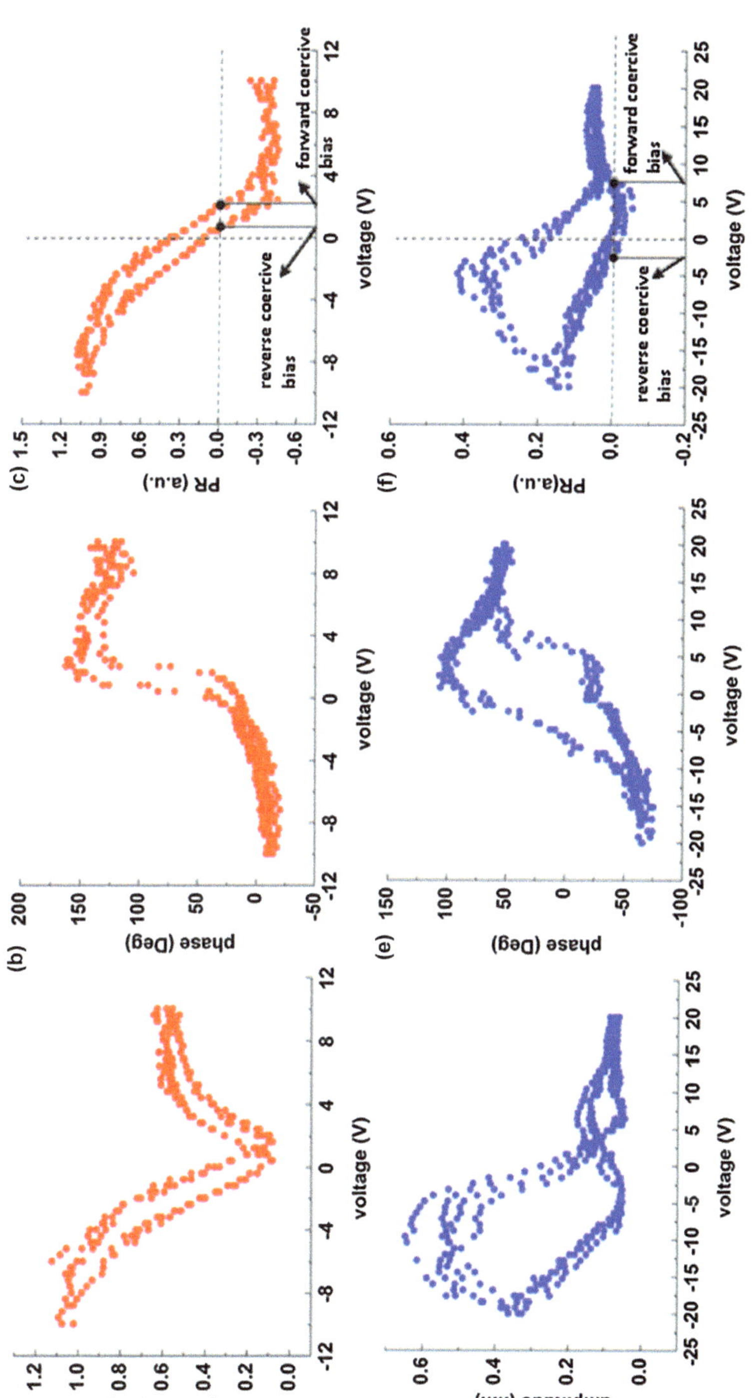

Figure 7.17 The PFS results of MOF (a–c) NUS-6-(Hf) and (d–f) NUS-6-(Zr). (a, d) Amplitude butterfly loops; (b, e) phase loops; (c, f) calculated PR loops.
Reproduced from ref. 44 with permission from the Royal Society of Chemistry.

7.5 Dipoles in MOF Lattices

The structural identification in the above ferroelectric MOF enables one to explore how to synthesize them with desirable functionalities. A ferroelectric MOF has two requirements, *i.e.* dipoles and a non-centrosymmetric lattice. In this section, dipoles in ferroelectric MOFs are discussed. In addition to the coordination bond and hydrogen bond, the host–guest molecular rotators and charge transfer complex within the lattice can lead to ferroelectrics.

7.5.1 Molecular Rotators

Amphidynamic material[171–173] is a kind of material that possesses structures of free volume compartments and features volume-conserving processes and correlated motions of two or more components.[174] One of the promising aspects is their host–guest rotation systems. The host–guest molecular rotation system inspired by new supramolecular ferroelectrics has long been investigated.[173] It is reported that the guests can exhibit dramatic molecular motion in the lattice. The system has proper energetic barriers that can lead to ferroelectricity.[68] For some MOF crystals, the free rotation of ligands around a static axis could be assumed to be bistable orientations.[174,175] Akutagawa *et al.* realized a stator-rotator host–guest molecular rotation using a complex of dibenzo[18]crown-6 with 4-fluoroanilinium.[176] The host–guest complexes of dibenzo[18]crown-6 with *m*-fluoroanilinium exhibited a ferroelectric response. The orientation of guest anilinium could be controlled by an external electric field, inducing its bipolar state stabilization, according to the site the fluorine atom occupied.[68] They also calculated the potential energy curve as a function of the rotation angle (φ) of the phenyl ring in the *m*-fluoroanilinium salt and proved that the presence of the fluorine increased the potential energy barrier to the 'flip-flop' motion. Garcia-Garibay *et al.* reported the amphidynamic character of crystalline MOF-5,[174] which comprised Zn_4O clusters bridged by 1,4-phenylenedicarboxylates.[174] Martineau *et al.* conducted advanced one- and two-dimensional high-field and ultrafast Magic-Angle Spinning (MAS) Nuclear Magnetic Resonance (NMR) measurements in tandem assisted by DFT calculations to fully explore the local environment and the long-range structure order of UiO-66(Zr) type MOFs functionalized by a series of polar groups such as –Br, –2OH, and $–NH_2$.[175] They found that the rotational motion of the organic linker requires overpassing an energy barrier that strongly depends on the functional groups. Both MOFs contain free rotation of ligands around a static axis and could present bistable orientations. Schurko *et al.* constructed a MOF (UWDM-1) with a mechanically interlocked molecule as part of its internal skeletal structure using a rotaxane as the organic linker and binuclear Cu(II) units as the nodes. They created a void inside the rigid framework in the as-synthesized material that allows the soft macrocyclic ring of the rotaxane to rotate rapidly, unimpeded by neighbouring molecular components.

The freely rotating soft macrocyclic ring incorporated into MOFs could be influenced by an electric field.[177] Moreover, Xiong *et al.* described a MOF-based inclusion compound, $[(CH_3)_2NH_2]_2$-$[KCo(CN)_6]$, which could be synthesized and characterized as a model compound with tunable dielectric properties. The angular motion of an unbound dimethylammonium in the lattice, which had a weak cooperative motion among adjacent dipoles, aroused an order–disorder phase transition of the system.[178] At the current stage, rotating polar ligands can be realized successfully,[179,180] but stabilizing their orientation is still difficult because of dipole–dipole interactions among neighbouring ligands.[68]

7.6 Two-dimensional (2D) MOFs

Most of the reported MOFs have three-dimensional (3D) structures in their microstructural scale. Compared to 3D MOFs, 2D MOF nanosheets have the advantages of both 2D layered materials and hybrid MOFs, which are considered to be the most promising materials for many applications. For example, 2D MOFs possess high specific surface areas that can present a large number of surface active sites for contacting with the electrolyte and electrode. Therefore, the large surface area can help to accelerate charge transfer at the catalytic interface, and the highly open structure exposes more interior atoms as highly accessible active sites.[181] These materials can be used as templates for supercapacitor applications.

7.6.1 Synthesis of 2D MOF Nanosheets

The direct synthesis of MOF nanosheets still remains difficult at the current stage. According to the limited information in the literature, there are two commonly adopted methods, *i.e.*, the top-down method (liquid or ultrasonic exfoliation of bulk MOFs)[182–185] and the bottom-up method (surfactant- or seed-modified growth).[41,186–189]

7.6.1.1 Top-down Methods

Two-dimensional MOF nanosheets are commonly prepared by top-down methods through exfoliation of layered MOFs.[190] The exfoliation can be realized by using various methods, including solvent intercalation,[191] ball-milling,[182] middle layer tailoring,[192,193] and interlayer ligand breaking,[194] and so on. For example, the low-cost layered MOFs M-BDC and M-BIM (M = Mn, Zn, Co, and Cu; BDC = 1,4-dicarboxybenzene; BIM = benzimidazole) have been successfully manufactured by using top-down methods.[195–197] However, top-down methods have limitations. The nanosheets can be obtained relatively satisfactorily, but it is difficult to control their thickness so that the generated nanostructures are non-uniform.

7.6.1.2 Bottom-up Methods

In top-down methods, the structural deterioration and morphology changes because the breaking of the coordination bonds during the exfoliation process has damaged the quality of the 2D MOF nanosheets.[198] In order to overcome the limitations in the top-down synthesis process, many researchers have tried to use the bottom-up method to synthesize high-quality 2D MOF nanosheets directly. The key point of the bottom-up synthesis method is to grow layered MOFs strictly in a direction perpendicular to the layers.[199] Makiura *et al.* have used liquid/liquid or liquid/air interfaces to control the growth of MOF nanosheets.[200] Rodenas *et al.* have synthesized Cu-BDC MOF nanosheets with a thickness of approximately 25 nm by using a three-layer strategy interface reaction to control the growth of the nanosheets.[186] Recently, Zhang and colleagues have developed a surfactant-assisted method to prepare a series of M-TCPP MOF nanosheets (M = Zn, Cd, Cu, and Co, and TCPP = tetrakis (4-carboxyphenyl) porphyrin).[91,201] Despite all these advancements, the bottom-up processes require a difficult experimental setup in general. Therefore, a better way of synthesizing 2D metal–organic nanosheets with uniform and controllable thickness in an efficient way is to combine both the top-down and bottom-up methods.[202]

7.6.2 2D Ferroelectric MOFs

Tang *et al.* constructed acentric MOFs, which belong to the 10 polar point groups (see Figure 7.1), by using an acentric organic ligand as a building block.[203] They reported the ferroelectric properties of two 2D MOFs, *i.e.*, $\{[Mn(tib)_2(H_2O)_4]SO_4\}_n$ ($C_{10}H_{16}N_4O_8SMn$) and $\{[Co(tib)_2(H_2O)_4]SO_4\}_n$ ($C_{10}H_{16}N_4O_8SMn$), which had a *R3c* space group and (6, 3) honeycomb networks. The metal ion Mn(II)/Co atom is coordinated by four H_2O molecules and two N atoms of imidazole from different tib ligands. The geometry of the coordination bond is a highly distorted octahedron. The ferroelectric *P–E* loops (Figure 7.18) indicate that both $\{[Mn(tib)_2(H_2O)_4]SO_4\}_n$ (Figure 7.18a) and $\{[Co(tib)_2(H_2O)_4]SO_4\}_n$ crystals (Figure 7.18b) are typical ferroelectrics with remnant polarization (P_r) values of *ca.* 0.586 μC cm^{-2} and *ca.* 0.208 μC cm^{-2}, and a coercive field (E_c) of 2000 V cm^{-1} and 2600 V cm^{-1}, respectively.

Ye *et al.* reported the ferroelectric and magnetic properties of 2D MOF $(Lig)_2Tb(H_2O)_2(ClO_4)$ that had a laminar framework (Figure 7.19a). They further revealed the ferroelectricity of its analogue, $(Lig)_2Tb(D_2O)_2(ClO_4)$ and confirmed its ferroelectric deuterium effect.[204] An X-ray crystal structure experiment showed that the MOF $(Lig)_2Tb(H_2O)_2(ClO_4)$ had a local coordination environment formed by the Tb ion coordinated with six O atoms from four different lactate ions and two water molecules, which could be best described as a slightly distorted square antiprism or dodecahedron. Figure 7.19b shows the *P–E* loop of $(Lig)_2Tb(D_2O)_2(ClO_4)$, which has a ferroelectric characteristic. It has a remanent polarization (P_r) of *ca.*

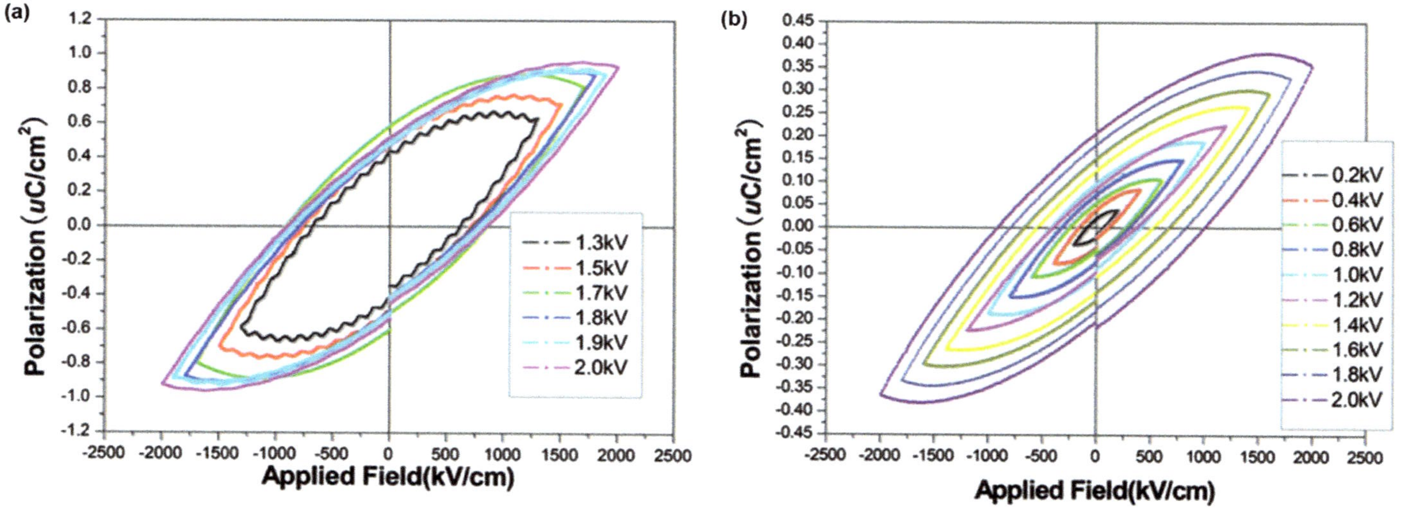

Figure 7.18 The *P–E* loops of (a) $\{[Mn(tib)_2(H_2O)_4]SO_4\}_n$; (b) $\{[Co(tib)_2(H_2O)_4]SO_4\}_n$ obtained using a RT6000 ferroelectric tester at room temperature at different voltages.
Reproduced from ref. 203 with permission from the Royal Society of Chemistry.

0.25 μC cm^{-2}, a coercive field (E_c) of 0.2 kV cm^{-1}, and a saturated polarization (P_s) of *ca.* 1.0 μC cm^{-2}. The saturated polarization of $(Lig)_2Tb(D_2O)_2(ClO_4)$ is found to be smaller than those of the typical ferroelectrics KDP (P_s = 5.0 μC cm^{-2}) and triglycine sulfate (TGS, P_s = 3.0 μC cm^{-2}), but larger than that of $NaKC_4H_4O_6 \cdot 4H_2O$ (Rochelle salt, P_s = 0.25 μC cm^{-2}).

Want *et al.* investigated the ferroelectric behaviour of a lanthanide-based MOF, $[Nd(C_4H_5O_6)(C_4H_4O_6)][3H_2O]$, which comprised a coordination system formed by a Nd ion coordinated by nine oxygen atoms that had a distorted capped square antiprismatic geometry (Figure 7.20). Dielectric studies revealed that the dielectric constant curve showed two anomalies at 295 K and 185 K. The former anomaly was demonstrated to be caused by a paraelectric–ferroelectric transition, and the latter was attributed to the freezing of the hydroxyl groups.[205]

7.7 Mechanical Properties of MOFs

Studies on the mechanical properties of MOFs remain scarce and far from incisive.[206,207] As a matter of fact, the mechanical properties of MOFs are quite crucial under practical applications, especially in the presence of stress, shear flow, *etc.*[208,209] Moreover, a better understanding of the mechanical properties is beneficial to design and synthesize MOF materials with high mechanical strength that can survive structural distortion and external mechanical stress or strains.[41,206]

An early measurement of the mechanical properties of MOFs was conducted on MOF-5 single crystals with a crystal size larger than 100 μm using the nanoindentation technique.[210,211] Nanoindentation has been successfully adopted in many studies of mechanical properties of large MOF single crystals[212–222] or thin films.[223,224] Since most MOFs are in the form of discrete particles, ranging from tens of nanometers to hundreds of micrometers,[225] and single crystals large enough for nanoindentation are difficult to prepare,[207] classic mechanical tests including the tensile test[226] and nanoindentation are not preferable choices. In fact, most of the hydrothermally and chemically stable MOFs are micro- or nano-particles ranging from 20 nm to 1 μm, such as UiO-66(Zr)-type or UiO-66(Hf)-type MOFs,[227–233] MILs,[234–236] and others. Therefore, effective nanomechanical test techniques for MOFs are in urgent demand.

Tan *et al.* quantitively measured the elastic moduli of HKUST-1[237] thin films *via* bimodal Amplitude Modulated-Frequency Modulated (AM-FM) Atomic Force Microscopy (AFM) and found that the values of elastic modulus were in the range of 3–6 GPa.[238] Recently, Sun *et al.* adopted the AM-FM technique to successfully measure the elastic moduli of a series of UiO-66-type MOFs[239] nanocrystals including UiO-66, UiO-66-NH_2, UiO-66-$(OH)_2$, UiO-66-$(COOH)_2$, and UiO-66-$(F)_4$.[240] By using the AM-FM technique, the microstructure and elasticity of materials can be obtained simultaneously. Figures 7.21 and 7.22 show the AM-FM results on UiO-66(Hf)-type and UiO-66(Zr)-type nanocrystals.

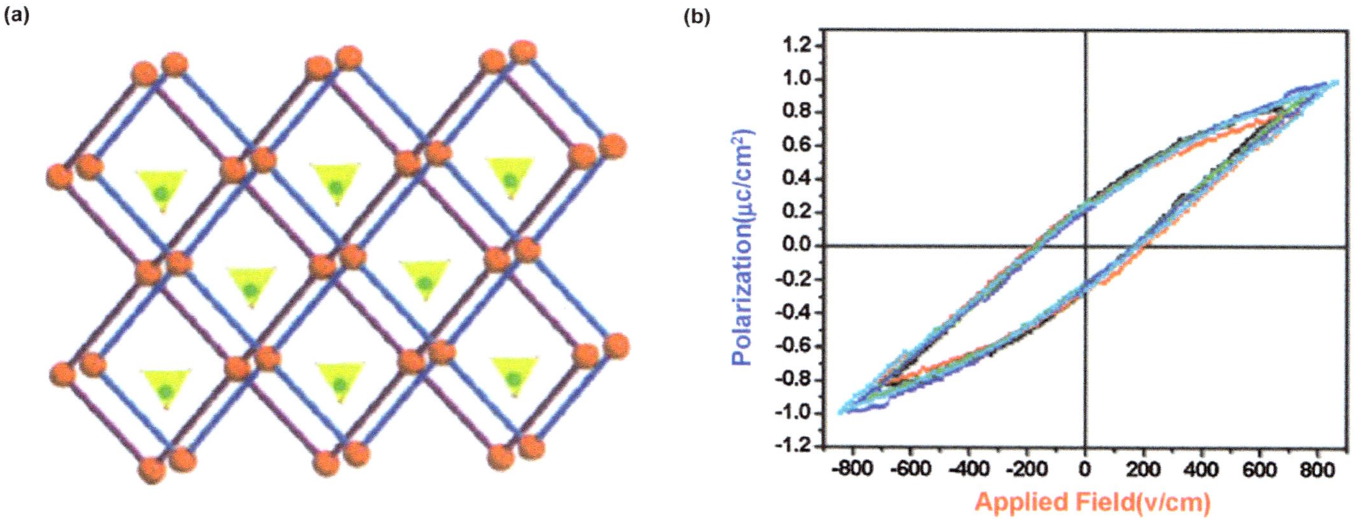

Figure 7.19 (a) The 2D laminar framework of MOF $(Lig)_2Tb(H_2O)_2(ClO_4)$ (the ball represents the Tb ion, and straight line denotes the lactate ligand). (b) The *P–E* loop of $(Lig)_2Tb(H_2O)_2(ClO_4)$ measured by a RT6000 ferroelectric tester at room temperature but at different voltage.
Reproduced from ref. 204 with permission from American Chemical Society, Copyright 2008.

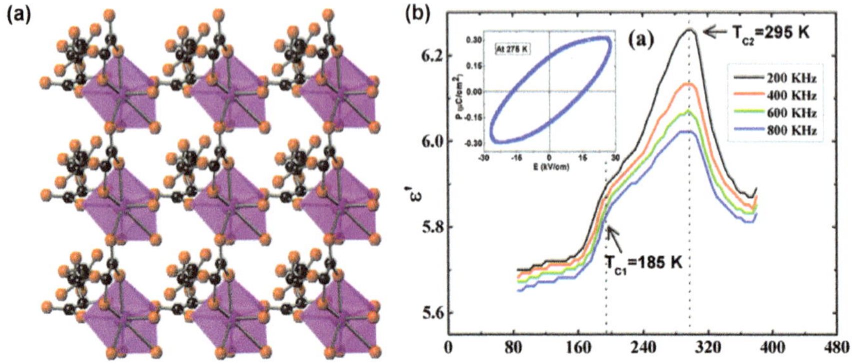

Figure 7.20 (a) Extended lamellar (2D) structure of $[Nd(C_4H_5O_6)(C_4H_4O_6)][3H_2O]$ (H atoms have been omitted for clarity). (b) Temperature dependence of dielectric constant ε_r; the inset shows a dielectric hysteresis loop at 275K.
Reproduced from ref. 205 with permission from AIP Publishing.

The topography images (Figures 7.21a–e and 7.22a–e) showed that the particle size of both UiO-66(Hf)-type MOFs and UiO-66(Zr)-type MOFs are approximately 100 ∼ 500 nm, which is consistent with the crystal size reported in the literature.[231,241,242] Both UiO-66-(Hf) and UiO-66-(Zr) particles have an octahedral shape reflected by a regular quasi-diamond shape in 2D images (see Figures 7.21a and 7.22a), while the others have a spherical shape (Figures 7.21b–e and 7.22b–e). According to elastic modulus mappings (Figures 7.21f–j and 7.22f–j), the elastic moduli of UiO-66(Hf), UiO-66(Hf)-$(OH)_2$, UiO-66(Hf)-NH_2, UiO-66(Hf)-$(COOH)_2$ and UiO-66(Hf)-$(F)_4$ are 27.1–63.8 GPa, 26.2–179 GPa, 17.4–70.6 GPa, 5.64–140 GPa and 28.4–232 GPa, respectively. The values are 22.3–45.7 GPa, 17.7–117 GPa, 7.57–75.6 GPa, 6.93–94.1 GPa and 69–210 GPa for UiO-66(Zr), UiO-66(Zr)-$(OH)_2$, UiO-66(Zr)-NH_2, UiO-66(Zr)-$(COOH)_2$ and UiO-66(Zr)-$(F)_4$, respectively. The large range of elasticity may be associated with the structural anisotropy of the MOF nanoparticles.

Bimodal AFM refers to the simultaneous excitation of first and higher eigenmodes of the probe cantilever that can quickly achieve mapping morphology as well as the elastic property of materials.[243] The excitation frequency is usually chosen at the resonance frequency of each eigenmode. Choosing different eigenmode vibrations can provide different properties with different sensitivity.[244] Normally, the first eigenmode amplitude or frequency is used as feedback on the sample surface topography imaging, while the second or higher eigenmode amplitude, phase and frequency shifts can be changed freely without a feedback loop.[240] In such way, bimodal AFM can be adopted to analyse the mechanical and electro-magnetic properties of materials.[244]

In the AM-AM mode, quantitative imaging can be obtained by the second eigenmode. Under the condition that the second mode amplitude is much

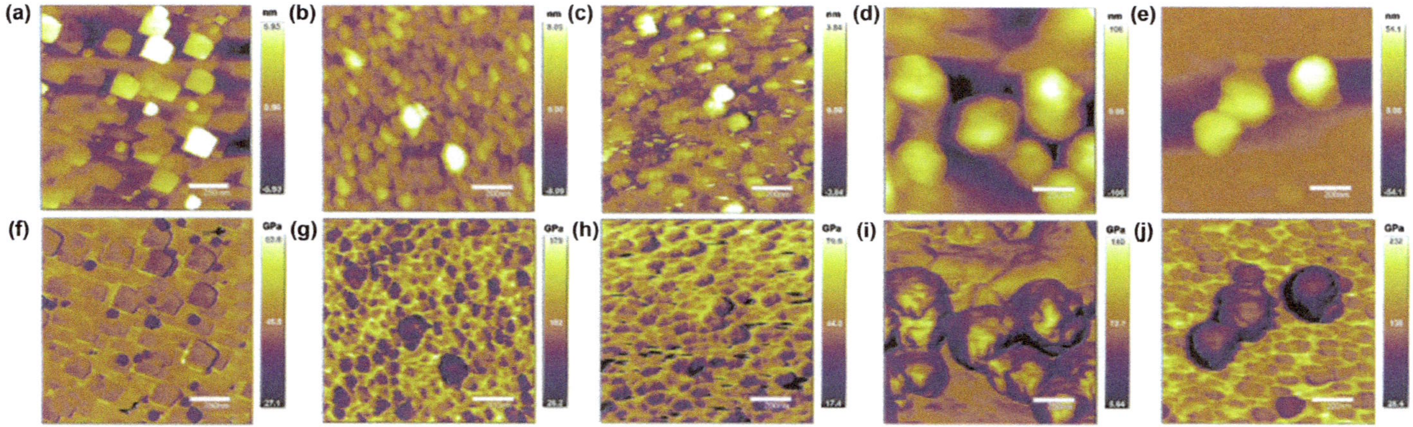

Figure 7.21 AM-FM results: Topography images of (a) UiO-66(Hf), (b) UiO-66(Hf)-$(OH)_2$, (c) UiO-66(Hf)-NH_2, (d) UiO-66(Hf)-$(COOH)_2$, (e) UiO-66(Hf)-$(F)_4$. Elastic modulus images of (f) UiO-66(Hf), (g) UiO-66(Hf)-$(OH)_2$, (h) UiO-66(Hf)-NH_2, (i) UiO-66(Hf)-$(COOH)_2$, (j) UiO-66(Hf)-$(F)_4$.
Reproduced from ref. 240 with permission from American Chemical Society, Copyright 2017.

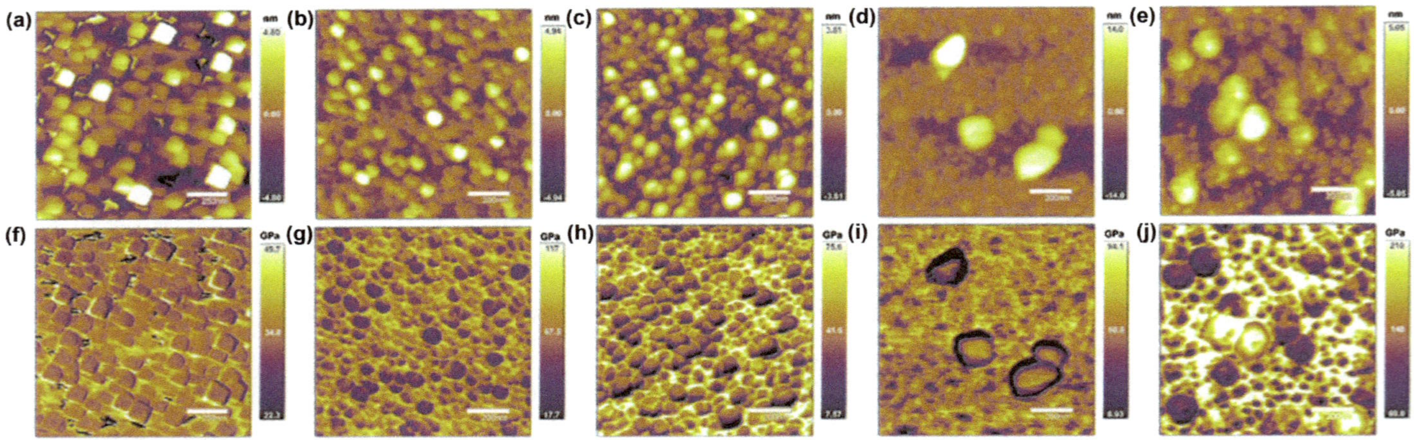

Figure 7.22 AM-FM results: Topography images of (a) UiO-66(Zr), (b) UiO-66(Zr)-$(OH)_2$, (c) UiO-66(Zr)-NH_2, (d) UiO-66(Zr)-$(COOH)_2$, (e) UiO-66(Zr)-$(F)_4$. Elastic modulus images of (f) UiO-66(Zr), (g) UiO-66(Zr)-$(OH)_2$, (h) UiO-66(Zr)-NH_2, (i) UiO-66(Zr)-$(COOH)_2$, (j) UiO-66(Zr)-$(F)_4$.
Reproduced from ref. 240 with permission from American Chemical Society, Copyright 2017.

smaller than the first mode amplitude, the tip–sample contact stiffness is derived from

$$k_{ts} = C\frac{k_2 A_{02} \cos \phi_2}{Q_2 A_2}, \tag{7.11}$$

Where A_2, A_{02} and Q_2 are the spring constant, free amplitude and Q factor of the second eigenmode. C is the correction factor that can be calculated from the tip–sample distance curve and frequency offset curve.[245]

Compared to the AM-AM technique, AM-FM and FM-FM techniques that use frequency shifts to calculate elastic properties are more direct. The relationship between the tip–sample contact stiffness and second eigenmode frequency shift is

$$k_{ts} \approx 2k_2 \Delta f_2 / f_2^0, \tag{7.12}$$

where k_2, f_2^0 and Δf_2 are the spring constant, second eigenmode resonant frequency and second eigenmode resonant frequency shift. From the Hertz contact theory,[50] the tip–sample contact stiffness can also be written as

$$k_{ts} = 2a_c E^*. \tag{7.13}$$

Where a_c is the tip–sample contact radius. Combining eqn (7.12) and (7.13), we have:

$$E^* = k_2 \Delta f_2 / a_c f_2^0 = C_2 \Delta f_2. \tag{7.14}$$

In general situation, the tip–sample contact radius is hard to determine, so a reference material is required. The modulus of the reference material should be approximately the same as that of the tested sample, also other testing parameters on both reference material and tested samples should be kept the same. For example, one should use the same tip for probing both the reference material and testing sample. Once the relationship between the frequency shifts and equivalent modulus of the reference sample is determined, the correction factor C_2 can be derived.

7.8 Conclusion

In this chapter, we discuss ferroelectric MOFs, which have been burgeoning recently for their promising applications in sensing, data communication, spin-crossover, and signal processing, *etc.* These ferroelectric MOFs possess versatile structural diversity and controlled functionalities according to the chemical modification of organic ligands and judicious choice of components. We also review the multiple experimental techniques (dielectric constant measurement, SHG measurement, Sawyer–Tower circuit, XRD and nanoscale spectroscopic measurement) and simulation methods (Monte Carlo method based on the Ginzburg–Landau model, *ab initio* supported model simulation) as well as the successive observations of the physical ferroelectric phenomena of MOF (2D and 3D) reported in the literature.

Two requirements should be fulfilled, *i.e.* the dipole moment and non-centrosymmetric lattice being ferroelectric. The ferroelectricity can originate from either the intrinsic polar structure or the encapsulation of small polar guest molecules in the MOF lattice. Moreover, we emphasize the significance of mechanical robustness of MOF in real applications. The goal of this chapter is to unravel the structural functionalities within artificially formed MOFs and offer an insight into the construction of promising hybrid ferroelectrics.

Acknowledgements

We acknowledge support from the Ministry of Education (Singapore) through the National University of Singapore under the Academic Research Funding (AcRF) R-265-000-495-112 and R-265-000-596-112. One of the authors (Y.S.) also thanks the postgraduate scholarship from the National University of Singapore.

References

1. W. Zhang and R.-G. Xiong, *Chem. Rev.*, 2012, **112**(2), 1163.
2. R. Cuppens, P. K. Larsen and G. A. C. M. Spierings, *Microelectron. Eng.*, 1992, **19**, 245.
3. A. I. Khan, K. Chatterjee, B. Wang, S. Drapcho, L. You, C. Serrao, S. R. Bakaul, R. Ramesh and S. Salahuddin, *Nat. Mater.*, 2014, **14**(2), 182.
4. L. Cima and D. Remiens, *Sens. Actuators, A*, 2005, **119**, 120.
5. U. Kushnir and O. Rabinovitch, *J. Intell. Mater. Syst. Struct.*, 2008, **19**(9), 1077.
6. Y. Ivry, N. Wang and C. Durkan, *Appl. Phys. Lett.*, 2014, **104**, 133505.
7. R. Watton, *Ferroelectrics*, 1989, **91**(1), 87.
8. N. Georgia, T. Despoina and E. K. Emmanouil, *J. Opt. A: Pure Appl. Opt.*, 2005, 7, 82.
9. R. Whatmore, *Springer Handbook of Electronic and Photonic Materials*, Springer, Boston, MA, 2006.
10. G. Busch, *Ferroelectrics*, 1987, **74**(1), 267.
11. W. S. Yun, J. J. Urban, Q. Gu and H. Park, *Nano Lett.*, 2002, **2**(5), 447.
12. M. E. Lines and A. M. Glass, *Principles and Applications of Ferroelectrics and Related Materials*, Oxford University Press, Oxford, 1977.
13. Y. Luo, I. Szafraniak-Wiza, N. Zakharov, N. Valanoor, M. Steinhart, R. Wehrspohn, J. H. Wendorff, R. Ramesh and M. Alexe, *Appl. Phys. Lett.*, 2003, **83**(3), 440.
14. I. Inbar and R. E. Cohen, *Ferroelectrics*, 1997, **194**(1), 83.
15. K. Heiji, *Jpn. J. Appl. Phys.*, 1969, **8**, 975.
16. M. Owczarek, K. A. Hujsak, D. P. Ferris, A. Prokofjevs, I. Majerz, P. Szklarz, H. Zhang, A. A. Sarjeant, C. L. Stern, R. Jakubas, S. Hong, V. P. Dravid and J. F. Stoddart, *Nat. Commun.*, 2016, 7, 13108.
17. F. Du, H. Zhang, C. Tian and S. Du, *Cryst. Growth Des.*, 2013, **13**(4), 1736.

18. S. Horiuchi, Y. Tokunaga, G. Giovannetti, S. Picozzi, H. Itoh, R. Shimano, R. Kumai and Y. Tokura, *Nature*, 2010, **463**(7282), 789.
19. D.-W. Fu, H.-L. Cai, Y. Liu, Q. Ye, W. Zhang, Y. Zhang, X.-Y. Chen, G. Giovannetti, M. Capone, J. Li and R.-G. Xiong, *Science*, 2013, **339**(6118), 425.
20. H.-Y. Ye, Q. Zhou, X. Niu, W.-Q. Liao, D.-W. Fu, Y. Zhang, Y.-M. You, J. Wang, Z.-N. Chen and R.-G. Xiong, *J. Am. Chem. Soc.*, 2015, **137**(40), 13148.
21. G. Férey, C. Serre, T. Devic, G. Maurin, H. Jobic, P. L. Llewellyn, G. De Weireld, A. Vimont, M. Daturi and J. S. Chang, *Chem. Soc. Rev.*, 2011, **40**(2), 550.
22. T. R. Cook, Y.-R. Zheng and P. J. Stang, *Chem. Rev.*, 2013, **113**(1), 734.
23. H.-C. Zhou and S. Kitagawa, *Chem. Soc. Rev.*, 2014, **43**(16), 5415.
24. H. Furukawa, K. E. Cordova, M. O'Keeffe and O. M. Yaghi, *Science*, 2013, **341**(6149), 1230444.
25. A. J. Howarth, Y. Liu, P. Li, Z. Li, T. C. Wang, J. T. Hupp and O. K. Farha, *Nat. Rev. Mater.*, 2016, **1**, 15018.
26. O. M. Yaghi, *J. Am. Chem. Soc.*, 2016, **138**(48), 15507.
27. D. J. Tranchemontagne, J. L. Mendoza-Cortes, M. O'Keeffe and O. M. Yaghi, *Chem. Soc. Rev.*, 2009, **38**(5), 1257.
28. M. Dincă and F. Léonard, *MRS Bull.*, 2016, **41**(11), 854.
29. H.-C. Zhou, J. R. Long and O. M. Yaghi, *Chem. Rev.*, 2012, **112**(2), 673.
30. E. Barea, C. Montoro and J. A. R. Navarro, *Chem. Soc. Rev.*, 2014, **43**(16), 5419.
31. K. Sumida, D. L. Rogow, J. A. Mason, T. M. McDonald, E. D. Bloch, Z. R. Herm, T.-H. Bae and J. R. Long, *Chem. Rev.*, 2012, **112**(2), 724.
32. J. Lee, O. K. Farha, J. Roberts, K. A. Scheidt, S. T. Nguyen and J. T. Hupp, *Chem. Soc. Rev.*, 2009, **38**(5), 1450.
33. J. Gascon, A. Corma, F. Kapteijn and F. X. Llabrés i Xamena, *ACS Catal.*, 2014, **4**(2), 361.
34. L. E. Kreno, K. Leong, O. K. Farha, M. Allendorf, R. P. Van Duyne and J. T. Hupp, *Chem. Rev.*, 2012, **112**(2), 1105.
35. P. Horcajada, R. Gref, T. Baati, P. K. Allan, G. Maurin, P. Couvreur, G. Férey, R. E. Morris and C. Serre, *Chem. Rev.*, 2012, **112**(2), 1232.
36. R. Medishetty and J. J. Vittal, in *Metal-Organic Frameworks for Photonics Applications*, ed. B. Chen and G. Qian, Springer, Berlin, Heidelberg, 2014, pp. 105.
37. K. K. Tanabe, C. A. Allen and S. M. Cohen, *Angew. Chem., Int. Ed.*, 2010, **49**(50), 9730.
38. M. Gutierrez, B. Cohen, F. Sanchez and A. Douhal, *Phys. Chem. Chem. Phys.*, 2016, **18**(40), 27761.
39. M. S. Denny Jr., J. C. Moreton, L. Benz and S. M. Cohen, *Nat. Rev. Mater.*, 2016, **1**(12), 16078.
40. A. A. Talin, A. Centrone, A. C. Ford, M. E. Foster, V. Stavila, P. Haney, R. A. Kinney, V. Szalai, F. El Gabaly, H. P. Yoon, F. Léonard and M. D. Allendorf, *Science*, 2014, **343**(6166), 66.

41. L. Zhu, X.-Q. Liu, H.-L. Jiang and L.-B. Sun, *Chem. Rev.*, 2017, **117**(12), 8129.
42. J. Della Rocca, D. Liu and W. Lin, *Acc. Chem. Res.*, 2011, **44**(10), 957.
43. R. Ramesh, *Nature*, 2009, **461**(7268), 1218.
44. Y. Sun, Z. Hu, D. Zhao and K. Zeng, *Nanoscale*, 2017, **9**(33), 12163.
45. S. L. Miller and P. J. McWhorter, *J. Appl. Phys.*, 1992, **72**(12), 5999.
46. B. H. Park, B. S. Kang, S. D. Bu, T. W. Noh, J. Lee and W. Jo, *Nature*, 1999, **401**, 682.
47. Q. Zhou, K. H. Lam, H. Zheng, W. Qiu and K. K. Shung, *Prog. Mater. Sci.*, 2014, **66**, 87.
48. K. Uchino, *Ceram. Int.*, 1995, **21**(5), 309.
49. A. Amin, *Ferroelectrics*, 1989, **98**(1), 129.
50. J. R. Barber and M. Ciavarella, *Int. J. Solids Struct.*, 2000, **37**, 29.
51. A. Kabir, A. Jamil, Y. Zhang and T. S. Kalkur, Voltage controlled oscillators with ferroelectric capacitors, 2008.
52. J. W. Flocken, R. A. Guenther, J. R. Hardy and L. L. Boyer, *Ferroelectrics*, 1992, **135**(1), 309.
53. L. K. Anderson, *Ferroelectrics*, 1974, 7(1), 55.
54. K. M. Johnson and G. Moddel, *Appl. Opt.*, 1989, **28**(22), 4888.
55. G. W. Taylor, *Ferroelectrics*, 1983, **49**(1), 17.
56. K. Okada, *J. Phys. Soc. Jpn.*, 1969, **27**, 420.
57. H. B. Cui, K. Takahashi, Y. Okano, H. Kobayashi, Z. Wang and A. Kobayashi, *Angew. Chem., Int. Ed.*, 2005, **44**(40), 6508.
58. H. Cui, Z. Wang, K. Takahashi, Y. Okano, H. Kobayashi and A. Kobayashi, *J. Am. Chem. Soc.*, 2006, **128**(47), 15074.
59. H. Cui, B. Zhou, L. S. Long, Y. Okano, H. Kobayashi and A. Kobayashi, *Angew. Chem., Int. Ed.*, 2008, **47**(18), 3376.
60. M. Guo, H.-L. Cai and R.-G. Xiong, *Inorg. Chem. Commun.*, 2010, **13**(12), 1590.
61. K.-L. Hu, M. Kurmoo, Z. Wang and S. Gao, *Chem. – Eur. J.*, 2009, **15**(44), 12050.
62. P. Jain, V. Ramachandran, R. J. Clark, H. D. Zhou, B. H. Toby, N. S. Dalal, H. W. Kroto and A. K. Cheetham, *J. Am. Chem. Soc.*, 2009, **131**(38), 13625.
63. W. Zhang, Y. Cai, R. G. Xiong, H. Yoshikawa and K. Awaga, *Angew. Chem., Int. Ed.*, 2010, **49**(37), 6608.
64. A. Stroppa, P. Jain, P. Barone, M. Marsman, J. M. Perez-Mato, A. K. Cheetham, H. W. Kroto and S. Picozzi, *Angew. Chem., Int. Ed.*, 2011, **50**, 5847.
65. A. Stroppa, P. Barone, P. Jain, J. M. Perez-Mato and S. Picozzi, *Adv. Mater.*, 2013, **25**(16), 2284.
66. D. Di Sante, A. Stroppa, P. Jain and S. Picozzi, *J. Am. Chem. Soc.*, 2013, **135**(48), 18126.
67. L. Pan, G. Liu, H. Li, S. Meng, L. Han, J. Shang, B. Chen, A. E. Platero-Prats, W. Lu, X. Zou and R.-W. Li, *J. Am. Chem. Soc.*, 2014, **136**(5), 17477.

68. A. S. Tayi, A. Kaeser, M. Matsumoto, T. Aida and S. I. Stupp, *Nat. Chem.*, 2015, 7(4), 281.
69. K. Kadau, T. C. Germann, P. S. Lomdahl and B. L. Holian, *Science*, 2002, **296**(5573), 1681.
70. E. Collet, M.-H. Lemée-Cailleau, M. Buron-Le Cointe, H. Cailleau, M. Wulff, T. Luty, S.-Y. Koshihara, M. Meyer, L. Toupet, P. Rabiller and S. Techert, *Science*, 2003, **300**(5619), 612.
71. I. T. T. Mitsui, E. Nakamura, *An Introduction to the Physics of Ferroelectrics*, Gordon and Breach Science Publishers, New York, 1976.
72. V. N. S. Y. A. Izyumov, *Phase Transitions and Crystal Symmetry*, Kluwer Publishers, Dordrecht, The Netherlands, 1990.
73. M. T. Wharmby, S. Henke, T. D. Bennett, S. R. Bajpe, I. Schwedler, S. P. Thompson, F. Gozzo, P. Simoncic, C. Mellot-Draznieks, H. Tao, Y. Yue and A. K. Cheetham, *Angew. Chem., Int. Ed.*, 2015, **54**(22), 6447.
74. R. Yadav, D. Swain, H. L. Bhat and S. Elizabeth, *J. Appl. Phys.*, 2016, **119**(6), 064103.
75. C. C. Neacsu, B. B. van Aken, M. Fiebig and M. B. Raschke, *Phys. Rev. B*, 2009, **79**(10), 100107.
76. S. A. Denev, T. T. A. Lummen, E. Barnes, A. Kumar, V. Gopalan and D. J. Green, *J. Am. Ceram. Soc.*, 2011, **94**(9), 2699.
77. Y. Uesu, H. Yokota, S. Kawado and N. Kato, *Recent Progress on the SHG Interference Microscope and Its Applications*, 2007.
78. C. Sheppard, J. Gannaway, R. Kompfner and D. Walsh, *IEEE J. Quantum Electron.*, 1977, **13**(9), 912.
79. J. N. Gannaway and C. J. R. Sheppard, *Opt. Quantum Electron.*, 1978, **10**(5), 435.
80. M. Flörsheimer, R. Paschotta, U. Kubitscheck, C. Brillert, D. Hofmann, L. Heuer, G. Schreiber, C. Verbeek, W. Sohler and H. Fuchs, *Appl. Phys. B*, 1998, **67**(5), 593.
81. W. R. Zipfel, R. M. Williams and W. W. Webb, *Nat. Biotechnol.*, 2003, **21**(11), 1369.
82. Y. R. Shen, *Nature*, 1989, **337**, 519.
83. R. Carriles, D. N. Schafer, K. E. Sheetz, J. J. Field, R. Cisek, V. Barzda, A. W. Sylvester and J. A. Squier, *Rev. Sci. Instrum.*, 2009, **80**(8), 081101.
84. M. Fiebig, V. V. Pavlov and R. V. Pisarev, *J. Opt. Soc. Am. B*, 2005, **22**(1), 96.
85. P. Jain, A. Stroppa, D. Nabok, A. Marino, A. Rubano, D. Paparo, M. Matsubara, H. Nakotte, M. Fiebig, S. Picozzi, E. S. Choi, A. K. Cheetham, C. Draxl, N. S. Dalal and V. S. Zapf, *Npj Quantum Mater.*, 2016, **1**, 16012.
86. D. Meier, M. Maringer, T. Lottermoser, P. Becker, L. Bohatý and M. Fiebig, *Phys. Rev. Lett.*, 2009, **102**(10), 107202.
87. T. Lottermoser, D. Meier, R. V. Pisarev and M. Fiebig, *Phys. Rev. B*, 2009, **80**(10), 100101.

88. T. Günter, A. Rubano, D. Paparo, M. Lilienblum, L. Marrucci, F. Miletto Granozio, U. Scotti di Uccio, R. Jany, C. Richter, J. Mannhart and M. Fiebig, *Phys. Rev. B*, 2012, **86**(23), 235418.
89. A. Rubano, T. Günter, M. Lilienblum, C. Aruta, F. Miletto Granozio, U. Scotti di Uccio, L. Marrucci, D. Paparo and M. Fiebig, *Appl. Surf. Sci.*, 2015, **327**, 413.
90. K. Markey, T. Putzeys, P. Horcajada, T. Devic, N. Guillou, M. Wubbenhorst, S. V. Cleuvenbergen, T. Verbiest, D. E. De Vos and M. A. van der Veen, *Dalton Trans.*, 2016, **45**(10), 4401.
91. W.-W. Zhou, B. Wei, F.-W. Wang, W.-Y. Fang, D.-F. Liu, Y.-J. Wei, M. Xu, X. Zhao and W. Zhao, *RSC Adv.*, 2015, **5**, 100956.
92. G. Ornelas-Arciniega, J. Reyes-Gómez and A. G. Castellanos-Guzman, *A new modification to the Sawyer-Tower ferroelectric hysteresis loop tracer*, 1998.
93. S. Zhang, *Electric-Field Control of Magnetization and Electronic Transport in Ferromagnetic/ Ferroelectric Heterostructures*, 2014.
94. Q. Ye, Y.-M. Song, G.-X. Wang, K. Chen, D.-W. Fu, P. W. Hong Chan, J.-S. Zhu, S. D. Huang and R.-G. Xiong, *J. Am. Chem. Soc.*, 2006, **128**(20), 6554.
95. A. K. Gupta, D. De, R. Katoch, A. Garg and P. K. Bharadwaj, *Inorg. Chem.*, 2017, **56**(8), 4697.
96. D. W. Fu, W. Zhang, H. L. Cai, Y. Zhang, J. Z. Ge, R. G. Xiong, S. D. Huang and T. Nakamura, *Angew. Chem., Int. Ed.*, 2011, **50**(50), 11947.
97. A. J. Bradley and A. H. Jay, *Proc. Phys. Soc.*, 1932, **44**(5), 563.
98. M. P. R. R. Saravanan, *Metal and Alloy Bonding – An Experimental Analysis: Charge Density in Metals and Alloys*, Springer, London, 2011.
99. J. Wooldridge, S. Ryding, S. Brown, T. L. Burnett, M. G. Cain, R. Cernik, R. Hino, M. Stewart and P. Thompson, *J. Synchrotron Radiat.*, 2012, **19**(5), 710.
100. G. Binnig and H. Rohrer, in *Proceedings of the 17th International Conference on the Physics of Semiconductors: San Francisco, California, USA August 6–10, 1984*, ed. J. D. Chadi and W. A. Harrison, Springer, New York, NY, 1985, pp. 3.
101. G. Binnig, C. F. Quate and C. Gerber, *Phys. Rev. Lett.*, 1986, **56**(9), 930.
102. A. L. Kholkin, S. V. Kalinin, A. Roelofs and A. Gruverman, in *Scanning Probe Microscopy: Electrical and Electromechanical Phenomena at the Nanoscale*, ed. S. Kalinin and A. Gruverman, Springer, New York, NY, 2007, pp. 173.
103. A. Gannepalli, D. G. Yablon, A. H. Tsou and R. Proksch, *Nanotechnology*, 2011, **22**(35), 355705.
104. V. K. Sergei, N. M. Anna, C. Long Qing and J. R. Brian, *Rep. Prog. Phys.*, 2010, **73**, 056502.
105. S. V. Kalinin, B. J. Rodriguez, S. Jesse, J. Shin, A. P. Baddorf, P. Gupta, H. Jain, D. B. Williams and A. Gruverman, *Microsc. Microanal.*, 2006, **12**(3), 206.

106. A. Gannepalli, D. G. Yablon, A. H. Tsou and R. Proksch, *Nanotechnology*, 2013, **24**, 159501.
107. J. M. Taylor, T. Komatsu, S. Dekura, K. Otsubo, M. Takata and H. Kitagawa, *J. Am. Chem. Soc.*, 2015, **137**(35), 11498.
108. Z. Hu, Y. Peng, Y. Gao, Y. Qian, S. Ying, D. Yuan, S. Horike, N. Ogiwara, R. Babarao, Y. Wang, N. Yan and D. Zhao, *Chem. Mater.*, 2016, **28**(8), 2659.
109. H.-N. Lin, S.-H. Chen, S.-T. Ho, P.-R. Chen and I.-N. Lin, *J. Vac. Sci. Technol., B*, 2003, **21**, 916.
110. Y. Zhang, L. Chen, J. Zeng, K. Zhou and D. Zhang, *Mater. Sci. Eng., C*, 2014, **39**, 143.
111. J. C. Tan, T. D. Bennett and A. K. Cheetham, *Proc. Natl. Acad. Sci. U. S. A.*, 2010, **107**(22), 9938.
112. J. M. Liu, X. Wang, H. L. W. Chan and C. L. Choy, *Mater. Sci. Eng., B*, 2004, **113**, 67.
113. Y. Lei, Y. Chen and J. D. Lee, *Comput. Mater. Sci.*, 2007, **41**(2), 242.
114. M. Sepliarsky, R. L. Migoni and M. G. Stachiotti, *Comput. Mater. Sci.*, 1998, **10**(1–4), 51.
115. W. Zhong, D. Vanderbilt and K. M. Rabe, *Phys. Rev. B*, 1995, **52**(9), 6301.
116. C. M. Bedoya-Hincapié, E. Restrepo-Parra, J. C. Riaño Rojas, J. J. Olaya-Flores and J. E. Alfonso, *Comput. Mater. Sci.*, 2012, **62**, 131.
117. T. Hang, W. Zhang, H.-Y. Ye and R.-G. Xiong, *Chem. Soc. Rev.*, 2011, **40**(7), 3577.
118. J. VandeVondele, M. Krack, F. Mohamed, M. Parrinello, T. Chassaing and J. Hutter, *Comput. Phys. Commun.*, 2005, **167**(2), 103.
119. H.-L. Hu and L.-Q. Chen, *Mater. Sci. Eng., A*, 1997, **238**(1), 182.
120. Y. Wei and C. Long-Qing, *J. Am. Ceram. Soc.*, 1995, **78**(9), 2554.
121. S. Nambu and D. A. Sagala, *Phys. Rev. B*, 1994, **50**(9), 5838.
122. B. G. Potter Jr., B. A. Tuttle and V. Tikare, *AIP Conf. Proc.*, 2000, **535**, 173.
123. J. Schmidt, J. VandeVondele, I. F. W. Kuo, D. Sebastiani, J. I. Siepmann, J. Hutter and C. J. Mundy, *J. Phys. Chem. B*, 2009, **113**(35), 11959.
124. D. J. Evans and G. P. Morriss, *Phys. Lett. A*, 1983, **98**(8–9), 433.
125. A. Ghoufi and G. Maurin, *J. Phys. Chem. C*, 2010, **114**(14), 6496.
126. R. G. Parr, S. R. Gadre and L. J. Bartolotti, *Proc. Natl. Acad. Sci. U. S. A.*, 1979, **76**(6), 2522.
127. R. G. Parr, *Density Functional Theory of Atoms and Molecules*, Dordrecht, 1980.
128. M. T. Carroll, R. F. W. Bader and S. H. Vosko, *J. Phys. B: At. Mol. Phys.*, 1987, **20**, 3599.
129. W. Kohn and L. J. Sham, *Phys. Rev.*, 1965, **140**(4A), A1133.
130. K. Rahmanizadeh, *Ferroelectric Materials with Interfaces: First Principles Calculations*, Master of Science Dissertation, 2011.
131. R. E. Cohen and H. Krakauer, *Phys. Rev. B*, 1990, **42**(10), 6416.
132. R. E. Cohen, *Nature*, 1992, **358**, 136.
133. R. E. Cohen and H. Krakauer, *Ferroelectrics*, 1992, **136**(1), 65.

134. R. E. Cohen and H. Krakauer, *Ferroelectrics*, 1990, **111**(1), 57.
135. R. D. King-smith and D. Vanderbilt, *Ferroelectrics*, 1992, **136**, 85.
136. Z.-X. Chen, Y. Chen and Y.-S. Jiang, *J. Phys. Chem. B*, 2001, **105**(24), 5766.
137. P. Ghosez, J. P. Michenaud and X. Gonze, *Phys. Rev. B*, 1998, **58**(10), 6224.
138. W. Zhong, R. D. King-Smith and D. Vanderbilt, *Phys. Rev. Lett.*, 1994, **72**(22), 3618.
139. R. Amritendu, M. Somdutta, G. Rajeev, A. Sushil, P. Rajendra and G. Ashish, *J. Phys.: Condens. Matter*, 2011, **23**, 325902.
140. C.-Z. Wang, R. Yu and H. Krakauer, *Phys. Rev. B*, 1996, **54**(16), 11161.
141. J. P. Perdew, K. Burke and M. Ernzerhof, *Phys. Rev. Lett.*, 1996, 77(18), 3865.
142. B. G. Lippert, J. H. Parrinello and A. Michele, *Mol. Phys.*, 1997, **92**(3), 477.
143. S. Grimme, *J. Comput. Chem.*, 2006, **27**(15), 1787.
144. S. Goedecker, M. Teter and J. Hutter, *Phys. Rev. B*, 1996, **54**(3), 1703.
145. C. Hartwigsen, S. Goedecker and J. Hutter, *Phys. Rev. B*, 1998, **58**(7), 3641.
146. S. D. Cox, T. E. Gier, G. D. Stucky and J. Bierlein, *J. Am. Chem. Soc.*, 1988, **110**(9), 2986.
147. F. Marlow, J. Kornatowski, B. Reck, I. Leike, K. Hoffman and J. Caro, *Mol. Cryst. Liq. Cryst. Sci. Technol., Sect. A*, 1996, **276**, 295.
148. A. M. Glass and M. E. Lines, *Principles and Applications of Ferroelectrics and Related Materials*, Oxford University Press, 1977.
149. A. Kamal and V. D. V. M. Ann, *Eur. J. Inorg. Chem.*, 2016, **2016**(27), 4332.
150. P. Jain, N. S. Dalal, B. H. Toby, H. W. Kroto and A. K. Cheetham, *J. Am. Chem. Soc.*, 2008, **130**(32), 10450.
151. M. Sánchez-Andújar, S. Presedo, S. Yáñez-Vilar, S. Castro-García, J. Shamir and M. A. Señarís-Rodríguez, *Inorg. Chem.*, 2010, **49**(4), 1510.
152. G.-C. Xu, X.-M. Ma, L. Zhang, Z.-M. Wang and S. Gao, *J. Am. Chem. Soc.*, 2010, **132**(28), 9588.
153. M. Maczka, A. Ciupa, A. Gagor, A. Sieradzki, A. Pikul and M. Ptak, *J. Mater. Chem. C*, 2016, **4**, 1186.
154. H.-X. Zhao, X.-J. Kong, H. Li, Y.-C. Jin, L.-S. Long, X. C. Zeng, R.-B. Huang and L.-S. Zheng, *Proc. Natl. Acad. Sci. U. S. A.*, 2011, **108**(9), 3481.
155. H. K. Chae, D. Y. Siberio-Pérez, J. Kim, Y. Go, M. Eddaoudi, A. J. Matzger, M. O'Keeffe and O. M. Yaghi, *Nature*, 2004, **427**(6974), 523.
156. G. J. Halder, C. J. Kepert, B. Moubaraki, K. S. Murray and J. D. Cashion, *Science*, 2002, **298**(5599), 1762.
157. C. Heng-Bo, T. Kazuyuki, O. Yoshinori, K. Hayao, W. Zheming and K. Akiko, *Angew. Chem., Int. Ed.*, 2005, **44**(40), 6508.
158. Z. Wang, B. Zhang, H. Fujiwara, H. Kobayashi and M. Kurmoo, *Chem. Commun.*, 2004, 416–417.
159. A. Yadav, A. K. Srivastava, P. Kulkarni, P. Divya, A. Steiner, B. Praveenkumar and R. Boomishankar, *J. Mater. Chem. C*, 2017, **5**, 10624.

160. R. Schroeder, L. A. Majewski and M. Grell, *Adv. Mater.*, 2004, **16**(7), 633–636.
161. G. Arlt and H. Neumann, *Ferroelectrics*, 1988, **87**(1), 109.
162. W. Zhang, H.-Y. Ye and R.-G. Xiong, *Coordin. Chem. Rev.*, 2009, **253**(23–24), 2980.
163. B. Chu, X. Zhou, K. Ren, B. Neese, M. Lin, Q. Wang, F. Bauer and Q. M. Zhang, *Science*, 2006, **313**(5785), 334.
164. Z. Zhang and T. C. M. Chung, *Macromolecules*, 2007, **40**(26), 9391.
165. Z. Zhang, Q. Meng and T. C. M. Chung, *Polymer*, 2009, **50**(2), 707.
166. Y. Wang, X. Zhou, Q. Chen, B. Chu and Q. Zhang, *IEEE Trans. Dielectr. Electr. Insul.*, 2010, **17**(4), 1036.
167. X. Hao, *J. Adv. Dielectr.*, 2013, **3**(1), 1330001.
168. M. H. Beyzavi, R. C. Klet, S. Tussupbayev, J. Borycz, N. A. Vermeulen, C. J. Cramer, J. F. Stoddart, J. T. Hupp and O. K. Farha, *J. Am. Chem. Soc.*, 2014, **136**(45), 15861.
169. W. Zheng, K. H. Bowen, J. Li, I. Dabkowska and M. Gutowski, *J. Phys. Chem. A*, 2005, **109**(50), 11521.
170. H.-B. Cui, K. Takahashi, Y. Okano, H. Kobayashi, Z. Wang and A. Kobayashi, *Angew. Chem., Int. Ed.*, 2005, **44**(40), 6508.
171. M. A. Garcia-Garibay, *Proc. Natl. Acad. Sci. U. S. A.*, 2005, **102**(31), 10771.
172. S. D. Karlen and M. A. Garcia-Garibay, in *Molecular Machines*, ed. T. R. Kelly, Springer, Berlin, Heidelberg, 2005, pp. 179.
173. T.-A. V. Khuong, J. E. Nuñez, C. E. Godinez and M. A. Garcia-Garibay, *Acc. Chem. Res.*, 2006, **39**(6), 413.
174. S. L. Gould, D. Tranchemontagne, O. M. Yaghi and M. A. Garcia-Garibay, *J. Am. Chem. Soc.*, 2008, **130**(11), 3246.
175. S. Devautour-Vinot, G. Maurin, C. Serre, P. Horcajada, D. Paula da Cunha, V. Guillerm, E. de Souza Costa, F. Taulelle and C. Martineau, *Chem. Mater.*, 2012, **24**(11), 2168.
176. T. Akutagawa, H. Koshinaka, D. Sato, S. Takeda, S.-I. Noro, H. Takahashi, R. Kumai, Y. Tokura and T. Nakamura, *Nat. Mater.*, 2009, **8**(4), 342.
177. V. N. Vukotic, K. J. Harris, K. Zhu, R. W. Schurko and S. J. Loeb, *Nat. Chem.*, 2012, **4**(6), 456.
178. W. Zhang, H.-Y. Ye, R. Graf, H. W. Spiess, Y.-F. Yao, R.-Q. Zhu and R.-G. Xiong, *J. Am. Chem. Soc.*, 2013, **135**(14), 5230.
179. R. D. Horansky, L. I. Clarke, J. C. Price, T.-A. V. Khuong, P. D. Jarowski and M. A. Garcia-Garibay, *Phys. Rev. B*, 2005, **72**(1), 014302.
180. E. B. Winston, P. J. Lowell, J. Vacek, J. Chocholousova, J. Michl and J. C. Price, *Phys. Chem. Chem. Phys.*, 2008, **10**(34), 5188.
181. L. Huang, X. Zhang, Y. Han, Q. Wang, Y. Fang and S. Dong, *J. Mater. Chem. A*, 2017, **5**(35), 18610.
182. Y. Peng, Y. Li, Y. Ban, H. Jin, W. Jiao, X. Liu and W. Yang, *Science*, 2014, **346**(6215), 1356.
183. J. Yin, Q. Fan, Y. Li, F. Cheng, P. Zhou, P. Xi and S. Sun, *J. Am. Chem. Soc.*, 2016, **138**(44), 14546.

184. S. Zhao, Y. Wang, J. Dong, C.-T. He, H. Yin, P. An, K. Zhao, X. Zhang, C. Gao, L. Zhang, J. Lv, J. Wang, J. Zhang, K. Abdul, N. Khan, Z. Wei, J. Zhang, S. Liu, H. Zhao and Z. Tang, *Nat. Energy*, 2016, **1**, 16184.
185. A. Gallego, C. Hermosa, O. Castillo, I. Berlanga, C. J. Gómez-García, E. Mateo-Martí, J. I. Martínez, F. Flores, C. Gómez-Navarro, J. Gómez-Herrero, S. Delgado and F. Zamora, *Adv. Mater.*, 2013, **25**(15), 2141.
186. T. Rodenas, I. Luz, G. Prieto, B. Seoane, H. Miro, A. Corma, F. Kapteijn, F. X. Llabrés i Xamena and J. Gascon, *Nat. Mater.*, 2015, **14**(1), 48.
187. M. Y. Jeon, D. Kim, P. Kumar, P. S. Lee, N. Rangnekar, P. Bai, M. Shete, B. Elyassi, H. S. Lee, K. Narasimharao, S. N. Basahel, S. Al-Thabaiti, W. Xu, H. J. Cho, E. O. Fetisov, R. Thyagarajan, R. F. DeJaco, W. Fan, K. A. Mkhoyan, J. I. Siepmann and M. Tsapatsis, *Nature*, 2017, **543**, 690.
188. W. Yixian, Z. Meiting, P. Jianfeng, C. Bo, C. Xiehong, H. Ying, T. Chaoliang, M. Qinglang, W. Shixin, Y. Yifu, L. Qipeng, C. Junze, Z. Wei, Y. Yibin and Z. Hua, *Adv. Mater.*, 2016, **28**(21), 4149.
189. Z. Meiting, W. Yixian, M. Qinglang, H. Ying, Z. Xiao, P. Jianfeng, Z. Zhicheng, L. Qipeng, Y. Yifu, X. Huan, Z. Yanli and Z. Hua, *Adv. Mater.*, 2015, **27**(45), 7372.
190. Z. Meiting, L. Qipeng, M. Qinglang and Z. Hua, *Small Methods*, 2017, **1**(1–2), 1600030.
191. H.-L. Liu, Y.-J. Chang, T. Fan and Z.-Y. Gu, *Chem. Commun.*, 2016, **52**, 12984.
192. M. Xu, S. Yuan, X.-Y. Chen, Y.-J. Chang, G. Day, Z.-Y. Gu and H.-C. Zhou, *J. Am. Chem. Soc.*, 2017, **139**(24), 8312.
193. M. J. Cliffe, E. Castillo-Martínez, Y. Wu, J. Lee, A. C. Forse, F. C. N. Firth, P. Z. Moghadam, D. Fairen-Jimenez, M. W. Gaultois, J. A. Hill, O. V. Magdysyuk, B. Slater, A. L. Goodwin and C. P. Grey, *J. Am. Chem. Soc.*, 2017, **139**(15), 5397.
194. Y. Ding, Y.-P. Chen, X. Zhang, L. Chen, Z. Dong, H.-L. Jiang, H. Xu and H.-C. Zhou, *J. Am. Chem. Soc.*, 2017, **139**(27), 9136.
195. Q.-F. Yang, X.-B. Cui, J.-H. Yu, J. Lu, X.-Y. Yu, X. Zhang, J.-Q. Xu, Q. Hou and T.-G. Wang, *CrystEngComm*, 2008, **10**, 1534.
196. B. Garai, A. Mallick, A. Das, R. Mukherjee and R. Banerjee, *Chem. – Eur. J.*, 2017, **23**, 7361.
197. S. Zhao, Y. Wang, J. Dong, C.-T. He, H. Yin, P. An, K. Zhao, X. Zhang, C. Gao, L. Zhang, J. Lv, J. Wang, J. Zhang, A. M. Khattak, N. A. Khan, Z. Wei, J. Zhang, S. Liu, H. Zhao and Z. Tang, *Nat. Energy*, 2016, **1**, 16184.
198. K. Zhao, S. Liu, G. Ye, Q. Gan, Z. Zhou and Z. He, *J. Mater. Chem. A*, 2018, **6**, 2166.
199. T. Rodenas, I. Luz, G. Prieto, B. Seoane, H. Miro, A. Corma, F. Kapteijn, F. X. Llabrés i Xamena and J. Gascon, *Nat. Mater.*, 2014, **14**(1), 48.
200. R. Makiura, S. Motoyama, Y. Umemura, H. Yamanaka, O. Sakata and H. Kitagawa, *Nat. Mater.*, 2010, **9**(7), 565.
201. Y. Huang, M. Zhao, S. Han, Z. Lai, J. Yang, C. Tan, Q. Ma, Q. Lu, J. Chen, X. Zhang, Z. Zhang, B. Li, B. Chen, Y. Zong and H. Zhang, *Adv. Mater.*, 2017, **29**(32), 1700102.

202. G. Bikash, M. Arijit, D. Anuja, M. Rabibrata and B. Rahul, *Chem. – Eur. J.*, 2017, **23**, 7361.
203. Y.-Z. Tang, Y.-M. Yu, Y.-H. Tan, J.-S. Wu, J.-B. Xiong and H.-R. Wen, *Dalton Trans.*, 2013, **42**, 10106.
204. Q. Ye, D.-W. Fu, H. Tian, R.-G. Xiong, P. W. H. Chan and S. D. Huang, *Inorg. Chem.*, 2008, **47**(3), 772.
205. B. Z. Ahmad and B. Want, *J. Appl. Phys.*, 2016, **119**(14), 144104.
206. J. C. Tan and A. K. Cheetham, *Chem. Soc. Rev.*, 2011, **40**(2), 1059.
207. H. Wu, T. Yildirim and W. Zhou, *J. Phys. Chem. Lett.*, 2013, **4**(6), 925.
208. G. D. Pirngruber and P. L. Llewellyn, *Metal-Organic Frameworks: Applications from Catalysis to Gas Storage*, Wiley-VCH Verlag GmbH & Co. KGaA, 2011.
209. M. E. Davis and R. J. Davis, *Fundamentals of Chemical Reaction Engineering*, McGraw-Hill Higher Education, New York, 2003.
210. D. F. Bahr, J. A. Reid, W. M. Mook, C. A. Bauer, R. Stumpf, A. J. Skulan, N. R. Moody, B. A. Simmons, M. M. Shindel and M. D. Allendorf, *Phys. Rev. B*, 2007, **76**(18), 184106.
211. A. Kuc, A. Enyashin and G. Seifert, *J. Phys. Chem. B*, 2007, **111**(28), 8179.
212. T. D. Bennett, J.-C. Tan, S. A. Moggach, R. Galvelis, C. Mellot-Draznieks, B. A. Reisner, A. Thirumurugan, D. R. Allan and A. K. Cheetham, *Chem. – Eur. J.*, 2010, **16**(35), 10684.
213. T. D. Bennett, A. L. Goodwin, M. T. Dove, D. A. Keen, M. G. Tucker, E. R. Barney, A. K. Soper, E. G. Bithell, J.-C. Tan and A. K. Cheetham, *Phys. Rev. Lett.*, 2010, **104**(11), 115503.
214. M. Kosa, J.-C. Tan, C. A. Merrill, M. Krack, A. K. Cheetham and M. Parrinello, *ChemPhysChem*, 2010, **11**(11), 2332.
215. J.-C. Tan, P. J. Saines, E. G. Bithell and A. K. Cheetham, *ACS Nano*, 2012, **6**(1), 615.
216. J.-C. Tan, B. Civalleri, C.-C. Lin, L. Valenzano, R. Galvelis, P.-F. Chen, T. D. Bennett, C. Mellot-Draznieks, C. M. Zicovich-Wilson and A. K. Cheetham, *Phys. Rev. Lett.*, 2012, **108**(9), 095502.
217. A. U. Ortiz, A. Boutin, A. H. Fuchs and F.-X. Coudert, *Phys. Rev. Lett.*, 2012, **109**, 195502.
218. W. Li, M. S. R. N. Kiran, J. L. Manson, J. A. Schlueter, A. Thirumurugan, U. Ramamurty and A. K. Cheetham, *Chem. Commun.*, 2013, **49**, 4471.
219. V. I. Hegde, J.-C. Tan, U. V. Waghmare and A. K. Cheetham, *J. Phys. Chem. Lett.*, 2013, **4**(20), 3377–3381.
220. W. Li, S. Henke and A. K. Cheetham, *APL Mater.*, 2014, **2**, 123902.
221. T. D. Bennett, J. Sotelo, J.-C. Tan and S. A. Moggach, *CrystEngComm*, 2015, **17**, 286.
222. C. Hermosa, B. R. Horrocks, J. I. Martinez, F. Liscio, J. Gomez-Herrero and F. Zamora, *Chem. Sci.*, 2015, **6**, 2553.
223. S. Bundschuh, O. Kraft, H. K. Arslan, H. Gliemann, P. G. Weidler and C. Wöll, *Appl. Phys. Lett.*, 2012, **101**, 101910.
224. B. Van de Voorde, R. Ameloot, I. Stassen, M. Everaert, D. De Vos and J.-C. Tan, *J. Mater. Chem. C*, 2013, **1**, 7716.

225. N. Stock and S. Biswas, *Chem. Rev.*, 2012, **112**, 933.
226. J. D. Lord and R. M. Morrell, *Metrologia*, 2010, **47**(2), S41.
227. M. H. Beyzavi, N. A. Vermeulen, A. J. Howarth, S. Tussupbayev, A. B. League, N. M. Schweitzer, J. R. Gallagher, A. E. Platero-Prats, N. Hafezi, A. A. Sarjeant, J. T. Miller, K. W. Chapman, J. F. Stoddart, C. J. Cramer, J. T. Hupp and O. K. Farha, *J. Am. Chem. Soc.*, 2015, **137**(42), 13624.
228. J. H. Cavka, S. Jakobsen, U. Olsbye, N. Guillou, C. Lamberti, S. Bordiga and K. P. Lillerud, *J. Am. Chem. Soc.*, 2008, **130**(42), 13850.
229. J. Canivet, A. Fateeva, Y. Guo, B. Coasne and D. Farrusseng, *Chem. Soc. Rev.*, 2014, **43**, 5594.
230. N. C. Burtch, H. Jasuja and K. S. Walton, *Chem. Rev.*, 2014, **114**(20), 10575.
231. Z. Hu, S. Faucher, Y. Zhuo, Y. Sun, S. Wang and D. Zhao, *Chem. - Eur. J.*, 2015, **21**(48), 17246.
232. N. Planas, J. E. Mondloch, S. Tussupbayev, J. Borycz, L. Gagliardi, J. T. Hupp, O. K. Farha and C. J. Cramer, *J. Phys. Chem. Lett.*, 2014, **5**(21), 3716.
233. J. E. Mondloch, M. J. Katz, W. C. Isley Iii, P. Ghosh, P. Liao, W. Bury, G. W. Wagner, M. G. Hall, J. B. DeCoste, G. W. Peterson, R. Q. Snurr, C. J. Cramer, J. T. Hupp and O. K. Farha, *Nat. Mater.*, 2015, **14**(5), 512.
234. G. Férey, C. Mellot-Draznieks, C. Serre, F. Millange, J. Dutour, S. Surblé and I. Margiolaki, *Science*, 2005, **309**(5743), 2040.
235. L. Bromberg, Y. Diao, H. M. Wu, S. A. Speakman and T. A. Hatton, *Chem. Mater.*, 2012, **24**(9), 1664.
236. L. Alaerts, C. E. A. Kirschhock, M. Maes, M. A. van der Veen, V. Finsy, A. Depla, J. A. Martens, G. V. Baron, P. A. Jacobs, J. F. M. Denayer and D. E. De Vos, *Angew. Chem., Int. Ed.*, 2007, **46**(23), 4293.
237. K.-Y. Andrew Lin and Y.-T. Hsieh, *J. Taiwan Inst. Chem. Eng.*, 2015, **50**, 223.
238. A. K. Chaudhari, I. Han and J. C. Tan, *Adv. Mater.*, 2015, **27**(3), 4438.
239. Z. Hu, Y. Peng, Z. Kang, Y. Qian and D. Zhao, *Inorg. Chem.*, 2015, **54**(10), 4862.
240. Y. Sun, Z. Hu, D. Zhao and K. Zeng, *ACS Appl. Mater. Interfaces*, 2017, **9**(37), 32202.
241. H. G. T. Nguyen, L. Mao, A. W. Peters, C. O. Audu, Z. J. Brown, O. K. Farha, J. T. Hupp and S. T. Nguyen, *Catal. Sci. Technol.*, 2015, **5**, 4444.
242. M. J. Katz, Z. J. Brown, Y. J. Colón, P. W. Siu, K. A. Scheidt, R. Q. Snurr, J. T. Hupp and O. K. Farha, *Chem. Commun.*, 2013, **49**(82), 9449.
243. G. Chawla and S. D. Solares, *Appl. Phys. Lett.*, 2011, **99**, 074103.
244. R. Garcia and E. T. Herruzo, *Nat. Nanotechnol.*, 2012, 7, 217.
245. D. Martinez-Martin, E. T. Herruzo, C. Dietz, J. Gomez-Herrero and R. Garcia, *Phys. Rev. Lett.*, 2011, **106**, 198101.

CHAPTER 8

Synthetic Techniques and Functionalization Approaches of 2D Transition Metal Dichalcogenides

GUIJIAN GUAN

Institute of Materials Research and Engineering, A*STAR, 2 Fusionopolis Way, 138634, Singapore
Email: guan_guijian@imre.a-star.edu.sg

8.1 Introduction

In recent years, the successful exploration and extensive investigation of graphene have aroused enormous research interest in the production and application of other two-dimensional (2D) nanomaterials with atomically thin thickness.[1] As one of the most promising 2D materials, transition-metal dichalcogenides (TMDs) in single and few layers have become a new class of nanomaterials for fundamental studies and outstanding applications because of their fascinating exotic physical and structural properties that arise from their increased surface-to-volume ratio, altered surface energy, and inherent confinement effect.[2] Given that M represents a transition-metal element such as Mo, W, Nb, Ta, Ti, Zr, Hf, *etc.*, and X represents a chalcogen element such as S, Se and Te, *etc.*, TMD materials can be expressed by the formula of a stoichiometric MX_2 compound, well-known as MoS_2, WS_2, $MoSe_2$. A single layer of MX_2 contains three atomic layers (X–M–X) with a

Smart Materials No. 34
Layered Materials for Energy Storage and Conversion
Edited by Dongsheng Geng, Yuan Cheng and Gang Zhang

Published by the Royal Society of Chemistry, www.rsc.org

thickness of 6–7 Å, where coordination octahedrons or trigonal prisms MX_6 are linked through shared edges. These hexagonal layers bind to each other through weak van der Waals interactions to form bulk MX_2,[3] similar to graphite comprised of a graphene layer.

Similar to graphite, the layered structure of TMD crystals makes most of the exfoliating strategies for graphene effective to produce TMD nanosheets through breaking the van der Waals forces between adjacent TMD layers in bulk, which are classified as top-down techniques (*i.e.*, the big picture breaks down into smaller segments). As shown in Chart 8.1, the top-down methods include mechanical cleavage, ion intercalation-assisted and mechanical force-assisted liquid exfoliation. The micromechanical cleavage method is the most straightforward route for obtaining ultrathin TMD nanosheets with high crystalline quality,[3] although it lacks sufficient scalability for large-scale production of TMD nanosheets. To address this concern, solution-based exfoliation methods were developed to the scalable production of TMD nanosheets in high yield, which are also more convenient and easier for engineering the chemical and physical properties of TMD nanosheets for achieving improved performance in applications.[3] Based on their exfoliation mechanisms, liquid exfoliation methods are divided into ion intercalation and mechanical force-assisted exfoliation. Ion intercalation exfoliation is able to effectively yield high concentrations of single-layer TMD nanosheets in solution, however, the introduction of small ions usually results in structural and electronic deformations of TMD nanosheets from their bulk. Alternatively, mechanical force-assisted liquid exfoliation, particularly direct sonication of TMD materials in solution with or without

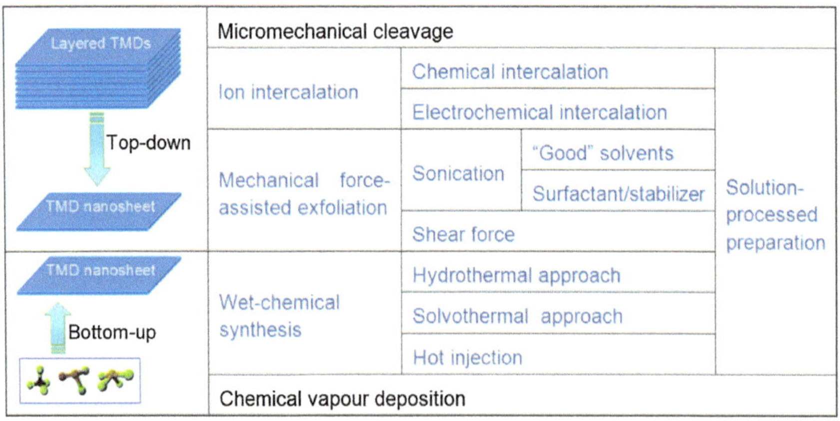

Chart 8.1 Overview of the synthetic techniques of TMD nanosheets. The top-down techniques include micromechanical cleavage, ion intercalation and mechanical force-assisted exfoliation; while the bottom-up techniques include wet-chemical synthesis and chemical vapour deposition. Among them, micromechanical cleavage and chemical vapour deposition are non-liquid methods, whereas all of the other approaches are solution-processed preparation methods.

additives, is a very simple technique to obtain a large quantity of TMD nanosheets dispersed in solution, which can further be purified to obtain single or few layer nanosheets *via* centrifugation at a special speed.

As opposed to the above top-down methods, TMD nanosheets can be also obtained *via* bottom-up methods, that is, nanosheets are produced *via* chemical reactions of certain precursors at given experimental conditions (see Chart 8.1). Among them, wet-chemical synthetic methods including the hydro/solvothermal approach and hot injection have been developed to produce a large amount of TMD nanosheets in solution by using metal salts and chalcogenide-containing chemicals as precursors.[4,5] An alternative to the liquid method, chemical vapour deposition (another bottom-up technique) is able to effectively grow high-quality TMD nanosheets on preselected substrates under high vacuum and high temperature,[6] which is very convenient to control the size and thickness of TMD nanosheets *via* the adjustment of deposition time and gas flow rate; but the rigid experimental requirements limit the wide practical applications of chemical vapour deposition.

In this chapter, different synthetic techniques towards ultrathin TMD nanosheets will be introduced systematically to realize their advantages and disadvantages. In the past few decades, there have been a large number of studies aimed at investigating the properties of TMD nanosheets or exploiting their promising applications. Although these works inevitably involved the preparation of TMD nanosheets, they will not be covered in this chapter. Herein, we will highlight the design principle, fabrication mechanism, typical process and special conditions for obtaining TMD nanosheets *via* different synthetic methods, followed by a discussion on the advances and limitations of each method. The recent exploitation to develop novel fabrication methods will also be demonstrated as typical examples. On the basis of various synthetic techniques, general strategies applied for the chemical/hybridized functionalization of TMD nanosheets will be presented in the following text. Finally, we propose the challenges and opportunities for exploiting TMD nanosheets in novel applications.

8.2 Micromechanical Cleavage

Micromechanical cleavage is used to weaken the van der Waals interaction between two adjacent layers in bulk crystals by applying mechanical force such as normal or shear force (see Figure 8.1A), but the in-plane covalent bonds are not broken. A universal method is use Scotch tape in micromechanical cleavage, which is shown in Figure 8.1B using the exfoliation of graphite as an example (the exfoliation of TMD nanosheets is carried out by following a similar procedure). First, the bulk crystal is attached to the adhesive surface of Scotch tape, while another adhesive surface is used to peel the bulk crystal into a thin flake.[7,8] After repeating the above process several times, a flake with an appropriate layer number is obtained. Then, the cleaved single- or few-layered flakes are transferred from the Scotch tape

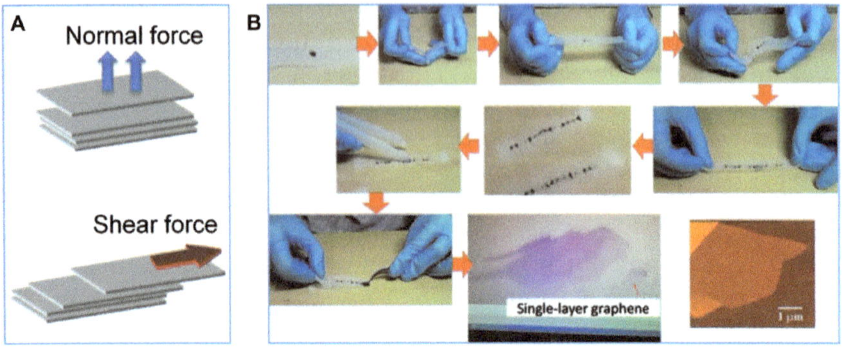

Figure 8.1 The mechanism and process for micromechanical cleavage of layered materials into ultrathin nanosheets: (A) two kinds of mechanical forces to yield ultrathin nanosheets, and (B) typical procedures in Scotch tape-based micromechanical cleavage. The bulk crystal is first attached to the adhesive on the Scotch tape and then peeled into a thin flake by using another adhesive surface, which can be repeated for obtaining an appropriate layer number. The freshly cleaved thin flake is further attached to a flat target surface to finally obtain single- or few-layered nanosheets on the used substrate after peeling off the Scotch tape. Reproduced from ref. 7 with permission from the Royal Society of Chemistry.

onto a flat SiO_2/Si substrate by peeling off the Scotch tape using plastic tweezers. In micromechanical cleavage, chemicals and chemical reactions are not involved, which means that it is a non-destructive method for keeping the "perfect" crystal quality of the exfoliated nanosheets. Meanwhile, the nanosheets have larger sizes (up to tens of micrometres) than those obtained using solution-processed methods. With a relatively large lateral size and excellent crystal quality, cleaved single- or few-layered 2D nanomaterials have been extensively investigated for the fundamental study of intrinsic physical, optical, and electronic properties, and the development of electronic and optoelectronic devices with high performance.[9–12]

For the exfoliation of TMD materials, mechanical cleavage is a traditional but efficient method to fabricate atomically thin TMD nanosheets on a re-prepared substrate. After successful cleavage of graphite into graphene *via* a micromechanical technique,[13] Novoselov and Geim further extended this technique to exfoliate other ultrathin 2D nanomaterials from their parent layered crystals, including TMD materials such as MoS_2 and $NbSe_2$.[14] In the past decade, the mechanical cleavage technique was extensively utilized to exfoliate single- and few-layered TMD nanosheets on SiO_2-coated Si substrates and quartz, including MoS_2, $MoSe_2$, $MoTe_2$, WS_2, WSe_2, TiS_2, TaS_2, $TaSe_2$, ReS_2, *etc.*[15,16] For example, Li *et al.* prepared single to five layers of MoS_2, TaS_2, and WSe_2 nanosheets on SiO_2 layer-coated Si substrates, and reliably identified the layer thickness by using optical microscopy.[17,18] For obtaining ultra-large monolayer TMD nanosheets, Javay and co-workers

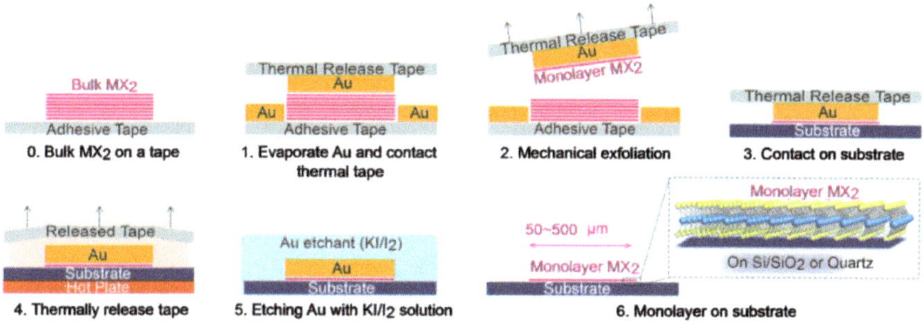

Figure 8.2 Schematic illustration of Au-mediated micromechanical exfoliation of monolayer MX_2 nanosheets with an ultra-large area. After being attached to adhesive tape, bulk MX_2 crystals are deposited with an evaporated Au film, followed by adhesion of a thermal release tape. Due to the stronger interaction between the topmost layer with the evaporated Au, the topmost layer is selectively peeled off by using the thermal release tape, which is then stuck onto the target substrate. After removal of the thermal tape and the etching of Au film, a large-area monolayer MX_2 nanosheet is obtained on the desired substrate.
Reproduced from ref. 19 with permission from John Wiley and Sons, © 2016 WILEY-VCH Verlag GmbH & Co. KGaA, Weinheim.

recently developed an Au-mediated mechanical exfoliation technique to significantly enlarge the area of exfoliated TMD nanosheets through evaporating Au films onto SiO_2/Si substrates (see Figure 8.2).[19] As demonstrated, the enlargement of the area of exfoliated nanosheets is a result of the enhanced contact between the TMD crystals and the substrate.

The micromechanical cleavage technique has many advantages as mentioned above, however, it is difficult to control the thickness and size/shape of the nanosheets because of the lack of precision, controllability, or repeatability from manually-operated exfoliation process by hand.[20,21] For the final products, the single-layered flakes always coexist on the substrate together with the few-layered nanosheets. Compared with the CVD growth and solution-based methods, the micromechanical cleavage rate is quite slow and the product is not scaled up (see the following text). Moreover, the substrate is essential for supporting the exfoliated 2D crystals, and so freestanding nanosheets cannot be produced *via* the micromechanical cleavage.

8.3 Ion Intercalation-assisted Liquid Exfoliation

Liquid phase exfoliation techniques were successfully used to produce large-scale and high-yield TMD nanosheets with single and few layers. Based on the intercalation chemistry of TMD materials (see Figure 8.3A),[22] ion intercalation-assisted liquid exfoliation was developed (see Figure 8.3B).[23] An ion-intercalated compound is first formed through intercalating small-sized cation ions into the interspacing of layered TMD crystals, leading to

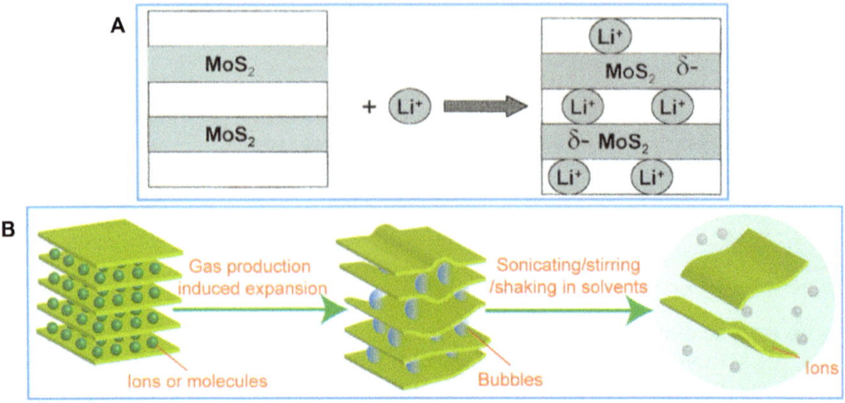

Figure 8.3 Schematic description of the ion intercalation-assisted liquid exfoliation. (A) The intercalation of lithium into layered MoS_2 crystals. Reprinted from *Coord. Chem. Rev.*, **224**, E. Benavente, Intercalation Chemistry of Molybdenum Disulfide, Pages 87–109, Copyright 2002, with permission from Elsevier.[22] (B) Ion intercalation-based exfoliation. Ions or molecules are intercalated between the layers to swell the crystal and weaken the interlayer attraction. Then, agitation can completely separate the layers, resulting in a solution of exfoliated nanosheets. Reproduced from ref. 23 with permission from the Royal Society of Chemistry.

significant expansion of the interspacing and weakening of the van der Waals interaction between the adjacent TMD layers in bulk crystals. The intercalated compound is subsequently exfoliated into single- or few-layer nanosheets under agitation (*e.g.*, ultrasonication, thermal). When alkaline metal ions such as Li^+ or Na^+ are used as intercalated ions (most cases), they can react with water to generate hydrogen gas, which can also help to separate the adjacent TMD layers during the sonication process and thus further promote the exfoliation efficiency.[24,25] Under the assistance of generated gas, the exfoliation of TMD nanosheets can be performed within a short time (*e.g.* 10 min), and then single- or few-layered nanosheets can be obtained at high yield after the removal of thick flakes *via* centrifugation. In the following section, we will introduce the recent developments on ion intercalation-assisted exfoliation of TMD nanosheets, which are classified as chemical or electrochemical intercalation in accordance with the intercalation motion/process.

8.3.1 Chemical Intercalation and Exfoliation

Layered TMD crystals can easily be intercalated chemically to form host–guest compounds with ions or molecules.[22] In particular, lithium-intercalated TMD compounds have been successfully developed and widely exploited to prepare single-layered TMD nanosheets of large scale.[20,25,26] In the lithium-intercalated exfoliation, organolithium reagents (*i.e.* *n*-butyllithium) are usually adopted to obtain the intercalated compounds *via* a reaction with TMD powders, followed by reacting with water to exfoliate the intercalated TMD bulk materials into single- or few-layered nanosheets.[24,25] Experimentally, layered TMD crystal is first intercalated with *n*-butyllithium (*n*-BuLi) in hexane solution under refluxed conditions for 3 days to form the Li intercalated TMD compound, which is then transferred into water to obtain the nanosheet suspension with the assistance of sonication.[24,25] It is worth noting that the Li intercalation process needs high temperature and long reaction times, which is a major disadvantage in the use of *n*-BuLi.[24,25,27] Recently, the Li intercalation process was successfully modified to address the aforementioned disadvantage *via* a versatile and rapid ultrasonication-enhanced lithium intercalation strategy.[28] The strategy made use of local high temperature and pressure provided by ultrasonication to activate the hexamer of *n*-BuLi and increase the mass-transfer of reactants at the solid–liquid interfaces. The extraordinary reaction conditions significantly increase the intercalation efficiency for TMD nanomaterials, including MoS_2, WS_2, and TiS_2. For MoS_2, the intercalation reaction time was reduced to 1–3 hours, and single-layer MoS_2 nanosheets were produced at a high yield of 17.2%, whereas it is 7.6% in the conventional intercalation method. Under the assistance of sonication, Rajamathi and co-workers replaced lithium-intercalated Li_xMS_2 (M = Mo and W) with NH_3/NH_4^+ to obtain stable MS_2 dispersions.[29] During the sonication process, ammonia gas was generated to promote the exfoliation of nanosheets, and the gas would finally escape from the solution under heating.

Besides *n*-BuLi, intercalation can also be implemented using other compounds as effective intercalators for the exfoliation of TMD nanosheets.[30,31] For example, Loh and co-workers achieved high yield exfoliation of MoS_2 nanosheets using metal naphthalenide (Li, Na, or K) as the intercalator after combining the hydrothermal method.[32] As shown in Figure 8.4, the MoS_2 crystals were first reacted with hydrazine (N_2H_4) and the subsequent decomposition of a N_2H_4 molecule greatly expanded the volume of the bulk crystals, which were then intercalated by metal naphthalenide and formed intercalated compounds. After sonication and purification, single-layered MoS_2 nanosheets with lateral sizes of up to 400 μm^2 were obtained with a high yield of 90%. Most recently, Daeneke *et al.* developed a hydrazine salt-based intercalation method to produce highly monodispersed sub-stoichiometric MoS_x bilayers.[33] They observed that sulphur loss and lattice deformations of MoS_2 were caused after reduction of bulk MoS_2 by hydrazine, which ultimately led to the delamination of MoS_2 layers and exfoliation of MoS_2 nanosheets. Alternatively, some common inorganic salts (NaCl, $CuCl_2$, *etc.*) can also be used as intercalators in the exfoliation of ultrathin TMD nanosheets from layered bulk TMD crystals, including MoS_2, $MoSe_2$, WS_2 and WSe_2.[34] Bulk TMD material was first dispersed in water containing saturated inorganic salts and the mixture was then heated to evaporate water at 100 °C, leading to the intercalation of cation ions into interlayer spacing.

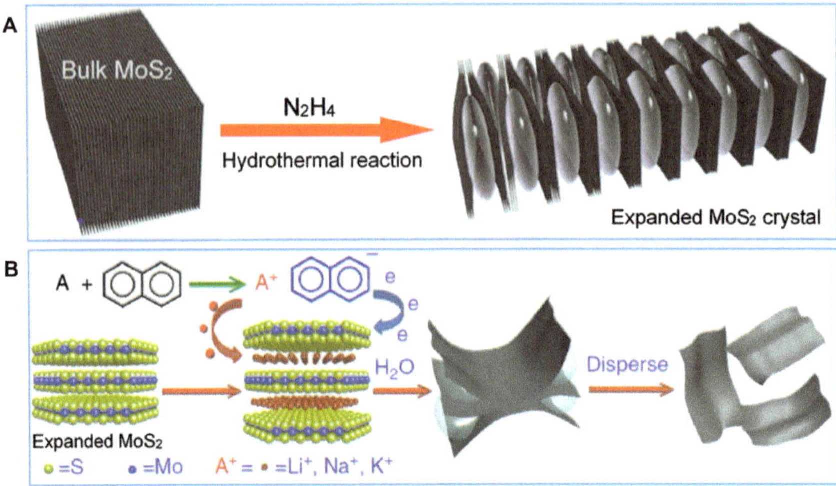

Figure 8.4 Schematic of fabrication processes of MoS_2 nanosheets using sodium naphthalenide as the intercalator. (A) Bulk MoS_2 is pre-exfoliated by the decomposition products of N_2H_4. (B) Pre-exfoliated MoS_2 reacts with $A^+C_{10}H_8^-$ to form an intercalation sample, and then exfoliates into single-layer sheets in water.
Reprinted from *Nat. Commun.*, **5**, J. Zhang, H. Zhang, S. Dong, Y. Liu, C. T. Nai, H. S. Shin, H. Y. Jeong, B. Liu and K. P. Loh, High Yield Exfoliation of Two-Dimensional Chalcogenides Using Sodium Naphthalenide, No. 2995, Copyright 2014, with permission from Springer Nature.[32]

The intercalated crystal was further sonicated in organic solvent to obtained single- or few-layered TMD nanosheets. As safer and cheaper alternatives, these inorganic salts are more suitable in practical productions of TMD nanosheets, compared to lithium reagents.

To date, a variety of ultrathin TMD nanosheets have been effectively produced *via* the chemical ion intercalation-assisted liquid exfoliation method.[15,24,25,28,31,32] Their layer number, lateral size, surface defects and concentration in solution can be tuned roughly by controlling the experimental conditions: such as initial concentration and particle size of bulk crystals, intercalating agents, intercalating temperature and time, and sonication situations. In addition, it was found that the ion intercalation into layered bulk crystals can induce the phase transformation of some TMDs (*e.g.*, MoS_2 and WS_2) from semiconducting hexagonal (2H) phase of bulk to a metallic octahedral (1T) phase of nanosheets. The phase engineering of TMDs will be useful for developing novel application of 2D TMDs.[35] In the ion-intercalation process, the experiment needs to be operated in a glovebox with extreme caution because the organometallic compounds used are highly explosive and very sensitive to moisture and oxygen. Moreover, it is difficult to achieve a sufficient but non-overwhelmed ion-intercalation due to the uncontrollable behaviour of the ion intercalation process.

8.3.2 Electrochemical Intercalation and Exfoliation

In a lithium ion battery system comprised of Li foil and layered materials, Li^+ ions are intercalated into the cathode (*i.e.* layered materials) during the discharge process when Li foil is used as the anode. Inspired by the intercalation process, an electrochemical method for intercalation was developed by Zhang and co-workers,[36] which similarly produced ultrathin nanosheets *via* the Li intercalation-assisted liquid exfoliation of TMD bulk crystals. As shown in Figure 8.5, instead of organometallic compounds, Li foil was used as the Li source and the electrochemical force drove the Li intercalation process. The intercalated degree of Li ions in the layered crystals could be monitored to realize sufficient but non-overwhelmed intercalation. Experimentally, TMD bulk crystals were coated on metal foils (*e.g.* Cu) to be used as cathodes in Li ion battery cells where Li foil was used as the anode.[36] Then, Li intercalated compounds were formed after intercalating Li ions into layered TMD crystals in the discharge process, which were further sonicated in water or ethanol to obtain suspensions of nanosheets. Before sonication, the intercalated compound on the electrodes should be washed with water

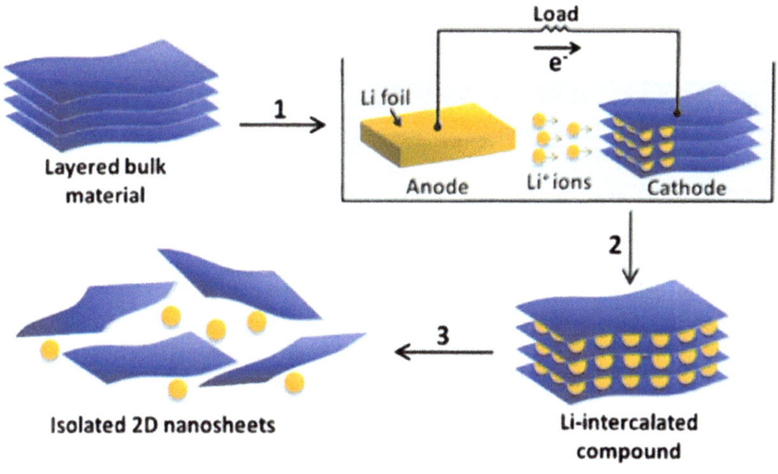

Figure 8.5 Schematic illustration of the electrochemical Li intercalation-assisted liquid exfoliation method for the preparation of single- or few-layer TMD nanosheets, which is carried out in a battery test system. The layered bulk material is incorporated in a test cell as a cathode (1) and lithium foil is used as an anode to provide lithium ions. The lithium intercalation process is conducted at a galvanostatic discharge (2). After completion of the lithium insertion, the intercalated compound (*e.g.* Li_xMoS_2) is washed with acetone to remove the residual electrolyte, and then ultrasonicated in water or ethanol to exfoliate and isolate the nanosheets (3).
Reproduced from ref. 36 with permission form John Wiley and Sons, Copyright © 2011 WILEY-VCH Verlag GmbH & Co. KGaA, Weinheim.

several times to remove the electrolyte. Finally, single- or few-layer nanosheets were produced at high yield after the centrifugation of nanosheet suspensions and purification of nanosheets. In this electrochemical method, the production yield of single-layered nanosheets is very high (*e.g.* >90% for MoS_2 and TaS_2 nanosheets), and the intercalated amount of Li ions and the intercalation speed can be controlled finely by the cut-off voltage and discharge current,[36,37] which is very important to achieve more efficient exfoliation of different layered bulk crystals.

As reported, electrochemical Li intercalation can be completed easily within 6 h during the discharge process at room temperature. Meanwhile, this method has been proven to be very general and effective for exfoliating a large number of other 2D nanosheets with few-layer thickness. The parameters for electrochemical intercalation were also studied systematically to further improve the quality of nanosheets.[38] Up to now, a variety of TMD nanosheets in single and few layers have been successfully exfoliated at high yield and large scale by Li intercalation in the electrochemical system.[36–38] Recently, the electrochemical intercalation method was also used for intercalating sodium ions in MoS_2 interlayers to realize the controversial phase transition from semiconductive 2H to metallic 1T phase.[39] In addition, the negatively charged ions-based electrochemical intercalation method was also employed successfully in Na_2SO_4 solution for exfoliating MoS_2 nanosheets (see Figure 8.6).[40] In this process, OH^-, O^{2-} and ${SO_4}^{2-}$ can be inserted into the MoS_2 layers instead of Li^+ ions. Moreover, the interlayers were greatly expanded *via* the formation of O_2 and/or SO_2 on the anode. After sonication, centrifugation and purification, MoS_2 flakes were obtained with a lateral size of 5–50 μm.

Compared to the chemical ion intercalation method, the assembly of battery cells makes the electrochemical intercalation method much more complicated. Meanwhile, to enhance the conductivity and quality of electrodes some additives were usually mixed with bulk TMD crystals for the fabrication of electrodes, such as activated carbon and polyvinylidenefluoride (PVDF), which may absorb on the exfoliated nanosheets and leave products that are undesirable for some specific applications.

8.4 Mechanical Force-assisted Liquid Exfoliation

In the micromechanical cleavage technique, mechanical force is applied on layered bulk crystals for exfoliating them into 2D nanomaterials at single or few layers. If mechanical forces occur on layered bulk crystals dispersed in liquid media, the layered bulk crystals are similarly exfoliated into ultrathin 2D nanosheets in liquid phase, which is called the mechanical force-assisted liquid exfoliation method. In this section, different types of mechanical forces will be introduced for the high-yield and large-scale production of ultrathin TMD nanosheets in liquid, including sonication-assisted and shear force-assisted liquid exfoliation.

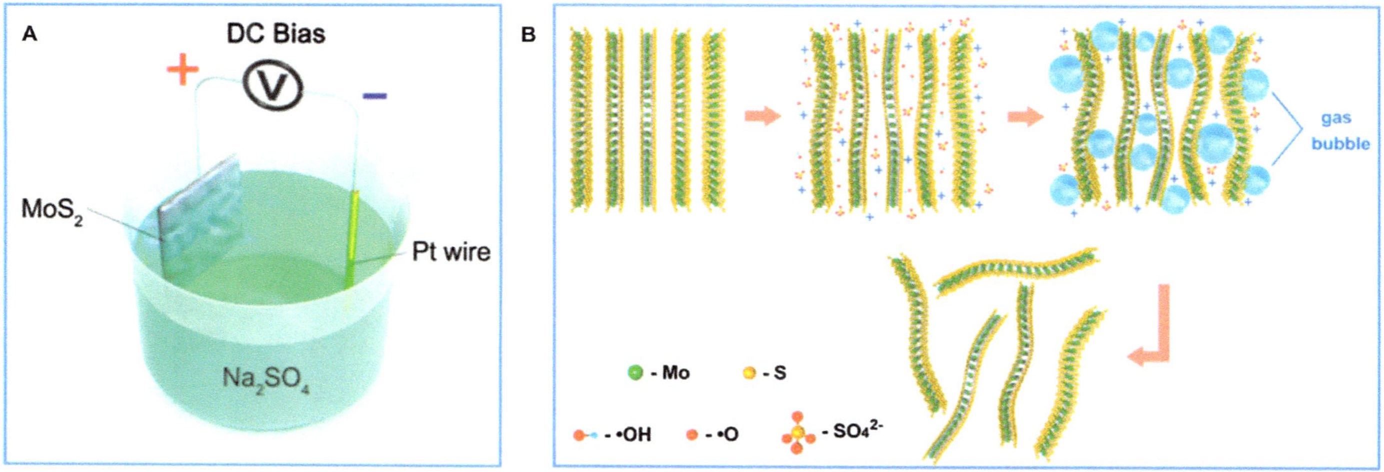

Figure 8.6 (A) Schematic illustration of the experimental setup for electrochemical exfoliation of bulk MoS_2 crystal in Na_2SO_4. (B) The mechanism of electrochemical exfoliation of bulk MoS_2 crystal in Na_2SO_4.
Reprinted with permission from *ACS Nano*, **8**, N. Liu, P. Kim, J. H. Kim, J. H. Ye, S. Kim and C. J. Lee, Large-Area Atomically Thin MoS_2 Nanosheets Prepared Using Electrochemical Exfoliation, 6902–6910, Copyright 2014 American Chemical Society.[40]

8.4.1 Sonication-assisted Liquid Exfoliation

In the exfoliation of ultrathin nanosheets from layered bulk crystals, sonication is the simplest and most common mechanical force in liquid media. Sonication waves can generate bubbles in the solution *via* inducement of liquid cavitation. When these bubbles collapse, microjets and shock waves pass through the layered bulk crystals dispersed in solution. Then, an intensive tensile stress is caused on the layered bulk crystals, which further leads to the exfoliation of thin layers of sheets from layered bulk crystals in solution. Modelling has shown that if the surface energy of the solvent is similar to that of the layered material (good solvent), the energy difference between the exfoliated and re-aggregated states will be very small to remove the driving force for re-aggregation (see Figure 8.7). Otherwise, re-aggregation and sedimentation of nanosheets will occur if the surface energy of the solvent is not similar to that of the layered material (poor solvent). For obtaining stable TMD nanosheets in the "poor" solvent, some special surfactant or stabilizer must be introduced to adjust the surface energy of solvent or stabilize the exfoliated nanosheets *via* the adsorption of a stabilizer (electrostatically or sterically stabilized). Generally, the surfactant or stabilizer can also improve the exfoliation efficiency significantly *via* the interaction with TMD bulk.

8.4.1.1 Exfoliation in a "Good" Solvent

To match the surface energy of the layered bulk crystal with the solvent system is the key factor for achieving efficient exfoliation of layered bulk crystals into ultrathin nanosheets, which significantly minimizes the energy

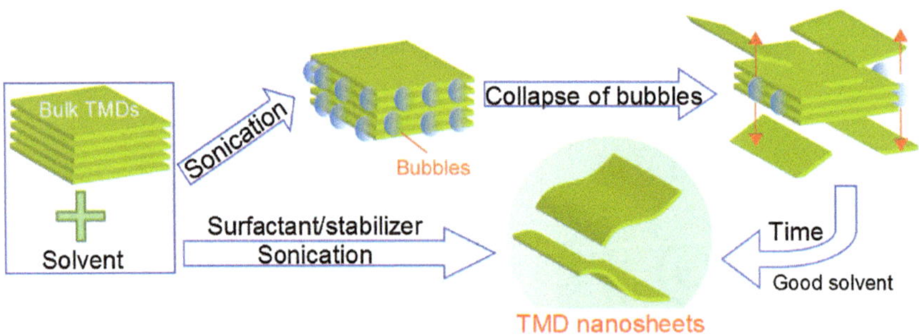

Figure 8.7 Sonication-assisted exfoliation of TMD nanosheets in liquid. The bulk TMD crystal is sonicated in a solvent, resulting in exfoliation and nanosheet formation. The exfoliated nanosheets are stabilized against re-aggregations in the good solvent due to appropriate surface energy. Alternatively, the introduction of a surfactant or stabilizer produces stable nanosheets in good or poor solvent *via* the adjustment of surface energy or the adsorption of the stabilizer.
Reproduced from ref. 23 with permission from the Royal Society of Chemistry.

of exfoliation for efficient exfoliation as indicated by both the experimental and theoretical results. Moreover, the solvent is also very important in stabilizing the exfoliated nanosheets for prohibiting their restacking and aggregating. Among the applicable solvents, *N*-methyl-pyrrolidone (NMP) is the most promising solvent to obtain a nanosheet suspension after sonication and purification. In 2008, Coleman's group first reported sonication-based exfoliation of graphene nanosheets from graphite in NMP.[41] In 2011, they extended this method to the facile exfoliation of other layered crystals and demonstrated that the sonication in NMP is an efficient and general liquid exfoliation method to produce few-layered nanosheets on a large scale.[42] Different types of layered compounds, including MoS_2, WS_2, $MoSe_2$, $NbSe_2$, $TaSe_2$, $NiTe_2$, $MoTe_2$, h-BN, and Bi_2Te_3 were successfully exfoliated into ultrathin nanosheets by using NMP as a good solvent, however, the concentration of TMD nanosheets is usually low ($<1\ mg\ mL^{-1}$). O'Neill *et al.* significantly improved the concentration of MoS_2 nanosheets in NMP ($\approx 40\ mg\ mL^{-1}$) by increasing the sonication time,[43] which also decreases the lateral size of 2D TMD nanosheets due to the scission induced by sonication. In 2016, a mild sonication bath was applied to produce luminescent suspensions of MoS_2 and WS_2 in NMP,[44] which were also able to be re-dispersed into water after centrifugation for further biological applications.

Besides NMP, an enormous number of exfoliation methods have been developed for the production of ultrathin TMD nanosheets by direct sonication of the parent crystals in different common solvents. A recent finding from Zecchina *et al.* revealed that intensive ultrasonication of MoS_2 in isopropyl alcohol (IPA) not only favoured exfoliation, but also induced fragmentation of layers, leading to the rupture of Mo–S–Mo bonds and the exposure of coordinatively unsaturated Mo and S species.[45] In another case, Zhang and co-workers demonstrated that a mixture of ethanol and water with appropriate proportion is effective for exfoliating and dispersing TMD nanosheets.[46] The mixture could convert these two "poor" solvents into "good" solvents in the exfoliation of 2D TMD nanosheets, providing more choice of proper solvent systems for effective exfoliation of TMD nanosheets and their novel applications. In order to distinguish the size effects on the intrinsic properties, they recently developed gradient centrifugation to separate flakes with different sizes (see Figure 8.8).[47] Briefly, the high concentration of TMD suspension was firstly centrifuged at a high rotation rate. The produced supernatant was preserved for later spectroscopic characterization. Meanwhile, the sedimentation was re-dispersed by sonication, and then centrifuged at a lower rate to obtain another batch of dispersed material and sedimentation. This process was repeated so that a series of suspensions with different size or thickness were obtained. The TMD nanosheets prepared by ultrasonication-assisted liquid exfoliation are usually in multilayer forms, whereas single-layered nanosheets are relatively few. Zhao and coworkers recently developed a simple and efficient strategy to synthesize single-layer MoS_2 nanosheets at high temperature *via* an improved oleum-treated liquid exfoliation,[48] and observed that the appropriate

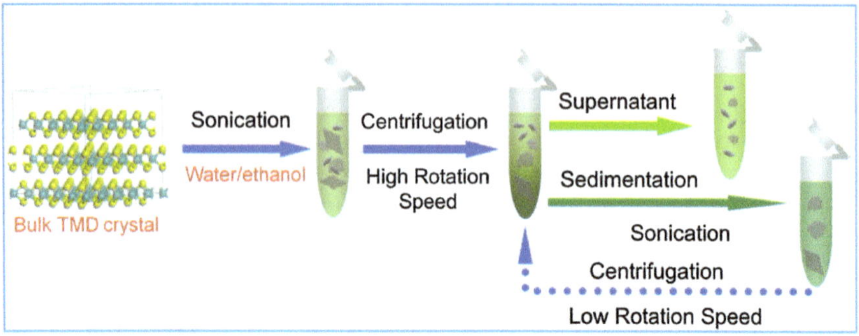

Figure 8.8 The gradient centrifugation procedure to separate TMD flakes with different sizes.
Reproduced from ref. 47 with permission from John Wiley and Sons, © 2014 WILEY-VCH Verlag GmbH & Co. KGaA, Weinheim.

grinding of powder before sonication can facilitate the exfoliation of MoS_2 nanosheets.

Owing to the large surface energy, pure water is always believed to be inefficient to directly exfoliate TMDs in liquid. Experimental observation also indicated pure water was insufficient for the exfoliation of nanosheets from layered bulk crystals at room temperature. Fortunately, a recent study has proven that when simply heating water to 60 °C, pure water will change into a promising solvent for the sonication-assisted exfoliation of TMD nanosheets.[49] At elevated temperature, water becomes effective for the exfoliation of nanosheets from layered TMD crystals during the sonication process, and the exfoliated nanosheets are stable *via* the presence of platelet surface charges, which are induced by edge functionalization or intrinsic polarity. Similarly, Forsberg *et al.* also reported the direct mechanical exfoliation of MoS_2 nanosheets in water,[50] although the concentration of the obtained nanosheets is only half the value obtained by exfoliation of MoS_2 nanosheets in organic solvents under the same conditions.

8.4.1.2 Exfoliation in the Presence of a Surfactant/Polymer or Stabilizer

As mentioned above, effective exfoliation of TMD nanosheets can only be carried out in the "good" solvents, which have a surface energy matching the layered TMD crystals. Due to the fact that the surface energy varies for different bulk crystals, it is difficult/impossible to find a suitable solvent for each layered crystal. Instead, the addition of surfactants/polymers or stabilizers is a very promising way for exfoliation of TMD bulk into ultrathin nanosheets *via* sonication of bulk crystals in solution.[51] On the one hand, surfactants/polymers can easily tune the surface tension of the aqueous solution to match the surface energy of layered bulk crystals, which is very important for achieving efficient exfoliation of layered materials.[23] On the other hand, the concentration, lateral size and layer number of the obtained nanosheets can be tuned roughly by controlling the shape of vessels, the solvent system, polymers' additive, and ultrasonic power/time/temperature. Alternatively, various stabilizers (*e.g.* biomolecules) can also be used to effectively exfoliate TMD nanosheets in solution under sonication, which benefits from the adsorption of biomolecules on the surface of TMD nanosheets and the electrostatic or steric stabilization of nanosheets. Over the past decade, ionic surfactants,[52] polymers,[53–55] and biomolecules[56–58] have extensively been reported to be used in the liquid exfoliation of TMD nanosheets from their bulky crystals. Table 8.1 shows the typical additives for exfoliating TMD nanosheets in solution under sonication and stabilizing the nanosheets after exfoliation.

The use of ionic surfactants sodium cholate was first reported by Coleman and co-workers to directly exfoliate bulk crystals into few-layered TMD nanosheets in water under ultrasonication.[52] In this process, *via* the van der Waals force, sodium cholate molecules would bind onto the exfoliated nanosheets and subsequently stabilize the nanosheets due to electrostatic repulsion. Similarly, polyvinylpyrrolidone (PVP) was used to exfoliate $MoSe_2$ nanosheets in water and was non-covalently modified onto nanosheets simultaneously to form hydrophilic PVP-coated $MoSe_2$ nanosheets.[53] It was proved that the PVP-coated nanosheets have high photothermal conversion efficiency and good biocompatibility. To further broaden the bio-applicability of 2D TMDs, the use of natural stabilizers, typically biomolecules, has also been demonstrated for the exfoliation and stabilization of TMD nanosheets. As one of the most important and interesting exploitations, protein bovine serum albumin (BSA) was found to be very effective for the exfoliation of various TMD nanosheets due to its rich functional groups.[56,61,62] In this field, Guan *et al.* first reported a facile method to exfoliate various layered materials into monolayer nanosheets, including MoS_2, WS_2, WSe_2, graphite, *etc.*, by using BSA as an effective exfoliating agent.[62] As indicated by theoretical simulations, the nonpolar groups in BSA

Table 8.1 Summary of the surfactant/polymers and biomolecules that have been reported to exfoliate TMD nanosheets under sonication and well disperse the 2D TMDs in solution.[30,59,60]

Surfactant/polymer	TMDs	Biomolecule	TMDs
Sodium cholate	MoS_2	Bovine serum albumin	MoS_2, WSe_2, WS_2
Tween 80	MoS_2	Silk fibroin	$MoSe_2$
Tween X-100	WS_2	Gelatin	MoS_2, WS_2
Sodium dodecylsulfate	MoS_2	DNA/RNA	WS_2, WSe_2
Gum arabic	MoS_2, WS_2	Cellulose	MoS_2
Pluronic P-123	MoS_2, WS_2	Chitosan	MoS_2
n-Dodecyl β-D-maltoside	MoS_2, WS_2	Tannic acid	MoS_2
Poly(vinylpyrrolidone)	$MoSe_2$	Hyaluronic acid	MoS_2
Polyvinylalcohol	MoS_2	Amino acid	MoS_2

can firmly bind on the surface of bulk crystals while the polar groups in BSA were exposed to water, which facilitates the exfoliation process under sonication to produce monolayer nanosheets in aqueous solution (see Figure 8.9). In this work, the exfoliation and functionalization of TMD nanosheets were realized by a one-step process and the obtained BSA-coated TMD nanosheets had high stability and good biocompatibility as suitable photothermal agents in biological systems.[63]

Compared to the micromechanical cleavage method, sonication-assisted liquid exfoliation provides a much higher production rate for various TMD nanosheets, which might be the most widely used approach for the high-yield and massive production of dispersed ultrathin nanosheets in solution due to its simple process and low cost. However, there are several disadvantages for the sonication-assisted liquid exfoliation method: (1) the yield of the single layer nanosheets is low; (2) the lateral size of the produced nanosheets is relatively small (typically 200–400 nm) because of the splitting of big nanosheets by the sonication force; (3) some defects may be generated on the exfoliated nanosheets during the sonication process, which will compromise some special properties of the exfoliated nanosheets; and (4) after using surfactant/polymer or stabilizer, the residual absorbers on the

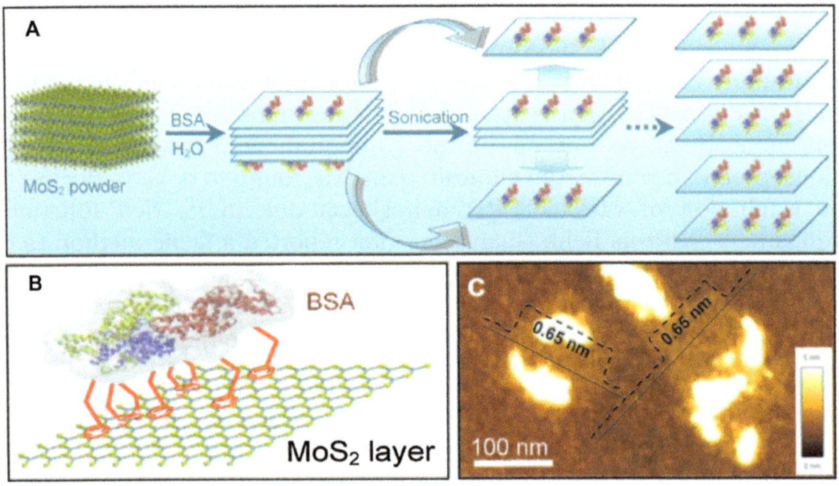

Figure 8.9 (A) Schematic for the BSA-induced exfoliation of single-layer MoS_2 nanosheets under sonication. (B) The simulated binding of BSA on a MoS_2 layer with benzene rings and disulfides, based on a calculated result. (C) AFM image for showing the thickness of exfoliated MoS_2 nanosheets and the binding of BSA.
Reprinted with permission from *J. Am. Chem. Soc.*, **137**, G. Guan, S. Zhang, S. Liu, Y. Cai, M. Low, C. P. Teng, I. Y. Phang, Y. Cheng, K. L. Duei, B. M. Srinivasan, Y. Zheng, Y. W. Zhang and M. Y. Han, Protein Induces Layer-by-Layer Exfoliation of Transition Metal Dichalcogenides, 6152–6155, Copyright 2015 American Chemical Society.[62]

exfoliated nanosheets will hinder the application of TMD nanosheets in some fields such as electrocatalysis, electronics, optoelectronics and energy storage.

8.4.2 Shear Force-assisted Liquid Exfoliation

In the exfoliation of graphene *via* sonication-assisted liquid exfoliation, the concentration of graphene nanosheets can be up to $\sim$1 mg mL^{-1} after optimizing various reaction conditions. However, the production yield and rate still cannot meet the requirements in industrial applications. For high yield and massive production of graphene flakes in solution, a high-shear rotor/stator mixer was used to develop the shear force-assisted liquid exfoliation method with a higher production rate.[64] Under high speed rotation, the mixer can generate high shear rates in liquid and trigger the exfoliation of layered bulk crystals in liquid by the shear forces. When treating the aqueous solution containing bulk TMD crystals and ionic surfactant by a shear force or a compression force, TMD nanosheets can similarly be exfoliated by the shear forces.[65,66] For example, Yao *et al.* demonstrated an efficient and scalable method for the fabrication of monolayer and few-layer MoS_2 in sodium dodecylsulfate (SDS) solution using combined low-energy ball milling and sonication.[65] Ball milling generates two forces on layered materials, shear force and compression force, which can cleave layered materials into ultrathin nanosheets from the top/bottom surfaces, and the edge of layered materials. Subsequent sonication would further break larger crystallites into smaller crystallites to fabricate nanosheets well dispersed in aqueous solutions at high concentrations. Another typical work was reported by Coleman and co-workers, to demonstrate the large-scale shear-exfoliation of MoS_2 nanosheets at the litre scale in aqueous surfactant solution using a kitchen blender.[66] They also suggested that large-scale shear exfoliation can be applied to the exfoliation of a wide range of layered compounds such as WS_2 and BN.

8.5 Wet-chemical Synthesis Method

In contrast to liquid-based exfoliation from layered bulk crystals to nanosheets (top-down), wet-chemical synthesis of nanosheets is a bottom-up process to liquid-phase produce a large amount of ultrathin TMD nanosheets at high yield,[67–69] which includes all of the synthetic methods *via* the chemical reaction of metal salts as precursors in the solution phase.[70] The wet-chemical synthesis is a controllable and reproducible method for the scalable production of TMD nanosheets in industrial applications. And, the lateral size and layer thickness of the nanosheets can be conveniently tuned by choosing the proper precursors and optimizing the reaction conditions. Moreover, the synthesized nanosheets can easily be dispersed in organic or aqueous media, which make them very suitable for various applications.[68] Based on the solvent used and the methods of heating, wet-chemical synthesis of 2D TMD nanomaterials can generally be classified into the hydrothermal method, the solvothermal method and hot-injection method.[71] The hydro and solvo thermal synthesis methods use water and an organic solvent, respectively, as the reaction medium, and a high temperature is applied for performing their reactions in a sealed vessel.[72] Generally, the reaction temperature used is higher than the boiling point of the solvents, leading to the auto-generation of a solvent at high pressure for promoting the reaction speed and improving the crystalline quality of the as-prepared products. In the synthesis of various TMD nanomaterials with different morphologies and crystallinity, wet-chemical protocols have been widely used and shown some specific advantages.[69] Herein, we will introduce the typical wet-chemical strategies for the production of ultrathin TMD nanosheets and discuss their advantages and limitations. With the description of a different strategy, some representative examples will also be given for understanding the formation process/mechanism of ultrathin TMD nanosheets in extreme conditions.

8.5.1 Hydrothermal Synthesis

The hydrothermal method, compared with the solvothermal approach, is more attractive owing to its simplicity and wide applicability.[71,73] As a representative example, defect-rich MoS_2 nanosheets at few layers were facilely synthesized at 220 °C for 18 h by using hexaammonium heptamolybdate tetrahydrate ($(NH_4)_6Mo_7O_{24}\cdot 4H_2O$) as a precursor with the assistance of excess thiourea.[74] In this process, thiourea not only acted as a reductant to reduce Mo(VI) to Mo(IV), but also acted as an additive to stabilize ultrathin MoS_2 nanosheets. Using the same precursors and a similar hydrothermal method, the oxygen-incorporated MoS_2 nanosheets were further prepared at 200 °C for 24 h,[75,76] and an enlarged interlayer spacing was observed as compared to the pristine 2H-MoS_2 (9.5 Å to 6.15 Å). Interestingly, a second solvothermal treatment can partially transform the 2H-MoS_2 nanosheets into the metallic 1T phase to show robust ferromagnetism at room temperature.[76] Different hydrothermal conditions (temperature, pH, time) were further investigated for revealing their effect on the structure of MoS_2 nanosheets by using $(NH_4)_6Mo_7O_{24}$ and $SC(NH_2)_2$ as reactants.[77] In this work, MoS_2 nanosheets with different structures (intercalated, intermediate and well crystalline) were synthesized by varying the conditions for crystallization. In anthracene hydrogenation, the measured results showed that the intermediate MoS_2 nanosheets exhibited the highest catalytic activity, with associated stability and abundant active sites.

Other precursors were also used to synthesize TMD nanosheets using hydrothermal methods. Typically, Chen *et al.* produced MoS_2 and $MoSe_2$ nanosheets *via* the chemical reaction of transition metal salts (Na_2MoO_4) with $Na_2S_2O_3$ (or Na_2SeSO_3) with the presence of hydrazine at ~140 °C.[78] The synthesized MoS_2 and $MoSe_2$ nanosheets had an average size of several nanometres. In another research, Na_2MoO_4 was used to synthesize MoS_2 in the solution-phase *via* reduction of Mo(VI) by L-cysteine, which acted as the sulfide source as well as the reducing agent.[79] Figure 8.10 shows the reaction conditions and experimental process in the hydrothermal method. In addition, Rao and co-workers demonstrated the preparation of few-layered MoS_2 nanosheets by reacting MoO_3 with KSCN in water at 180 °C.[67] Recently, Shi and co-workers used an "integrated" precursor $(NH_4)_2MoS_4$ containing both Mo and S elements to synthesize MoS_2 nanosheets in PEG solution *via* a one-pot route.[80] It was evidenced that PEG chains can strongly bind on the surface of the nanosheets to benefit the controllability of nanosheet size and enhance the stability and biocompatibility of nanosheets in drug-delivery systems, as well as photothermal therapy.

8.5.2 Solvothermal Synthesis

In contrast to the hydrothermal method, solvothermal synthesis uses high boiling point organic solvents (*e.g.* oleylamine) to replace water as the reaction medium in the preparation of 2D TMD nanomaterials. To date, this method is extensively utilized to synthesize TMD nanosheets for biomedical applications.[15,81,82] As is known, organic ligands are usually very important to improve the dispersibility of TMD nanomaterials and control their morphology and size. For example, Altavilla *et al.* performed the preparation of free-standing nanosheets of MoS_2 and WS_2 with high quality *via* the thermal decomposition of single-source precursors ($(NH_4)_2MS_4$, M = Mo or W) in oleylamine at 360 °C (see Figure 8.11),[83] where the layer numbers of TMD nanosheets were related to the reaction time. Mahler and co-workers developed a novel synthesis route for preparing WS_2 monolayers by the chemical reaction of WCl_6 and CS_2 at 320 °C under the assistance of oleylamine.[84] The experimental results showed that WS_2 monolayers with 1T structure were obtained by using tungsten precursors with low reactivity, while WS_2 monolayers with 2H structure were obtained by using precursors with higher reactivity. In addition, a small amount of hexamethyldisilazane (HMDS) was found to be beneficial for the generation of 2H-WS_2. Recently, metal ion-doped WS_2 and TiS_2 nanosheets were successfully synthesized through a quick reaction of metal chloride with oleylamine followed by the subsequent addition of sulfur powder.[85,86] Furthermore, uniform ultrathin ReS_2 nanosheets were also produced in oleylamine through the solvothermal reaction of $NaReO_4$ and sulfur,[87] which were further functionalized with PEG for strong NIR light absorption and high X-ray attenuation. It is noteworthy that the solvothermal synthesis of TMD nanosheets is usually carried out in a flask under argon, which is different to hydrothermal synthesis that occurs in a sealed autoclave.

Hydro/solvothermal methods are simple and low-cost strategies for the high-yield and scalable production of ultrathin 2D nanosheets in liquid. However, it is impossible to systematically observe the evolution of the reaction solution occurring in the sealed autoclave, leading to difficulties in figuring out the growth mechanism during the hydrothermal synthesis process and designing optimized experimental conditions for other

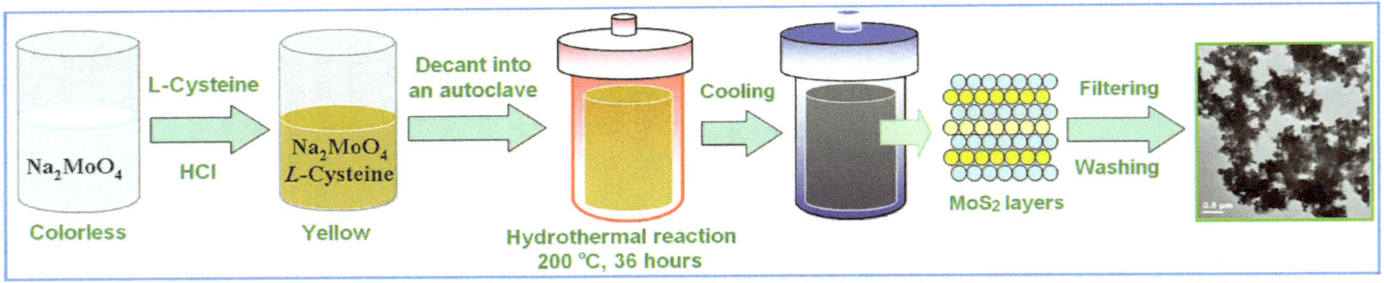

Figure 8.10 Preparation of MoS_2 using a hydrothermal method. $Na_2MoO_4{\cdot}2H_2O$ was dissolved in water and HCl was added to this solution to adjust the pH to 6.5, followed by the addition of L-cysteine. The mixture was transferred to a Teflon-lined, stainless steel autoclave and treated for 36 h at 200 °C. The mixture was allowed to cool and the black precipitate was collected to obtain MoS_2 nanosheets.
Reproduced from ref. 79 with permission from the Royal Society of Chemistry.

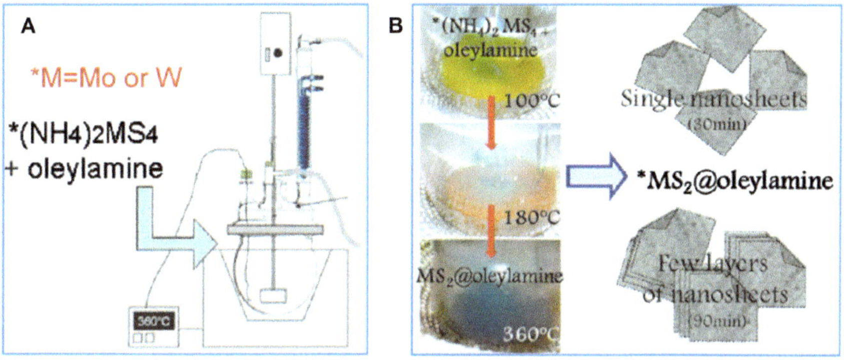

Figure 8.11 (A) Scheme of the apparatus for the synthesis of MS_2@oleylamine nanosheets by thermal decomposition of thio-salts of Mo and W in the presence of oleylamine. (B) Colour changes of the reaction mixture during the synthesis of WS_2@oleylamine after the decomposition. Reprinted with permission from *Chem. Mater.*, **23**, C. Altavilla, M. Sarno and P. Ciambelli, A Novel Wet Chemistry Approach for the Synthesis of Hybrid 2D Free-Floating Single or Multilayer Nanosheets of MS_2@oleylamine (M = Mo, W), 3879–3885, Copyright 2011 American Chemical Society.[83]

nanomaterial systems. Moreover, it is also hard to precisely control the morphology and structure of the obtained products in different laboratories or different batches by the same group. This is because a slight change of experimental conditions will have a great influence on the morphology and structure of the obtained products in hydro/solvothermal synthesis. For the solvothermal method, the ligands will absorb at the surface of TMD nanosheets, which are usually difficult to remove completely for further application. In addition, it is still a challenge to control the layer number of TMD nanosheets in the hydro/solvothermal synthesis method. Most of the synthesized 2D nanosheets were few-layer rather than single-layer. Recently, a modified solution-phase synthetic route was developed to synthesize group IV metal sulfides nanosheets in a single layer, which was called "diluted chalcogen continuous influx (DCCI)".[88] The interesting and important finding is that single layered nanosheets were obtained through the selective growth along the lateral directions by slow and continuous influx of chalcogen sources into the reaction system, whereas multi-layered nanosheets were synthesized through the growth in both the lateral and vertical directions by a burst influx of the chalcogen sources into the reaction system.

8.5.3 Hot-injection Method

In the hot-injection method, solutions of oleylamine or oleic acid are first heated to high temperature and then some highly reactive reactants are rapidly injected into the hot solution to drive a burst reaction by high initial supersaturation. This approach is very attractive for the production of various colloidal nanocrystals, which have a uniform shape/size, good monodispersity and high crystalline purity.[89] Recently, a variety of ultrathin inorganic nanosheets have been synthesized *via* the hot-injection method, including nanosheets of layered TMDs.[90,91] For example, Cheon *et al.* used the hot-injection method to synthesize single-layered $MoSe_2$ and WSe_2 nanosheets, with lateral sizes in the range of 200 to 400 nm.[92] One of their important findings is that the layer number of the nanosheets can be controlled by different capping ligands. When oleic acid was replaced with oleyl alcohol and oleylamine as the capping ligand, the nanosheet thickness could be tuned from single layer to 2–3 layers and 4–8 layers, respectively. Similarly, CS_2 and Se powders were also used as injection chalcogen precursors to synthesize other transition metal sulfide and selenide nanosheets.[88,93] In another interesting exploitation, Ozin *et al.* found that the crystal phase of WS_2 nanosheets could be tuned by simply changing the injecting precursors.[84] Typically, single-layered metallic 1T-WS_2 nanosheets with a round-shaped morphology and a diameter of ~100 nm were synthesized when rapidly adding CS_2 and WCl_6 precursors into hot oleylamine solution at 320 °C. In contrast to that, flower-like 2H-WS_2 nanosheets were obtained when a small amount of hexamethyldisilazane was introduced into the reaction solvent (Figure 8.12).

By using the hot injection approach, different 2D TMD nanomaterials were synthesized directly after optimizing various reaction conditions, which generally have a uniform size/shape, good monodispersity and high crystalline purity. However, it is not easy to completely remove long-chain

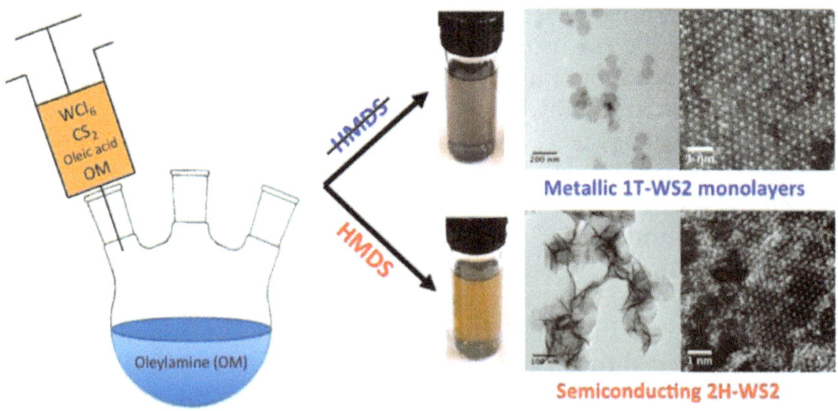

Figure 8.12 Schematic illustration of the synthesis of 2H and 1T WS_2 nanosheets by the hot-injection method, and TEM observations of the obtained products.
Reprinted with permission from *J. Am. Chem. Soc.*, **136**, B. Mahler, V. Hoepfner, K. Liao and G. A. Ozin, Colloidal Synthesis of 1T-WS2 and 2H-WS2 Nanosheets: Applications for Photocatalytic Hydrogen Evolution, 14121–14127, Copyright 2014, American Chemical Society.[84]

surfactants (*i.e.* oleylamine and oleic acid) used in the hot-injection process from the obtained TMD nanosheets, which normally limits the applications of TMD nanosheets for high performance. Moreover, a solvent with high boiling point is often used in the system, and thus the reaction temperature needs to reach 300 or 350 °C. The relatively high reaction temperature usually cannot be used for large-scale production because of the limits of the precursor injection. On the basis of these concerns, the solvothermal synthesis would be better suited for large-scale production of TMD nanosheets in the one-pot heating method.[83] All of the reactants are first mixed in the solvent with high boiling point at room temperature and then the mixture is heated to a high temperature to trigger a chemical reaction for obtaining the final products.

8.6 Chemical Vapour Deposition

Chemical vapour deposition (CVD) is applicable in the synthesis of materials *via* the chemical reaction occurring between precursors and one or more gases. This is usually carried out in a furnace chamber containing the reaction precursors and a properly selected substrate.[94] When the vapour precursors are poured into the chamber, a chemical reaction will occur under special experimental conditions and the reaction product will be deposited at the surface of the substrates to obtain high purity materials or thin films. As a very popular and traditional technique, the CVD method has widely been used in the growth of metal and inorganic thin films on various substrates over the last few decades. Recent research also proved that ultrathin 2D nanomaterials in single or few layers can similarly be produced on different substrates by the CVD method.[95,96] For example, a solution of $(NH_4)_2MoS_4$ was first coated onto insulating substrates and then few-layered MoS_2 nanosheets successfully grew on the substrates through thermal decomposition and subsequent sulfurization of $(NH_4)_2MoS_4$ in a sulphur vapour atmosphere at 1000 °C (see Figure 8.13).[97] In another work, Mo metal film, 1–5 nm in thickness, was first deposited on SiO_2 substrate by using an electronic beam evaporator, and then the large-area growth of few-layered MoS_2 nanosheets was performed *via* sulfurization of Mo film in sulphur vapour at 750 °C.[98] The experimental results indicated that the thickness and the size of the Mo film determined the thickness and size of the nanosheet, respectively, so, the thickness and the size of the MoS_2 nanosheets were conveniently tuned *via* controlling the size of the substrate and the thickness of the Mo film. Meanwhile, the annealing treatment in sulphur vapour proved to be very important for improving the quality of the crystal structure and the electrical performance of MoS_2 nanosheets.

Besides $(NH_4)_2MoS_4$ and Mo metal, other precursors containing a transition metal element, particularly transition metal oxides (*i.e.* MoO_3) were also used successfully in CVD approaches for the synthesis of ultrathin TMD nanosheets on substrates.[99,100] In 2012, Balendhran *et al.* reported the simultaneous thermal evaporation of MoO_3 nanoparticles and sulphur

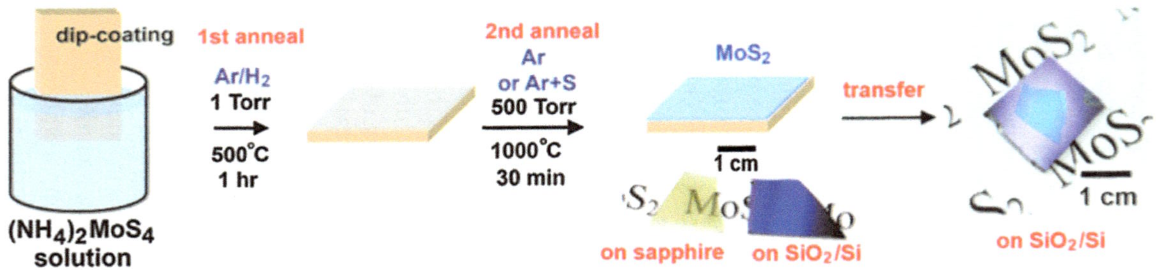

Figure 8.13 Schematic illustration of the two-step thermolysis process for the synthesis of thin MoS_2 layers on insulating substrates. The precursor $(NH_4)_2MoS_4$ was dip-coated on SiO_2/Si or sapphire substrates followed by a two-step annealing process. Reprinted with permission from *Nano Lett.*, **12**, K. K. Liu, W. Zhang, Y. H. Lee, Y. C. Lin, M. T. Chang, C. Y. Su, C. S. Chang, H. Li, Y. Shi, H. Zhang, C. S. Lai and L. J. Li, Growth of Large-Area and Highly Crystalline MoS_2 Thin Layers on Insulating Substrates, 1538–1544, Copyright 2012 American Chemical Society.[97]

powder at 830 °C to prepare monolayer MoS_2 nanosheets in a sulphur-rich environment.[101] By using the same precursors but a lower temperature of about 650 °C, Lee and co-workers directly synthesized MoS_2 nanosheets with a lateral size up to 2 mm on SiO_2/Si substrates.[99] Unexpectedly, non-uniform layer thickness was observed on the large area MoS_2 nanosheets, including single-layer, bilayer and few-layer. Recently, a two-step thermal process was reported for the synthesis of wafer-scaled MoS_2 thin films with a uniform thickness *via* the CVD method: MoO_3 materials were firstly reduced in a H_2/Ar atmosphere at 500 °C, and then the reduced intermediate product was sulfurized in the presence of sulphur at 1000 °C.[102] As a clear advantage in this method, it is very easy to transfer the synthesized MoS_2 nanosheets onto other substrates regardless of any materials. In another interesting work, Zhang and co-workers demonstrated a low-pressure CVD method for the growth of WS_2 flakes on sapphire substrate.[103] The size and the layer thickness (single or few layers) of the WS_2 nanosheets can be conveniently controlled by altering the growth temperature and the distance of the source from the substrate. In the optimized conditions, the as-synthesized WS_2 flakes exhibited the shape of a jagged triangle ~50 μm in size. When introducing a mixture of H_2 and Ar gases in the chamber, the shape of the WS_2 nanosheets became straight-edge triangles and had a single crystalline structure (Figure 8.14).

By varying the types of precursors, temperatures of thermal evaporation, flowing rate of gas, and material of substrates, a variety of ultrathin TMD nanosheets in single or few layers have been synthesized *via* the CVD technique, including MoS_2, WS_2, $MoSe_2$, WSe_2, ZrS_2, ReS_2, $MoTe_2$, *etc.*[15,97–103] Besides pure TMD nanosheets, some alloyed TMD nanosheets (*e.g.* $MoS_{2x}Se_{2(1-x)}$, $Mo_xW_{1-x}S_2$, and $WS_{2x}Se_{2(1-x)}$) have also been synthesized by using mixed precursors in the CVD technique, which also provide a novel way for tuning the bandgap of 2D nanomaterials because the bandgap is significantly dependent on the ratio of elemental compositions in alloyed TMD nanosheets.[15] In addition, the nanosheets of TMD

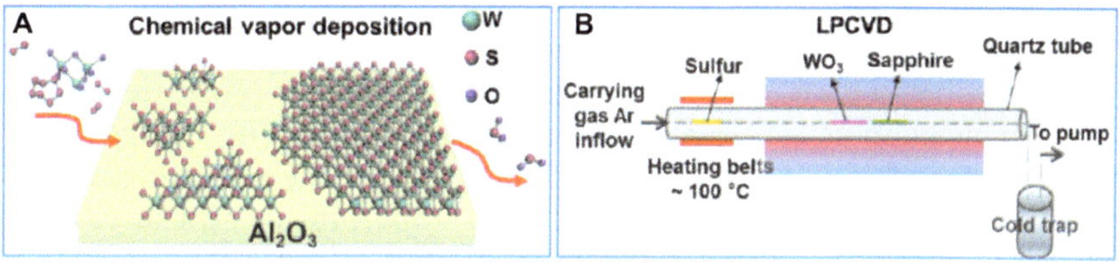

Figure 8.14 Low-pressure CVD synthesis of WS_2 on sapphire (Al_2O_3). (A) Schematic view of the related chemical reaction. (B) Experimental setup of the low-pressure CVD system.
Reprinted with permission from *ACS Nano*, 7, Y. Zhang, Y. Zhang, Q. Ji, J. Ju, H. Yuan, J. Shi, T. Gao, D. Ma, M. Liu, Y. Chen, X. Song, H. Y. Hwang, Y. Cui and Z. Liu, Controlled Growth of High-Quality Monolayer WS_2 Layers on Sapphire and Imaging Its Grain Boundary, 8963–8971, Copyright 2013 American Chemical Society.[103]

heteronanostructures (*e.g.* WSe_2–WS_2, $MoSe_2$–MoS_2, and $MoSe_2$–WSe_2) have also been synthesized using the CVD technique, and this was carried out by a two-step method: one TMD was first grown on the substrate and then another TMD was grown epitaxially on the existing TMD seed.[15] By further tuning the experimental conditions, the size and thickness of the TMD heteronanosheets can also be altered to obtain lateral and vertical heteronanostructures *via* controlling the lateral and vertical growth of TMDs. Because different TMD nanosheets exhibited different band gaps, lateral and vertical TMD heteronanosheets can be regarded as natural p–n junctions with atomic thickness for the construction of high-performance electronic devices.

As a non-liquid technique, micromechanical cleavage only obtains single- or few-layered TMD nanosheets at a low yield and low production rate. In contrast, the CVD technique is able to achieve scalable and massive production of high-purity 2D TMD nanomaterials with high crystal quality, which are promising candidates for the fabrication of high-performance electronic and optoelectronics devices. Meanwhile, the CVD technique has the highest level of control for the size, thickness and shape of ultrathin 2D nanomaterials among all the other synthetic methods. Therefore, the CVD technique is believed to be a promising method with the potential to synthesize ultrathin TMD nanosheets for industrial applications. However, as-synthesized 2D TMD nanomaterials are always deposited on special substrates by the CVD technique. In the next investigation and further applications, TMD nanomaterials need to be transferred carefully onto other substrates. Another disadvantage in the CVD method is its relatively high cost for producing ultrathin TMD nanosheets due to the usage of unique equipment and the harsh reaction conditions such as higher temperature and inert atmospheres.

8.7 Chemical Functionalization and Metal Hybridization of 2D TMD Nanosheets

With fascinating physical and chemical properties, single- and few-layered TMD nanosheets have exhibited numerous outstanding properties in some applications such as sensing, catalysis, and composite and energy storage. Proper surface functionalization can further render the required properties onto TMD nanosheets, which may play an important or critical role for a specific application. The surface functionalization not only can tune the properties of TMD nanosheets effectively, but can also broaden the scope of their applications significantly. Based on the interaction between ligands and TMDs, the functionalization of TMD nanosheets can be performed by physical and chemical routes. Among them, physical adsorption can be achieved during the exfoliation of TMD nanosheets, and the adsorbed ligands are generally both exfoliating agents and stabilizing agents. The related research works have been introduced as the sonication-based exfoliation of

TMD nanosheets under the assistance of a stabilizer. In this section, we will further demonstrate the chemical functionalization of TMD nanosheets after obtaining TMD nanosheets. Chemical functionalization of TMD nanosheets is usually realized through grafting organic molecules or atoms on the surface of TMD nanosheets *via* the formation of covalent bonds or coordinate bonds. Differing from the weak interaction in physical absorption, the strong attachment from chemical bonds between functional groups and TMD nanosheets is not dependent on surrounding environments (*i.e.* a very stable configuration). In addition, when the modified molecules have some unique functional groups, the functionalized TMD nanosheets can become chemical reactive 2D nanomaterials to largely facilitate their promising applications in more fields. In this section, we will introduce novel chemical routes to modify the surface of TMD nanosheets. Meanwhile, the hybridization of TMD nanosheets with metal nanoparticles is also summarized briefly.

8.7.1 Covalent Functionalization

The atoms on the basal plane are usually inert in chemical reaction due to their saturated nature, so covalent functionalization generally occurs on the edge atoms of TMD nanosheets with a higher reaction activity, particularly for the intact crystalline 2H-MoS_2 nanosheets obtained *via* non-intercalation methods. Typically, Leite *et al.* reported the covalent functionalization of 2H-MoS_2 nanosheets by mixing liquid-exfoliated 2H-MoS_2 nanosheets with polybutadiene in NMP/toluene.[104] The chemical functionalization only occurred on the edge sites of 2H-MoS_2 nanosheets, and the formation of a C–S bond (at 630 cm^{-1}) was verified by FTIR analysis. Similarly, Jeong *et al.* functionalized 2H-MoS_2 nanosheets on their edges with lipoic acid (LA) *via* a simple sonochemical aqueous dispersion procedure.[105]

On the basal planes, covalent functionalization of TMD nanosheets is intrinsically dependent of the presence of defects or vacancies. As mentioned above, Li intercalation-exfoliated TMD nanosheets exhibit 1T phase with structural and electronic deformations from their bulk, making them applicable for surface functionalization. So, Li intercalation-based TMD nanosheets are usually used to produce functionalized TMD nanosheets. For example, Backes and co-workers reported the successful functionalization of diazonium salts onto the basal plane of 1T-MoS_2 nanosheets *via* the quenching of residing surface negative charges (see Figure 8.15).[106] The extent of functionalization was controllable through varying the intercalating conditions, which could be up to 20 atoms% at an optimized condition. In contrast to the chemical functionalization on sulphur vacancies, Chhowalla and co-workers developed another effective grafting route for the covalent modification of functional groups to sulphur atoms in the basal plane of TMD nanosheets, which is based on electron transfer between 1T-TMD nanosheets and organohalides.[107]

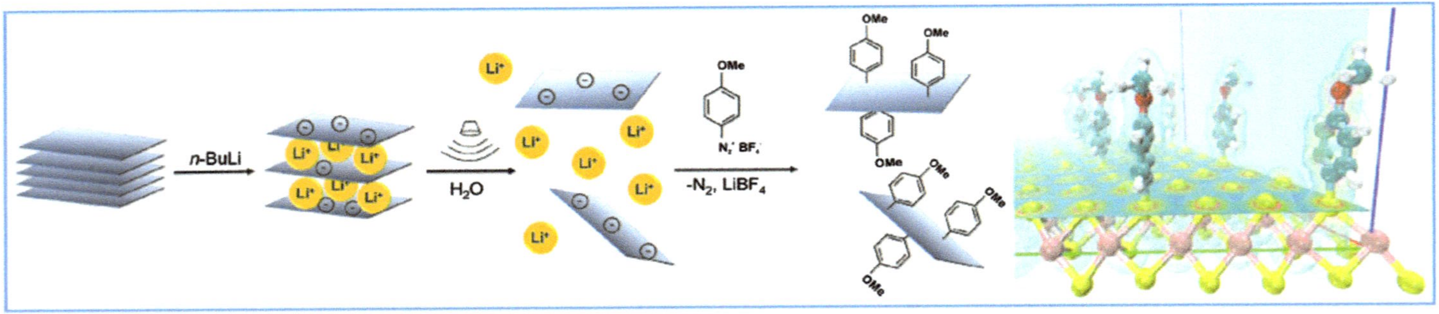

Figure 8.15 Schematic representation of the basal-plane functionalization of MoS_2. After intercalation with *n*-BuLi, the negatively-charged MoS_2 is dispersed in water by a mild bath-type sonication leading to an efficient exfoliation into individual sheets. The charges on the MoS_2 are quenched by the addition of 4-methoxyphenyldiazonium tetrafluoroborate obtaining functionalized MoS_2 nanosheets.
Reprinted with permission from *ACS Nano*, **9**, K. C. Knirsch, N. C. Berner, H. C. Nerl, C. S. Cucinotta, Z. Gholamvand, N. McEvoy, Z. Wang, I. Abramovic, P. Vecera, M. Halik, S. Sanvito, G. S. Duesberg, V. Nicolosi, F. Hauke, A. Hirsch, J. N. Coleman and C. Backes, Basal-Plane Functionalization of Chemically Exfoliated Molybdenum Disulfide by Diazonium Salts, 6018–6030, Copyright 2015 American Chemical Society.[106]

8.7.2 Coordinate Functionalization

In thiol chemistry, ligand conjugation has been demonstrated to be an effective and convenient method for the chemical functionalization of TMD nanosheets. For example, Chou and co-workers realized the surface functionalization of 1T-MoS_2 nanosheets by bifunctional thiolated polyethylene glycol (PEG) ligands. The PEG molecule has a thiol group at one end for ligand conjugation on 1T-MoS_2 nanosheets and other groups (carboxyl and hydroxyl, *etc.*) at another end to stabilize/functionalize the nanosheets.[108] As an important extension, various small bifunctional molecules containing thiol groups can also be modified onto the surface of 1T-MoS_2 nanosheets by a similar procedure, including l-cysteine, 1-thioglycerol and mercaptopropionic acid.[109] These bifunctional ligand-modified 1T-MoS_2 nanosheets exhibited high chemical reactivity, which could be further functionalized with other target elements by proper chemical reactions for more complex applications. Compared to a single thiol ligand, two sulphur atoms will have a much stronger affinity for binding on MoS_2 nanosheets, which was used to modify LA-conjugated PEG onto the surface of 1T-MoS_2 nanosheets.[110] Besides 1T-MoS_2 nanosheets, 2H-MoS_2 nanosheets were also used for surface functionalization *via* thiol conjugation. For example, dibenzothiophene (DBT) molecules were successfully introduced onto the corner defect sites in single-layer 2H-MoS_2 nanosheets obtained by the CVD method.[111] The strong interaction of DBT molecules with Mo atoms can be confirmed by the smaller Mo-S distance at the modified sites than the pristine 2H-MoS_2 nanosheets. Recently, McDonald and co-workers reported the first examples to demonstrate functionalization on basal plane S atoms of 2H-MoS_2 nanosheets with metal carboxylate salts.[112] As shown in Figure 8.16, these 2H-MoS_2 nanosheets exfoliated in liquid phase were first dispersed in IPA and then reacted with $M(OAc)_2$ to produce 2H-MoS_2-$M(OAc)_2$ nanosheets *via* the coordinating of metal carboxylate on basal plane S atoms. The final functionalized nanosheets exhibited good dispersity/stability and could be easily dispersed in various conventional solvents such as water, ethanol and acetone.

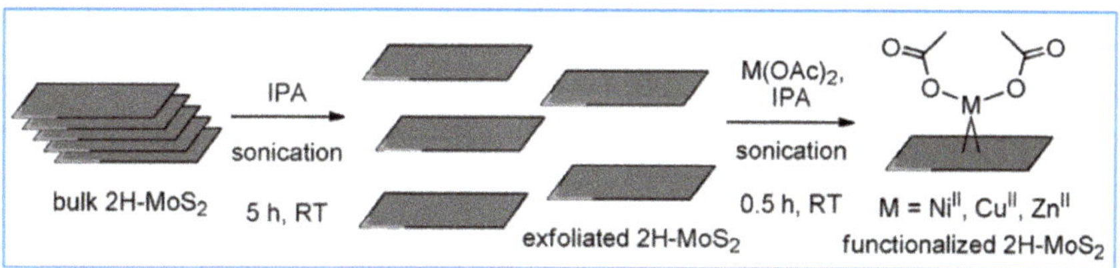

Figure 8.16 Schematic representation of the procedure for exfoliation and functionalization of 2H-MoS_2 nanosheets.
Reprinted from ref. 112 with permission from John Wiley and Sons, © 2015 WILEY-VCH Verlag GmbH & Co. KGaA, Weinheim.

Surface functionalization of TMD nanosheets could also be used to eliminate structure defects and traps for better electron transport. For example, Yu and co-workers reported the repairing of sulphur vacancies through the modification of (3-mercaptoprpyl) trimethoxysilane onto the surface of 2H-MoS_2 nanosheets under a mild annealing condition.[113] Cho and co-workers demonstrated the passivation of sulphur vacancies through the deposition of alkanethiol molecules onto 2H-MoS_2 nanosheets, and investigated the influence of thiol molecules on the physical and electrical properties of 2H-MoS_2 for use as field-effect transistors.[114]

8.7.3 Metal Hybridization of TMD Nanosheets

To incorporate novel functionalities for achieving synergistic effects, various metal and semiconductor nanoparticles were chemically grown/coupled on TMD nanosheets to prepare hybrid nanostructures with combined optical, electronic or magnetic properties. In particular, the hybridizations of MoS_2 nanosheets have been realized with metal NPs at the dangling bond-rich edges of non-liquid exfoliated MoS_2 nanosheets or the detective basal planes of Li-intercalated MoS_2 nanosheets *via* a reduction of metal salts under the assistance of surfactants/stabilizers. For example, Zhang's group first reported the epitaxial growth of noble metal nanomaterials on single-layer MoS_2 nanosheets by the proper choice of precursors, reducing agents and surfactants.[115] Typically, 5 nm-sized Pd NPs were epitaxially grown on the surface of MoS_2 nanosheets *via* the reduction of K_2PdCl_4 with ascorbic acid in the presence of PVP. 1–3 nm-sized Pt NPs were synthesized on MoS_2 nanosheets *via* the photochemical reduction of K_2PtCl_4 in the presence of sodium citrate. While silver triangular nanoplates were prepared on MoS_2 nanosheets in the presence of cetyl trimethylammonium bromide or PVP.

In recent years, other research groups have also started to report the preparation of metal nanoparticle-modified MoS_2 nanosheets. For example, Wang's group systematically investigated four noble metallic nanoparticles decorated on the surface of MoS_2 nanosheets *via* different synthesized methods, such as Au, Ag, Pt and Pd nanoparticles. They found that such noble metallic nanoparticles–MoS_2 nanocomposites showed higher catalytic activity compared with commercial Pt catalysts.[116] Su *et al.* successfully synthesized Au–Pt core–shell bimetallic nanoparticles-decorated MoS_2 nanocomposites, which exhibited better catalytic activity toward methanol oxidation than that of Au–MoS_2 and Pt–MoS_2 nanocomposites.[117] In addition, Kim *et al.* decorated MoS_2 nanosheets with Au nanoparticles by directly reducing Au(III) cations with MoS_2 nanosheets and the resulting Au–MoS_2 nanosheets showed a significantly enhanced electrocatalytic performance.[118] Although this method is very simple, the lack of a stabilizer often leads to aggregation of the products. Very recently, Guan *et al.* demonstrated a one-step process for exfoliation and functionalization of MoS_2 nanosheets by using BSA-caged Au_{25} clusters as both exfoliating and functionalizing agents.[119] As compared to the existing hybridizations of MoS_2

with metal NPs, the clusters-based functionalized process achieved the epitaxial growth of Au NPs on exfoliated MoS_2 nanosheets rather than the free growth in solution, and the control of Au NPs in size upon reaction with H_2O_2. Similarly, the strategy was further extended to other nanosheets with the use of different BSA-caged clusters (*e.g.* Ag clusters). Clearly, the introduction of protein-caged functional nanomaterials provides a new and effective strategy for the direct exfoliation and simultaneous functionalization of 2D materials towards their enhanced applications.

8.8 Conclusions and Outlook

In this chapter, a series of synthetic methods for producing single- and few-layered TMD nanosheets have been summarized, including micromechanical cleavage, ion intercalation, mechanical force-assisted exfoliation, wet-chemical synthesis and chemical vapour deposition. Among them, micromechanical cleavage and chemical vapour deposition are the traditional routes for obtaining high-quality nanosheets with a large size. However, the large-sized nanosheets are usually not suitable for biological applications due to their poor dispersity in water. In comparison, the liquid-based techniques are convenient and cost-effective approaches for the massive production of small-sized TMD nanosheets, which can be easily sorted and separated to produce nanosheets of desirable size and thickness, and also offer a rich platform for the modification and hybridization of nanosheets with functional molecules/ions or other nanomaterials in solution. Currently, the solution-based methods have been extensively used in laboratories for the preparation of TMD nanosheets, however, the quantity of TMD nanosheets is still relatively small for the requirement of large-scale commercial applications. Therefore, developing reliable methods to scale up the production of TMD nanosheets, especially the single-layer nanosheets, is still a great challenge.

As mentioned above, liquid-phase exfoliated TMD nanosheets offer many opportunities for the functionalization and hybridization of nanosheets with a variety of molecules or other nanomaterials, for example, organic and biomolecules, metals, metal oxides, metal chalcogenides, and carbonaceous materials. The functionalized or hybridized TMD nanosheets do show great potential in various promising applications, such as sensing, energy storage and conversion, electronic and bio-applications. An important work is the epitaxial growth of metal NPs on the surface of MoS_2 nanosheets for preparing MoS_2–NP composites, which show higher electrocatalytic activity toward the hydrogen evolution reaction (HER) due to the exposure of high index facets induced by the epitaxial effect. However, most of the nanomaterials were just simply hybridized with TMD nanosheets without much concern for their hybridization manner, such as the growth orientation, surface exposure and component distribution, or the interaction of each component. These features of the hybrid nanomaterial largely determine their performance in specific applications. Therefore, a great challenge in the hybridization of solution-processed MoS_2 nanosheets is still the highly controllable construction of hybrid nanomaterials with desired structural features for better applications.

Recent research has also started to realize that the peripheral edge sites of nanosheets rather than the basal planes often act as active centres in enhanced applications. For instance, the peripheral edge sites of MoS_2 and $MOOH_x$ (M = Ni/Fe, Co, Mn) nanosheets have been demonstrated to render high activities towards HER and the oxygen evolution reaction, respectively.[120–123] It has been shown that the edges of MoS_2 are the active sites for HER and the vertically-aligned MoS_2 nanosheets with rich edges have better HER catalytic performance. As such, it is of key importance to expose more active edge sites *via* structural manipulation for enhancing the outstanding performance of 2D nanomaterials. In the light of current research achievements, there are still many opportunities in these promising fields. Very recently, we reported a surface-mediated chemical dissolution for the convenient creation of holes on atomically thin MoS_2 nanosheets that are partially coated with BSA.[124] The resulting porous nanosheets exhibit not only greatly increased bandgaps but also substantially enhanced photocurrents together with much better performance during selective adsorption and photocatalytic degradation of substances. The facile production of nanosheets with holes may offer an effective route to exploit the exotic properties and novel applications of geometrically constructed 2D TMD nanomaterials.

References

1. M. Xu, T. Liang, M. Shi and H. Chen, *Chem. Rev.*, 2013, **113**(5), 3766.
2. C. Tan and H. Zhang, *Chem. Soc. Rev.*, 2015, **44**(9), 2713.
3. X. Zhang, Z. Lai, C. Tan and H. Zhang, *Angew. Chem., Int. Ed.*, 2016, **55**(31), 8816.
4. H. Matte, A. Gomathi, A. K. Manna, D. J. Late, R. Datta, S. K. Pati and C. N. R. Rao, *Angew. Chem., Int. Ed.*, 2010, **49**(24), 4059.
5. P. Wang, H. Sun, Y. Ji, W. Li and X. Wang, *Adv. Mater.*, 2014, **26**(6), 964.
6. Y. M. Shi, W. Zhou, A. Y. Lu, W. J. Fang, Y. H. Lee, A. L. Hsu, S. M. Kim, K. K. Kim, H. Y. Yang, L. J. Li, J. C. Idrobo and J. Kong, *Nano Lett.*, 2012, **12**(6), 2784.
7. M. Yi and Z. Shen, *J. Mater. Chem. A*, 2015, **3**(22), 11700.
8. H. Li, J. Wu, Z. Y. Yin and H. Zhang, *Acc. Chem. Res.*, 2014, **47**(4), 1067.
9. H. Li, Z. Yin, Q. He, H. Li, X. Huang, G. Lu, D. W. Fam, A. I. Tok, Q. Zhang and H. Zhang, *Small*, 2012, **8**(1), 63.
10. D. Sarkar, W. Liu, X. Xie, A. C. Anselmo, S. Mitragotri and K. Banerjee, *ACS Nano*, 2014, **8**(4), 3992.
11. Z. Zeng, Z. Yin, X. Huang, H. Li, Q. He, G. Lu, F. Boey and H. Zhang, *Angew. Chem., Int. Ed.*, 2011, **50**(47), 11093.
12. B. Radisavljevic, M. B. Whitwick and A. Kis, *ACS Nano*, 2011, **5**(12), 9934.

13. K. S. Novoselov, A. K. Geim, S. V. Morozov, D. Jiang, Y. Zhang, S. V. Dubonos, I. V. Grigorieva and A. A. Firsov, *Science*, 2004, **306**(5696), 666.
14. K. S. Novoselov, D. Jiang, F. Schedin, T. J. Booth, V. V. Khotkevich, S. V. Morozov and A. K. Geim, *Proc. Natl. Acad. Sci. U. S. A.*, 2005, **102**(30), 10451.
15. C. Tan, X. Cao, X. I. Wu, Q. He, J. Yang, X. Zhang, J. Chen, W. Zhao, S. Han, G. H. Nam, M. Sindoro and H. Zhang, *Chem. Rev.*, 2017, **117**(9), 6225.
16. H. Li, G. Lu, Y. Wang, Z. Yin, C. Cong, Q. He, L. Wang, F. Ding, T. Yu and H. Zhang, *Small*, 2013, **9**(11), 1974.
17. H. Li, J. Wu, X. Huang, G. Lu, J. Yang, X. Lu, Q. Xiong and H. Zhang, *ACS Nano*, 2013, **7**(11), 10344.
18. H. Li, J. Wu, Z. Y. Yin and H. Zhang, *Acc. Chem. Res.*, 2014, **47**(4), 1067.
19. S. B. Desai, S. R. Madhvapathy, M. Amani, D. Kiriya, M. Hettick, M. Tosun, Y. Zhou, M. Dubey, J. W. Ager III, D. Chrzan and A. Javey, *Adv. Mater.*, 2016, **28**(21), 4053.
20. Q. H. Wang, K. Kalantar-Zadeh, A. Kis, J. N. Coleman and M. S. Strano, *Nat. Nanotechnol.*, 2012, **7**(11), 699.
21. Z. Zeng, T. Sun, J. Zhu, X. Huang, Z. Yin, G. Lu, Z. Fan, Q. Yan, H. H. Hng and H. Zhang, *Angew. Chem., Int. Ed.*, 2012, **51**(36), 9052.
22. E. Benavente, *Coord. Chem. Rev.*, 2002, **224**(1–2), 87.
23. X. Cai, Y. Luo, B. Liu and H. M. Cheng, *Chem. Soc. Rev.*, 2018, **47**(16), 6224.
24. M. B. Dines, *Mater. Res. Bull.*, 1975, **10**(4), 287.
25. P. Joensen, R. F. Frindt and S. R. Morrison, *Mater. Res. Bull.*, 1986, **21**(4), 457.
26. G. Eda, H. Yamaguchi, D. Voiry, T. Fujita, M. Chen and M. Chhowalla, *Nano Lett.*, 2011, **11**(12), 5111.
27. M. Chhowalla, H. S. Shin, G. Eda, L. J. Li, K. P. Loh and H. Zhang, *Nat. Chem.*, 2013, **5**(4), 263.
28. L. Yuwen, H. Yu, X. Yang, J. Zhou, Q. Zhang, Y. Zhang, Z. Luo, S. Su and L. Wang, *Chem. Commun.*, 2016, **52**(3), 529.
29. M. Chhowalla, H. S. Shin, G. Eda, L. J. Li, K. P. Loh and H. Zhang, *Nat. Chem.*, 2013, **5**(4), 263.
30. E. D. Grayfer, M. N. Kozlova and V. E. Fedorov, *Adv. Colloid Interface Sci.*, 2017, **245**, 40.
31. T. Fujita, Y. Ito, Y. Tan, H. Yamaguchi, D. Hojo, A. Hirata, D. Voiry, M. Chhowalla and M. Chen, *Nanoscale*, 2014, **6**(21), 12458.
32. J. Zhang, H. Zhang, S. Dong, Y. Liu, C. T. Nai, H. S. Shin, H. Y. Jeong, B. Liu and K. P. Loh, *Nat. Commun.*, 2014, **5**, 2995.
33. T. Daeneke, R. M. Clark, B. J. Carey, J. Z. Ou, B. Weber, M. S. Fuhrer, M. Bhaskaran and K. Kalantar-zadeh, *Nanoscale*, 2016, **8**(33), 15252.
34. L. Niu, K. Li, H. Zhen, Y. S. Chui, W. Zhang, F. Yan and Z. Zheng, *Small*, 2014, **10**(22), 4651.

35. D. Voiry, A. Mohite and M. Chhowalla, *Chem. Soc. Rev.*, 2015, **44**(9), 2702.
36. Z. Zeng, Z. Yin, X. Huang, H. Li, Q. He, G. Lu, F. Boey and H. Zhang, *Angew. Chem., Int. Ed.*, 2011, **50**(47), 11093.
37. Z. Y. Zeng, C. L. Tan, X. Huang, S. Y. Bao and H. Zhang, *Energy Environ. Sci.*, 2014, 7(2), 797.
38. Z. Y. Zeng, T. Sun, J. X. Zhu, X. Huang, Z. Y. Yin, G. Lu, Z. X. Fan, Q. Y. Yan, H. H. Hng and H. Zhang, *Angew. Chem., Int. Ed.*, 2012, **51**(36), 9052.
39. E. L. Chng, Z. Sofer and M. Pumera, *Nanoscale*, 2014, **6**(23), 14412.
40. N. Liu, P. Kim, J. H. Kim, J. H. Ye, S. Kim and C. J. Lee, *ACS Nano*, 2014, **8**(7), 6902.
41. Y. Hernandez, V. Nicolosi, M. Lotya, F. M. Blighe, Z. Sun, S. De, I. T. McGovern, B. Holland, M. Byrne, Y. K. Gun'Ko, J. J. Boland, P. Niraj, G. Duesberg, S. Krishnamurthy, R. Goodhue, J. Hutchison, V. Scardaci, A. C. Ferrari, J. N. Coleman, *Nat. Nanotechnol.*, 2008, **3**(9), 563.
42. J. N. Coleman, M. Lotya, A. O'Neill, S. D. Bergin, P. J. King, U. Khan, K. Young, A. Gaucher, S. De, R. J. Smith, I. V. Shvets, S. K. Arora, G. Stanton, H. Y. Kim, K. Lee, G. T. Kim, G. S. Duesberg, T. Hallam, J. J. Boland, J. J. Wang, J. F. Donegan, J. C. Grunlan, G. Moriarty, A. Shmeliov, R. J. Nicholls, J. M. Perkins, E. M. Grieveson, K. Theuwissen, D. W. McComb, P. D. Nellist and V. Nicolosi, *Science*, 2011, **331**(6017), 568.
43. A. O'Neill, U. Khan and J. N. Coleman, *Chem. Mater.*, 2012, **24**(12), 2414.
44. M. M. Bernal, L. Alvarez, E. Giovanelli, A. Arnaiz, L. Ruiz-Gonzalez, S. Casado, D. Granados, A. M. Pizarro, A. Castellanos-Gomez and E. M. Perez, *2D Mater.*, 2016, **3**, 035014.
45. L. Muscuso, S. Cravanzola, F. Cesano, D. Scarano and A. Zecchina, *J. Phys. Chem. C*, 2015, **119**(7), 3791.
46. K. G. Zhou, N. N. Mao, H. X. Wang, Y. Peng and H. L. Zhang, *Angew. Chem., Int. Ed.*, 2011, **50**(46), 10839.
47. K. G. Zhou, M. Zhao, M. J. Chang, Q. Wang, X. Z. Wu, Y. Song and H. L. Zhang, *Small*, 2015, **11**(6), 694.
48. W. Yin, L. Yan, J. Yu, G. Tian, L. Zhou, X. Zheng, X. Zhang, Y. Yong, J. Li, Z. Gu and Y. Zhao, *ACS Nano*, 2014, **8**(7), 6922.
49. J. Kim, S. Kwon, D. H. Cho, B. Kang, H. Kwon, Y. Kim, S. O. Park, G. Y. Jung, E. Shin, W. G. Kim, H. Lee, G. H. Ryu, M. Choi, T. H. Kim, J. Oh, S. Park, S. K. Kwak, S. W. Yoon, D. Byun, Z. Lee and C. Lee, *Nat. Commun.*, 2016, **6**, 8294.
50. V. Forsberg, R. Zhang, J. Backstrom, C. Dahlstrom, B. Andres, M. Norgren, M. Andersson, M. Hummelgard and H. Olin, *PLoS One*, 2016, **11**(4), e0154522.
51. A. Ciesielskia and P. Samorì, *Chem. Soc. Rev.*, 2014, **43**(1), 381.
52. R. J. Smith, P. J. King, M. Lotya, C. Wirtz, U. Khan, S. De, A. O'Neill, G. S. Duesberg, J. C. Grunlan, G. Moriarty, J. Chen, J. Wang, A. I. Minett, V. Nicolosi and J. N. Coleman, *Adv. Mater.*, 2011, **23**(34), 3944.

53. Z. Y. Lei, W. C. Zhu, S. J. Xu, J. Ding, J. X. Wan and P. Y. Wu, *ACS Appl. Mater. Interfaces*, 2016, **8**(32), 20900.
54. Z. Lei, Y. Zhou and P. Wu, *Small*, 2016, **12**(23), 3112.
55. L. Zong, M. Li and C. Li, *Adv. Mater.*, 2017, **29**(10), 1604691.
56. X. Jia, J. Bai, Z. Ma and X. Jiang, *J. Mater. Chem. B*, 2017, **5**(2), 269.
57. G. S. Bang, S. Cho, N. Son, G. W. Shim, B. K. Cho and S. Y. Choi, *ACS Appl. Mater. Interfaces*, 2016, **8**(3), 1943.
58. E. Satheeshkumar, A. Bandyopadhyay, M. B. Sreedhara, S. K. Pati, C. N. R. Rao and M. Yoshimura, *ChemNanoMat*, 2017, **3**(3), 172.
59. S. Zhu, L. Gong, J. Xie, Z. Gu and Y. Zhao, *Small Methods*, 2017, **1**(12), 1700220.
60. H. Tao, Y. Zhang, Y. Gao, Z. Sun, C. Yanb and J. Texter, *Phys. Chem. Chem. Phys.*, 2017, **19**(2), 921.
61. R. Deng, H. Yi, F. Fan, L. Fu, Y. Zeng, Y. Wang, Y. Li, Y. Liu, S. Jib and Y. Su, *RSC Adv.*, 2016, **6**(80), 77083.
62. G. Guan, S. Zhang, S. Liu, Y. Cai, M. Low, C. P. Teng, I. Y. Phang, Y. Cheng, K. L. Duei, B. M. Srinivasan, Y. Zheng, Y. W. Zhang and M. Y. Han, *J. Am. Chem. Soc.*, 2015, **137**(19), 6152.
63. G. Yang, C. Zhu, D. Du, J. Zhu and Y. Lin, *Nanoscale*, 2015, **7**(34), 14217.
64. K. R. Paton, E. Varrla, C. Backes, R. J. Smith, U. Khan, A. O'Neill, C. Boland, M. Lotya, O. M. Istrate, P. King, T. Higgins, S. Barwich, P. May, P. Puczkarski, I. Ahmed, M. Moebius, H. Pettersson, E. Long, J. Coelho, S. E. O'Brien, E. K. McGuire, B. M. Sanchez, G. S. Duesberg, N. McEvoy, T. J. Pennycook, C. Downing, A. Crossley, V. Nicolosi and J. N. Coleman, *Nat. Mater.*, 2014, **13**(6), 624.
65. Y. G. Yao, Z. Y. Lin, Z. Li, X. J. Song, K. S. Moon and C. P. Wong, *J. Mater. Chem.*, 2012, **22**(27), 13494.
66. E. Varrla, C. Backes, K. R. Paton, A. Harvey, Z. Gholamvand, J. McCauley and J. N. Coleman, *Chem. Mater.*, 2015, **27**(3), 1129.
67. H. S. S. Ramakrishna Matte, A. Gomathi, A. K. Manna, D. J. Late, R. Datta, S. K. Pati and C. N. R. Rao, *Angew. Chem., Int. Ed.*, 2010, **49**(24), 4059.
68. J. H. Han, S. Lee and J. Cheon, *Chem. Soc. Rev.*, 2013, **42**(7), 2581.
69. C. L. Tan and H. Zhang, *Nat. Commun.*, 2015, **6**, 7873.
70. Z. B. Zhuang, Q. Peng and Y. D. Li, *Chem. Soc. Rev.*, 2011, **40**(11), 5492.
71. B. L. Cushing, V. L. Kolesnichenko and C. J. O'Connor, *Chem. Rev.*, 2004, **104**(9), 3893.
72. W. Shi, S. Song and H. Zhang, *Chem. Soc. Rev.*, 2013, **42**(13), 5714.
73. C. N. R. Rao, H. S. S. Ramakrishna Matte and U. Maitra, *Angew. Chem., Int. Ed.*, 2013, **52**(50), 13162.
74. J. Xie, H. Zhang, S. Li, R. Wang, X. Sun, M. Zhou, J. Zhou, X. W. Lou and Y. Xie, *Adv. Mater.*, 2013, **25**(40), 5807.
75. J. F. Xie, J. J. Zhang, S. Li, F. Grote, X. D. Zhang, H. Zhang, R. X. Wang, Y. Lei, B. C. Pan and Y. Xie, *J. Am. Chem. Soc.*, 2013, **135**(47), 17881.

76. L. Cai, J. F. He, Q. H. Liu, T. Yao, L. Chen, W. S. Yan, F. C. Hu, Y. Jiang, Y. D. Zhao, T. D. Hu, Z. Shun and S. Wei, *J. Am. Chem. Soc.*, 2015, **137**(7), 2622.
77. M. Li, D. E. Wang, J. H. Li, Z. D. Pan, H. J. Ma, Y. X. Jiang and Z. J. Tian, *RSC Adv.*, 2016, **6**(75), 71534.
78. X. H. Chen and R. Fan, *Chem. Mater.*, 2001, **13**(3), 802.
79. X. Guo, Y. Wang, F. Wu, Y. Ni and S. Kokot, *Analyst*, 2015, **140**(4), 1119.
80. S. Wang, K. Li, Y. Chen, H. Chen, M. Ma, J. Feng, Q. Zhao and J. Shi, *Biomaterials*, 2015, **39**, 206.
81. X. Li, J. Shan, W. Zhang, S. Su, L. Yuwen and L. Wang, *Small*, 2017, **13**(5), 1602660.
82. Y. Chen, C. L. Tan, H. Zhang and L. Z. Wang, *Chem. Soc. Rev.*, 2015, **44**(9), 2681.
83. C. Altavilla, M. Sarno and P. Ciambelli, *Chem. Mater.*, 2011, **23**(17), 3879.
84. B. Mahler, V. Hoepfner, K. Liao and G. A. Ozin, *J. Am. Chem. Soc.*, 2014, **136**(40), 14121.
85. X. Qian, S. Shen, T. Liu, L. Cheng and Z. Liu, *Nanoscale*, 2015, **7**(14), 6380.
86. L. Cheng, C. Yuan, S. Shen, X. Yi, H. Gong, K. Yang and Z. Liu, *ACS Nano*, 2015, **9**(11), 11090.
87. S. Shen, Y. Chao, Z. Dong, G. Wang, X. Yi, G. Song, K. Yang, Z. Liu and L. Cheng, *Adv. Funct. Mater.*, 2017, **27**(28), 1700250.
88. D. Yoo, M. Kim, S. Jeong, J. Han and J. Cheon, *J. Am. Chem. Soc.*, 2014, **136**(42), 14670.
89. C. B. Murray, D. J. Norris and M. G. Bawendi, *J. Am. Chem. Soc.*, 1993, **115**(19), 8706.
90. P. D. Antunez, D. H. Webber and R. L. Brutchey, *Chem. Mater.*, 2013, **25**(12), 2385.
91. W. Y. Wu, S. Chakrabortty, C. K. L. Chang, A. Guchhait, M. Lin and Y. Chan, *Chem. Mater.*, 2014, **26**(21), 6120.
92. W. Jung, S. Lee, D. Yoo, S. Jeong, P. Miro, A. Kuc, T. Heine and J. Cheon, *J. Am. Chem. Soc.*, 2015, **137**(23), 7266.
93. S. Jeong, J. H. Han, J. T. Jang, J. W. Seo, J. G. Kim and J. Cheon, *J. Am. Chem. Soc.*, 2011, **133**(37), 14500.
94. J. Yu, J. Li, W. Zhang and H. Chang, *Chem. Sci.*, 2015, **6**(12), 6705.
95. Q. Ji, Y. Zhang, Y. Zhang and Z. Liu, *Chem. Soc. Rev.*, 2015, **44**(9), 2587.
96. Y. Shi, H. Li and L. J. Li, *Chem. Soc. Rev.*, 2015, **44**(9), 2744.
97. K. K. Liu, W. Zhang, Y. H. Lee, Y. C. Lin, M. T. Chang, C. Y. Su, C. S. Chang, H. Li, Y. Shi, H. Zhang, C. S. Lai and L. J. Li, *Nano Lett.*, 2012, **12**(3), 1538.
98. Y. Zhan, Z. Liu, S. Najmaei, P. M. Ajayan and J. Lou, *Small*, 2012, **8**(7), 966.
99. Y. H. Lee, X. Q. Zhang, W. Zhang, M. T. Chang, C. T. Lin, K. D. Chang, Y. C. Yu, J. T. W. Wang, C. S. Chang, L. J. Li and T. W. Lin, *Adv. Mater.*, 2012, **24**(17), 2320.

100. S. Najmaei, Z. Liu, W. Zhou, X. Zou, G. Shi, S. Lei, B. I. Yakobson, J. C. Idrobo, P. M. Ajayan and J. Lou, *Nat. Mater.*, 2013, **12**(8), 754.
101. S. Balendhran, J. Z. Ou, M. Bhaskaran, S. Sriram, S. Ippolito, Z. Vasic, E. Kats, S. Bhargava, S. Zhuiykov and K. Kalantar-Zadeh, *Nanoscale*, 2012, **4**(2), 461.
102. Y. C. Lin, W. Zhang, J. K. Huang, K. K. Liu, Y. H. Lee, C. T. Liang, C. W. Chu and L. J. Li, *Nanoscale*, 2012, **4**(20), 6637.
103. Y. Zhang, Y. Zhang, Q. Ji, J. Ju, H. Yuan, J. Shi, T. Gao, D. Ma, M. Liu, Y. Chen, X. Song, H. Y. Hwang, Y. Cui and Z. Liu, *ACS Nano*, 2013, 7(10), 8963.
104. R. H. Goncalves, R. Fiel, M. R. Soares, W. H. Schreiner, C. M. Silva and E. R. Leite, *Chem. – Eur. J.*, 2015, **21**(44), 15583.
105. M. Jeong, S. Kim and S. Y. Ju, *RSC Adv.*, 2016, **6**(43), 36248.
106. K. C. Knirsch, N. C. Berner, H. C. Nerl, C. S. Cucinotta, Z. Gholamvand, N. McEvoy, Z. Wang, I. Abramovic, P. Vecera, M. Halik, S. Sanvito, G. S. Duesberg, V. Nicolosi, F. Hauke, A. Hirsch, J. N. Coleman and C. Backes, *ACS Nano*, 2015, **9**(6), 6018.
107. D. Voiry, A. Goswami, R. Kappera, C. de Carvalho Castro e Silva, D. Kaplan, T. Fujita, M. Chen, T. Asefa and M. Chhowalla, *Nat. Chem.*, 2015, 7, 45.
108. S. S. Chou, M. De, J. Kim, S. Byun, C. Dykstra, J. Yu, J. Huang and V. P. Dravid, *J. Am. Chem. Soc.*, 2013, **135**(12), 4584.
109. L. Zhou, B. Z. He, Y. Yang and Y. G. He, *RSC Adv.*, 2014, **4**(11), 32570.
110. T. Liu, C. Wang, X. Gu, H. Gong, L. Cheng, X. Shi, L. Feng, B. Sun and Z. Liu, *Adv. Mater.*, 2014, **26**(21), 3433.
111. A. Tuxen, J. Kibsgaard, H. Gobel, E. Laegsgaard, H. Topsoe, J. V. Lauritsen and F. Besenbacher, *ACS Nano*, 2010, **4**(8), 4677.
112. C. Backes, N. C. Berner, X. Chen, P. Lafargue, P. LaPlace, M. Freeley, G. S. Duesberg, J. N. Coleman and A. R. McDonald, *Angew. Chem., Int. Ed.*, 2015, **54**(9), 2638.
113. Z. Yu, Y. Pan, Y. Shen, Z. Wang, Z. Y. Ong, T. Xu, R. Xin, L. Pan, B. Wang, L. Sun, J. Wang, G. Zhang, Y. W. Zhang, Y. Shi and X. Wang, *Nat. Commun.*, 2014, **5**, 5290.
114. K. Cho, M. Min, T. Y. Kim, H. Jeong, J. Pak, J. K. Kim, J. Jang, S. J. Yun, Y. H. Lee, W. K. Hong and T. Lee, *ACS Nano*, 2015, **9**(8), 8044.
115. X. Huang, Z. Zeng, S. Bao, M. Wang, X. Qi, Z. Fan and H. Zhang, *Nat. Commun.*, 2013, **4**, 1444.
116. L. Yuwen, F. Xu, B. Xue, Z. Luo, Q. Zhang, B. Bao, S. Su, L. Weng, W. Huang and L. Wang, *Nanoscale*, 2014, **6**(11), 5762.
117. S. Su, C. Zhang, L. Yuwen, X. Liu, L. Wang, C. Fan and L. Wang, *Nanoscale*, 2016, **8**(1), 602.
118. S. Su, C. Zhang, L. Yuwen, J. Chao, X. Zuo, X. Liu, C. Song, C. Fan and L. Wang, *ACS Appl. Mater. Interfaces*, 2014, **6**(21), 18735.
119. G. Guan, S. Liu, Y. Cheng, Y. H. Zhang and M. Y. Han, *Nanoscale*, 2018, **10**(23), 10911.

120. T. F. Jaramillo, K. P. Jørgensen, J. Bonde, J. H. Nielsen, S. Horch and I. Chorkendorff, *Science*, 2007, **317**(5834), 100.
121. B. Hinnemann, P. G. Moses, J. Bonde, K. P. Jørgensen, J. H. Nielsen, S. Horch, I. Chorkendorff and J. K. Nørskov, *J. Am. Chem. Soc.*, 2005, **127**(15), 5308.
122. M. W. Louie and A. T. Bell, *J. Am. Chem. Soc.*, 2013, **135**(33), 12329.
123. C. C. McCrory, S. Jung, J. C. Peters and T. F. Jaramillo, *J. Am. Chem. Soc.*, 2013, **135**(45), 16977.
124. G. Guan, M. Wu, Y. Cai, S. Liu, Y. Cheng, S. Y. Tee, Y. W. Zhang and M. Y. Han, *Chem. Mater.*, 2018, **30**(15), 5108.

CHAPTER 9

High Electrocatalytic Performance of Two-dimensional Layered MoS_2-based Materials for the Hydrogen Evolution Reaction

SUBRAMANIAM JAYABAL AND DONGSHENG GENG*

Beijing Advanced Innovation Center for Materials Genome Engineering, Center for Green Innovation, School of Mathematics and Physics, University of Science and Technology Beijing, Beijing 100083, China
*Email: dgeng@ustb.edu.cn

9.1 Introduction

Two-dimensional (2D) materials are a form of layer-type materials whose structures and properties are determined by their chemical environment, and which are only one or a few atomic layers thick. Systematic investigation of graphene-based 2D materials has revealed a number of unusual physical, chemical, and electronic properties.[1,2] Transition metal dichalcogenides (TMDs) are a type of 2D layered materials with diverse properties for fundamental research and emerging applications.[1,3,4] TMDs belong to a type of MX_2, where M indicates the transition metal of the groups IV–X (Ti, Zr, Hf, V, Nb, Ta, Mo, W, Tc, Re, Pd or Pt) and X indicates a chalcogen atom (S, Se or Te). Depending upon their unique electronic properties, TMDs act as a

Smart Materials No. 34
Layered Materials for Energy Storage and Conversion
Edited by Dongsheng Geng, Yuan Cheng and Gang Zhang

Published by the Royal Society of Chemistry, www.rsc.org

semiconductor (*e.g.*, MoS_2 and WS_2), insulator (*e.g.*, HfS_2), metal (*e.g.*, NbS_2 and VSe_2), and semimetal (*e.g.*, WTe_2 and $TiSe_2$).[1] Molybdenum disulfide (MoS_2) is a class of TMDs that is formed by a graphene-like hexagonal arrangement of Mo and S atoms bonded together in sandwiched S–Mo–S layered units.[1] The S–Mo–S layered structures are arranged in a stack on top of each other by weak non-covalent interactions to form a MoS_2 crystalline structure.[1] Therefore, single- or few-layer MoS_2 nanosheets can be prepared by direct exfoliation from the bulk material. MoS_2 in its layered structure features a number of unexpected properties depending upon the coordination environment, which differentiates it from bulk MoS_2.[1,5]

9.1.1 Why 2D Layered MoS_2 Nanosheets?

When the thickness of bulk MoS_2 decreases to a monolayer, transition of an indirect-to-direct bandgap arises, and strong optical properties ensue. Moreover, MoS_2 nanosheets can be designed with optimized thickness and size, with heteroatoms doping and hybridized/composed with active conducting materials for unusual interfacial properties. Furthermore, the electronic property of MoS_2 nanosheets is strongly determined by the Mo metal coordination and its d-electrons, which make it a promising application in catalysis and electronic/optoelectronic devices. Besides, the edges and corners of MoS_2 nanosheets possess unsaturated coordination and dangling bonds, which remain essential in surface chemistry applications. Subsequently, the unique properties of MoS_2 nanosheets provide openings for innovative research in a variety of fields.[1,3,5–8] There have been a large number of studies investigating the design of 2D layered MoS_2-based materials and their uses in the fields of catalysis,[9–11] electronics,[11–13] sensors[14,15] and energy.[9,16] Figure 9.1 shows the various uses of the 2D layered MoS_2-based materials that have emerged in recent years.[17–19]

Electrocatalysis processes play an important role in the energy-converting process, which involves the mutual conversion of renewable electricity and chemical energies.[20] Electrocatalytic energy storage and conversion consuming renewable power sources (*e.g.*, solar and wind energy) are considered to be one of the most effective and cleanest energy-converting pathways.[20] On the one hand, energy storage that involves an electrocatalytic hydrogen evolution reaction (HER) in the cathode and an oxygen evolution reaction (OER) in the anode can effectively induce the water-splitting process, and finally convert electrical energy into chemical energy (Figure 9.2).[20,21] On the other hand, the hydrogen oxidation reaction (HOR) and the oxygen reduction reaction (ORR) involved in the conversion of chemical energy to electrical energy behave as half-cell reactions in fuel cells or rechargeable Li–O_2 batteries.[20,21] Therefore, achieving high-performance energy-converting electrocatalysis and also developing efficient electrocatalysts for these above energy-converting pathways is urgent and essential for the future. Electrocatalytic water splitting is one of the most up-to-date energy conversion processes for converting renewable electrical energy into

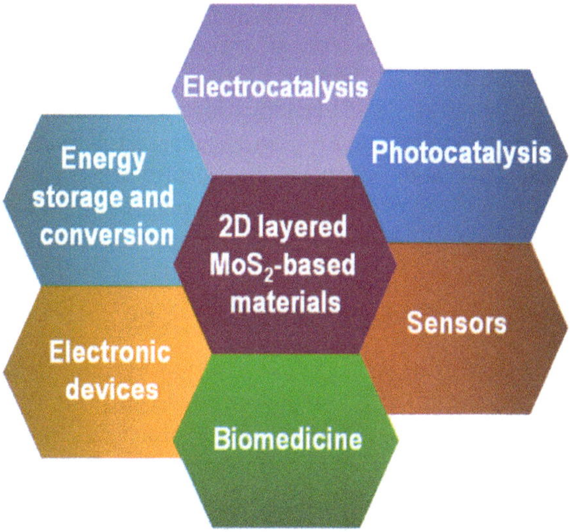

Figure 9.1 Pictorial illustration of the emerging applications of 2D layered MoS_2-based materials in various fields.

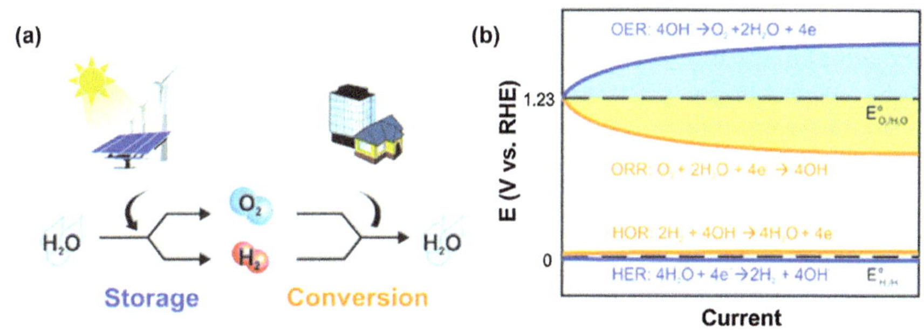

Figure 9.2 (a) Hydrogen and oxygen cycles for energy storage and conversion. Water electrolysis comprised of a HER and OER can realize energy storage in chemical form. For energy conversion, an ORR and HOR are involved in the construction of a fuel cell. (b) Scheme of the overpotentials associated with the HER, OER, ORR, and HOR.
Reproduced from ref. 20 with permission from the Royal Society of Chemistry.

chemical energy. The HER plays an important role in energy conversion processes that achieve the transformation of renewable electricity into chemical energy (*e.g.*, H_2).

Given the facts already discussed, this chapter aims at presenting a complete and up-to-date summary for understanding the high-performance of 2D layered MoS_2-based materialsfor HER electrocatalysis, which focuses on the relationships between properties and structures. Then we will explore their structural features, interfacial design, and performance as electrocatalysts. Subsequently, we review recent approaches and updated advances for improving the electrocatalytic activity of MoS_2 nanosheets and their composite materials. We believe that the 2D layered MoS_2 nanosheets and their composite materials as electrocatalysts can fulfill the requirements of innovative electrocatalytic energy storage systems and future outcome challenges. Finally, we also comment on certain existing research accomplishments, and provide explanations for the persisting potential challenges and upcoming opportunities in this exciting and rapidly developing field.

9.2 Properties of MoS_2 Nanosheets

The properties of MoS_2 are strongly determined by their crystalline phases, dimensional structures, and chemical compositions.[1,7] Bulk MoS_2 can be in one of three phases, namely 1T-MoS_2, 2H-MoS_2, and 3R-MoS_2, where the letters denote the tetragonal, hexagonal, and rhombohedral symmetries, respectively, and the digits show the number of S–Mo–S layers in a unit cell.[1,24] Each Mo atom is covalently bonded to six surrounding S atoms forming a trigonal prismatic coordination sphere in both phases of the 2H-MoS_2 and 3R-MoS_2 structures (Figure 9.3a).[3] In contrast, the 1T-MoS_2 metastable crystalline phase forms a distorted octahedron coordination around one Mo atom (Figure 9.3a).[25,26] The stable S–Mo–S monolayer includes two hexagonal planes of S atoms and an intermediate sandwiched hexagonal plane of Mo atoms. The oxidation states of Mo and S atoms in the MoS_2 have +4 and −2. Two polymorphs may exist for the phase coordination of the MoS_2 monolayer (trigonal D_{3h} group with a hexagonal symmetry; and octahedral D_{3d} group with a tetragonal symmetry).[1,24,27] Due to these prominent structural differences, the two phases display entirely diverse properties.

The electronic structure of MoS_2 greatly depends on the Mo metal coordination and non-bonding electrons in the d orbitals of Mo–S bonds found within the gap between the bonding and antibonding bands.[1,3] On the one hand, the non-bonding d orbitals in the trigonal prismatic coordinated 2H-MoS_2 are completely occupied and this contributes to the semiconducting behavior; whereas on the other hand, the non-bonding d orbitals are partially occupied in the octahedrally coordinated 1T-MoS_2 and this leads to a metallic nature. The filling of non-bonding electrons in the d orbitals for the MoS_2 monolayer greatly determines its symmetry.[1,3,28] Thus, depending upon the occupation of non-bonding electrons in the d orbital, MoS_2 undergoes a transition of electronic property from semiconductor to conductor.

Figure 9.3b displays the calculated electronic band structures attained for bulk, quadrilayer, bilayer, and a monolayer of 2H-MoS_2.[22] The direct excitonic transitions of the MoS_2 monolayer occurring at high energies at the K point barely changes with layer thickness, but the indirect bandgap of bulk MoS_2 increases monotonically as the number of layers decreases. When the thickness of the layer is reduced, the indirect bandgap becomes larger, while

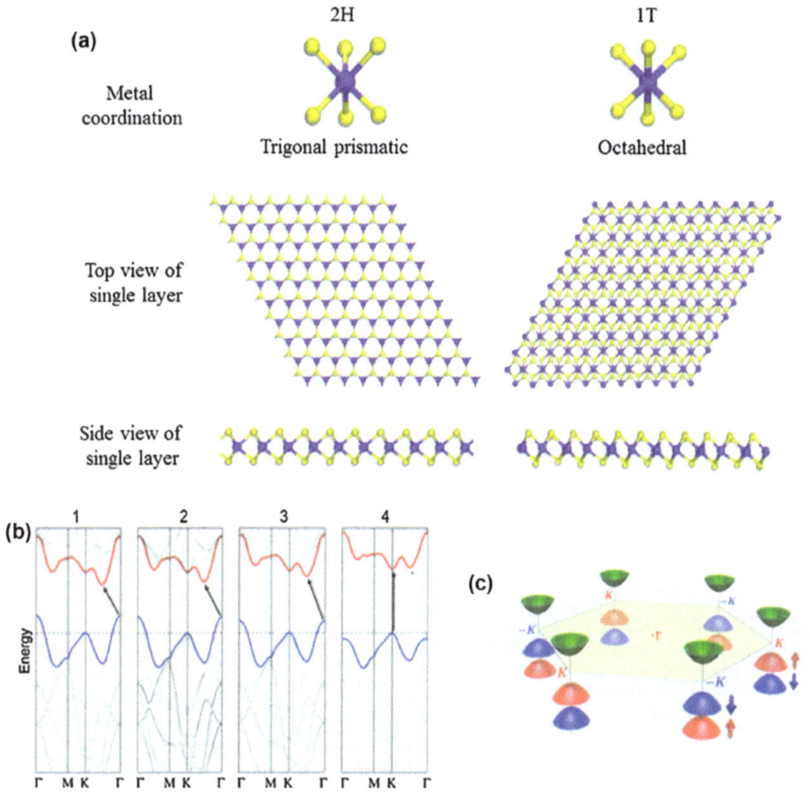

Figure 9.3 (a) Schematic metal coordination of a MoS$_2$ single layer in 2H and 1T forms. (b) Calculated band structures of MoS$_2$ in various forms: (1) bulk, (2) quadrilayer, (3) bilayer and (4) monolayer. (c) Schematic diagram of the band structure at the band edges located at the K points.
(a) Reproduced from ref. 3 with permission from the Royal Society of Chemistry. (b) Reproduced from ref. 22 with permission from American Chemical Society, Copyright 2010. (c) Reproduced from ref. 23, with permission from American Physical Society, https://doi.org/10.1103/PhysRevLett.108.196802, Copyright 2012.

the direct excitonic transition barely changes. Consequently, the indirect transition energy becomes very high in the MoS$_2$ monolayer, which induces the indirect-to-direct bandgap semiconductor transition that occurs from a bulk to a monolayer of MoS$_2$ due to quantum confinement effects.[1] Figure 9.3c displays a schematic illustration of the MoS$_2$ electronic band structure at the band edges located at the K and −K points showing six valleys and opposite spin–orbit splitting of the valence band. The blue and red surfaces indicate a spin–orbit split valence band maximum and the green surfaces indicate the conduction band minimum or the valleys. Due to the strong spin–orbit coupling, these valleys split at the valence band. The time-reversal symmetry of the MoS$_2$ monolayer that needs a splitting of the spin at different valleys must be opposite (blue and red as shown in Figure 9.3c), which leads to a unique situation where the spin and valley degrees of freedom are coupled.[23]

9.3 MoS_2-based Materials

Although MoS_2 acts as a good catalyst, its poor intrinsic conductivity limits the kinetics of electrocatalytic reactions and also the parallel stacking of MoS_2 layers confine the transport of the electron between the active sites and electrode materials.[3] These concerns can be fixed by constructing supportive catalysts composed of conductive materials uniting with the MoS_2 nanosheets. By means of the large surface area, it is believed that MoS_2 nanosheets can behave as model building blocks for the synthesis of MoS_2-based hybrids/composites/heterostructures.[19] Consequently, MoS_2 nanosheets can be combined with other active materials to conquer the limitations of individual constituents, which has been a topic of exhaustive studies in recent years.[17–19,29–31] Subsequently, a great variety of active materials have been explored to combine with MoS_2 nanosheets such as metals,[32,33] metal oxides,[34,35] polymers,[36,37] graphene,[38,39] carbon nanotubes,[40,41] heteroatoms,[42,43] metal chalcogenides,[44,45] metal–organic frameworks,[46,47] and so on.[48–51] The resulting MoS_2-based materials can assist the electron transfer kinetics, decrease the contact resistivity and avoid aggregation.[52,53] Furthermore, the synergistic effect, probably from the combination of two or more constituents, might enhance the electrocatalytic activities of the MoS_2-based materials.[53]

9.4 MoS_2-based Electrocatalyst

When bulk MoS_2 is exfoliated into few- or mono-layers, the edge planes are more dominant at their surfaces over the basal planes, and the edge atom can either be a Mo or S, probably depending upon the growth conditions.[1,8] When a S–Mo–S layer is decreased in size horizontally, this provides the nanosheet with less-coordination at the edge, kink, and corner atoms, and also the surface is controlled by this edge, kink, and corner atoms rather than the basal atoms.[1,10,54] The catalytic activity of the MoS_2 nanosheets is assigned to the Mo-terminated $(10\bar{1}0)$ and S-terminated $(\bar{1}010)$ edge planes, that possess a divergent local structural arrangement compared with the catalytically inert (0001) basal planes (Figure 9.4). The Mo edges covered by S atoms are the most active sites, which possess less coordination and provide a sulfide-rich environment.[55,56] In addition, large numbers of exposed surface atoms present in the MoS_2 nanosheets are routinely responsible for interactions (oxidation or reduction) with various components and they can simply escape from their own lattice, which results in the creation of vacancies. These vacancies, together with the associated structural disorder existing in the MoS_2 nanosheet, can reduce the coordination number of the atoms at the surface and therefore lead to the creation of dangling bonds.[1,9]

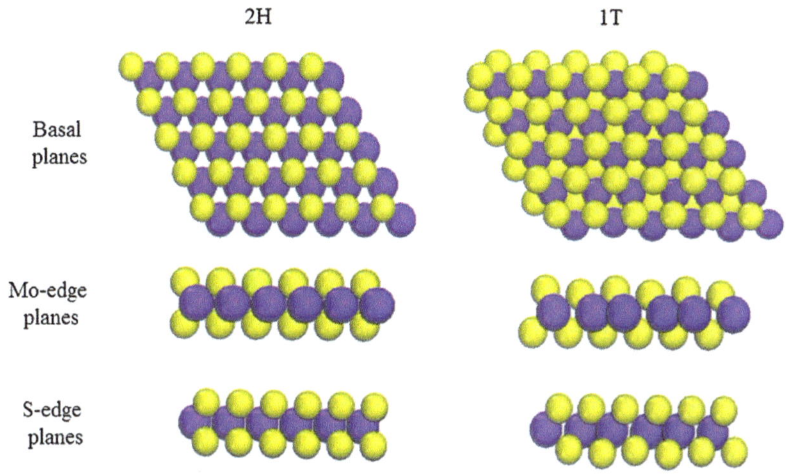

Figure 9.4 Ball model of 2H and 1T-MoS_2 nanosheets.
Reproduced from ref. 3 with permission form the Royal Society of Chemistry.

These vacancies cover an approach to alter the properties of MoS_2 nanosheets and thus improve the overall electrocatalytic performance. Moreover, these uncoordinated edges and dangling bonds, present at the edges and corners, that enable the reactive molecule (or substrate) to co-ordinate (or adsorb) the Mo or S atoms at the surface of the MoS_2 nanosheets more easily, therefore enhance the electrocatalytic performance.[57]

Usually, an effective electrocatalyst must have a very low overpotential, high current density, be relatively inexpensive and have excellent stability.[9,53,58] When compared with the noble metal catalysts, MoS_2 nanosheets are a promising non-precious catalyst because they are inexpensive, have high chemical stability and high electrocatalytic performance for the HER.[7,53] However, owing to their less active sites and poor charge transport property, the electrocatalytic performance of pure 2H-MoS_2 nanosheet is not ideal.[53,58] Consequently, improving the charge transport property and increasing the effective active sites in MoS_2 nanosheets has continued to be an essential problem for designing high-performance electrocatalysts by means of combining the MoS_2 nanosheets with conducting materials and doping other species to tune the basal planes.[3] The existence of uncoordinated atoms at the Mo edge and also the high specific surface area offers an efficient coupling or doping with active materials; this makes an efficient transfer of electrons through pre-concentration of the target molecules at the surface of the electrode, thus improving the electrocatalytic performance.[3]

9.5 Hydrogen Evolution Reaction

The increasing demand for energy and the subsequent effects on the planet have inspired significant attempts to improve ecological and environmentally-friendly energy alternatives for future generations.[9,53] Environmental problems affected by fossil fuels have attracted extensive attention and several efforts are currently ongoing to find inexpensive, carbon-less and abundant energy alternatives.[59] Hydrogen is one of the most tremendous molecular fuels and has the highest energy density, which makes it an excellent energy-carrier and also a green energy source for the substitution of conventional fossil fuels due to it being recyclable and environmentally-friendly.[60–62] Among various energy substitutes, the electrocatalytic splitting of water for the formation of hydrogen has been attracting considerable interest owing to the fact it is pollution-free and inexpensive.[58] Consequently, construction of high-performance HER electrocatalysts is pivotal for the improvement of environmentally-friendly energy for electrocatalytic applications.

The HER is a cathodic reaction in electrochemical water splitting. The electrochemical hydrogen evolution reaction (HER) is the reduction of protons or water to hydrogen molecules by applying external potential or electrochemical energy. The elementary steps and overall reactions in acidic and alkaline electrolytes for a HER are presented in Table 9.1.[9,53,58,63–65] The HER plays an essential part in the water electrolysis process, where molecular hydrogen is produced from the electrochemical reduction of H^+. There are two commonly accepted mechanisms in the HER process; namely, Volmer–Heyrovsky and Volmer–Tafel mechanisms. In the Volmer step, hydrogen adsorption is either from a hydronium ion (in acidic solutions) or a dissociated water molecule (in alkaline solutions) on the electrode surface as shown in Table 9.1. The final step includes either: recombination on the electrode surface from the two adsorbed hydrogen atoms to form molecular hydrogen (Tafel step), or bonding between the adsorbed hydrogen atom with the hydrated proton from the surface of the electrode directly by the transfer of an electron (Heyrovsky step). Therefore, catalysts possess a Gibbs free energy of hydrogen adsorption (ΔG_H) very close to zero that is beneficial at both extremely high or low values, and which disturbs the hydrogen ion adsorption and molecular hydrogen desorption effectively.[66–68] The water

Table 9.1 Mechanistic steps involved in the acidic and alkaline conditions for the HER process.

Acidic solution	
$2H_{(aq)}^+ + 2e^- \rightarrow H_{2(g)}$	
Volmer step	$H^+ + e^- \rightarrow H_{ads}$
Tafel step	$2H_{ads} \rightarrow H_2$
Heyrovsky step	$H^+ + H_{ads} + e^- \rightarrow H_2$
Alkaline solution	
$2H_2O + 2e^- \rightarrow H_{2(g)} + 2OH_{(aq)}^-$	
Volmer step	$2H_2O + 2e^- \rightarrow 2H_{ads} + 2OH^-$
Tafel step	$2H_{ads} \rightarrow H_2$
Heyrovsky step	$H_2O + H_{ads} + e^- \rightarrow H_2 + OH^-$

dissociation step in alkaline solutions is usually lower by two to three orders of an extent than in acidic solutions.[69–71] The increase in the potential of the resulting current at the cathode by one order of extent gives the Tafel slope and it is calculated by plotting the electrode potential *versus* the logarithm of the current density.[1] In addition, the rate determining step is also determined using the Tafel plot of the polarization curve obtained from cyclic voltammetry or linear sweep voltammetry.[58]

In the past few years, Pt-based materials have been extensively utilized as electrocatalysts for high electrocatalytic performance and low overpotential for a HER. Nevertheless, large-scale hydrogen production for practical applications is not preferable due to very low natural reserves and because it is relatively expensive.[3,72] Hence, the search for a substitute for HER electrocatalysts that is low cost, electrochemically stable, eco-friendly, and highly efficient is the main challenge in recent years.[3] Currently, many researchers have focused on replacing Pt with low-cost 2D layered MoS_2-based materials as efficient catalysts for the HER; and this has become a compelling new research field.[1,3] The HER process at 2H-MoS_2 takes place on the edge planes, which includes adsorption of H^+ on the S edge atoms, followed by diffusion of H^+ from S to Mo by forming molybdenum hydride (Mo–H), and formation of H_2 takes place finally from Mo–H and a hydronium ion (H_3O^+).[73] However, the HER process at 1T-MoS_2 mainly takes place on the basal planes, which includes adsorption of hydrogen on the S edges, followed by the adsorbed H with the attacked $(H_3O)^+$ proton and formation of H_2 finally takes place.[74] For both 2H-MoS_2 and 1T-MoS_2, the HER process primarily occurs *via* the Volmer–Heyrovsky reaction. Although the HER process occurs through the Volmer–Heyrovsky reaction for 1T-MoS_2 nanosheets, the reaction mechanism pathway is completely divergent from the 2H-MoS_2 nanosheets, in which the Heyrovsky process takes place between adsorbed H on S and H from the hydronium ion directly because the Mo–H formation on 1T-MoS_2 nanosheets is highly unfavorable. For pure MoS_2 nanosheets, the low number of catalytically active sites and their poor intrinsic conductivity limits their overall electrocatalytic performance.[53] As a result, the basic design principles for the high electrocatalytic performance of MoS_2-based materials include creating more active centers along with improving the charge transport during the electrocatalytic HER process.[3] Table 9.2 displays a detailed outline of the literature reporting MoS_2 nanosheets and their composite materials for HER performance. Noticeably, a great development in electrocatalytic HER activity has been accomplished for MoS_2 nanosheets in recent years by means of doping/engineering with various heteroatoms/composite materials.

9.5.1 Emerging Design Strategy for Improving the Electrocatalytic Performance of MoS_2

As mentioned in previous sections, pure MoS_2 nanosheets are sluggish for electrocatalysis due to the limited number of active sites and poor electron

Table 9.2 Comparison of electrocatalytic performances involving MoS_2-based materials for HER performance.

Catalyst	Electrolyte	Onset potential (mV)	Exchange current density ($mA\,cm^{-2}$)	Overpotential at 10 $mA\,cm^{-2}$ (mV)	Tafel slope ($mV\,dec^{-1}$)	Ref.
MoS_2 nanosheets	0.5 M H_2SO_4	—	3.89×10^{-2}	—	68	75
Mesoporous/1T-MoS_2 nanosheets	0.5 M H_2SO_4	—	1.58×10^{-2}	153	43	76
MoS_2 nanosheets/strained S vacancies	0.5 M H_2SO_4	—	—	170	60	66
Au-deposited MoS_2 nanosheets	0.5 M H_2SO_4	−100	—	—	50	77
O-incorporated MoS_2 nanosheets	0.5 M H_2SO_4	120	126.5	300	55	43
V-doped MoS_2	1 M H_2SO_4	−130	—	130	69	78
MoS_2/graphene nanosheets	0.5 M H_2SO_4	—	0.14	30	67.4	38
MoS_{2+x}/N-CNTs/CP	0.5 M H_2SO_4	−135	—	160	36	79
MoS_2/$CoSe_2$ hybrid	0.5 M H_2SO_4	−11	7.3×10^{-2}	68	36	80
C-doped MoS_2@graphene nanosheets	0.5 M H_2SO_4	−160	—	—	46	81
MoS_2/N-rGO	0.5 M H_2SO_4	−5	7.4×10^{-1}	56	41.3	82
Hierarchical MoS_2 nanosheets	0.5 M H_2SO_4	50	0.036	167	70	83
Pt-MoS_2 nanosheets	0.5 M H_2SO_4	—	—	—	40	32
P-doped 2H-MoS_2 nanosheets	0.5 M H_2SO_4	130	—	300	49	42
2D MoS_2 nanosheets	0.5 M H_2SO_4	−480	—	—	40	64
MoS_2/carbon aerogel hybrids	0.5 M H_2SO_4	−140	—	300	59	84
Au decorated MoS_2 nanosheets	0.5 M H_2SO_4	—	—	205	57	33
MoS_2-Au nanohybrids	0.5 M H_2SO_4	−17	—	66	40	85
MoS_2/macroporous carbon foam/N-doped graphene hybrids	0.5 M H_2SO_4	—	—	170	53	86

MoS_2 nanosheets with rich in-plane edges	0.5 M H_2SO_4	−87	—	—	41	87
Pt decorated MoS_2 nanosheets	0.5 M H_2SO_4	0	0.75	—	49	88
MoS_2/reduced graphene oxide	0.5 M H_2SO_4	—	—	140	41	39
Mo_2N-MoS_2 composites	0.5 M H_2SO_4	—	0.6×10^{-2}	190	59	89
Co-MoS_2/rGO hybrids	0.5 M H_2SO_4	—	2.9×10^{-2}	65	50	90
Metallic-MoS_2 nanosheets	0.5 M H_2SO_4	—	0.1	—	41	91
1T-2H MoS_2 nanosheets	0.5 M H_2SO_4	90	—	126	35	92
1T-MoS_2 nanosheets	1 M H_2SO_4	—	—	130	49	93
1T-MoS_2 nanosheets	0.5 M H_2SO_4	—	—	100	40	94
1T/2H MoS_2 nanosheets	0.5 M H_2SO_4	—	—	−234	46	95
MoS_2 nanosheets/N-doped carbon nanofibers	0.5 M H_2SO_4	—	2.4×10^{-2}	135	48	96
N-doped MoS_2 nanosheets	0.5 M H_2SO_4	—	—	−168	40.5	97
MoS_2/carbon black nanocomposites	0.5 M H_2SO_4	−78	3.6×10^{-2}	200	39	98
C-MoS_2 hierarchical spheres	0.5 M H_2SO_4	−103	—	159	56.1	99
MoS_2 nanosheets/graphene frameworks	0.5 M H_2SO_4	—	—	−121	46.3	100
MoS_2 thin films	0.5 M H_2SO_4	—	0.6×10^{-2}	—	140	101
Ni-doped MoS_2 nanosheets	1 M KOH	—	0.98	98	60	102
MoS_2/Ni_3S_2 heterostructures	1 M KOH	50	—	110	83	103
$Ni(OH)_2$/MoS_2 heterostructures	1 M KOH	20	—	80	60	104
MoS_2-Graphene-Ni networks	0.1 M KOH	−25	—	—	98	105
MoS_2@Ni/carbon cloth	1 M KOH	30	0.807	91	89	70
Ni-doped MoS_2/Ni foam	1 M NaOH	—	—	207	65	106

transport properties. Therefore, different strategies have been established to alter the electrocatalytic activity of MoS_2 nanosheets such as interfacial composition, thickness, defects, and surface properties. We outline the recent approaches (see Table 9.2), which focuses on the two main issues to enhance the performance of MoS_2 nanosheets, *i.e.*, (i) creating more active sites; (ii) improving the charge transport property.[7,53,58] Current strategies for increasing the effective active sites and improving electron transportation in electrocatalytic reactions include introducing edge engineering, defect engineering, strain engineering, doping engineering, interlayer engineering, interface engineering and phase engineering, which are further briefly summarized below.

9.5.1.1 *Engineering Edges*

When MoS_2 layers are reduced in the lateral dimension, it gives rise to exposed edges and corners of MoS_2 nanosheets, which can give high electrocatalytic HER performance by increasing the number of active sites.[7,24] In recent years, several potential synthetic approaches have been exploited in order to increase the number of active centers for improving the electrocatalytic performance of MoS_2 nanosheets. Specifically, Kibsgaard *et al.*[57] designed the surface of a MoS_2 structure to expose the edge and corner planes favorably for high-performance electrocatalysis through the combination of large-area contiguous thin films with a tunable thickness of highly ordered mesoporous design of MoS_2 (Figure 9.5(a)). The high surface curving of this mesostructured electrocatalyst exposed with a huge number of active edge sites, also possessing a high surface area, results in a high electrocatalytic performance in a HER. The results indicate that by morphological control of MoS_2 nanosheets at the nanoscale, the surface structure of the electrocatalyst at the atomic scale changed and especially exposed more catalytically active edge sites for improved performance.

9.5.1.2 *Engineering Defects*

Defects existing in MoS_2 materials are usually identified as the active centers for electrocatalytic reactions. The 2H-MoS_2 shows poor electrocatalytic activity due to the catalytically inert basal plane atoms on its surface.[57] However, increasing the effective active sites in MoS_2 nanosheets can be done by introducing defects.[108] Consequently, by introducing defects on the basal planes of the MoS_2 nanosheets the unique electronic, physical and chemical properties of the MoS_2 electrocatalysts can be activated. The main issues of defect engineering associated with the performance of the electrocatalysts are in the forms of active centers, for *e.g.*, uncoordinated atoms, pores, kinks, and terraces, which can make basal planes with more active sites for high electrocatalytic performance. Moreover, the number of defects present in the MoS_2 nanosheets can also have a certain effect on the electrocatalytic performance.[109] As a result, Ar or O_2 plasma has been widely

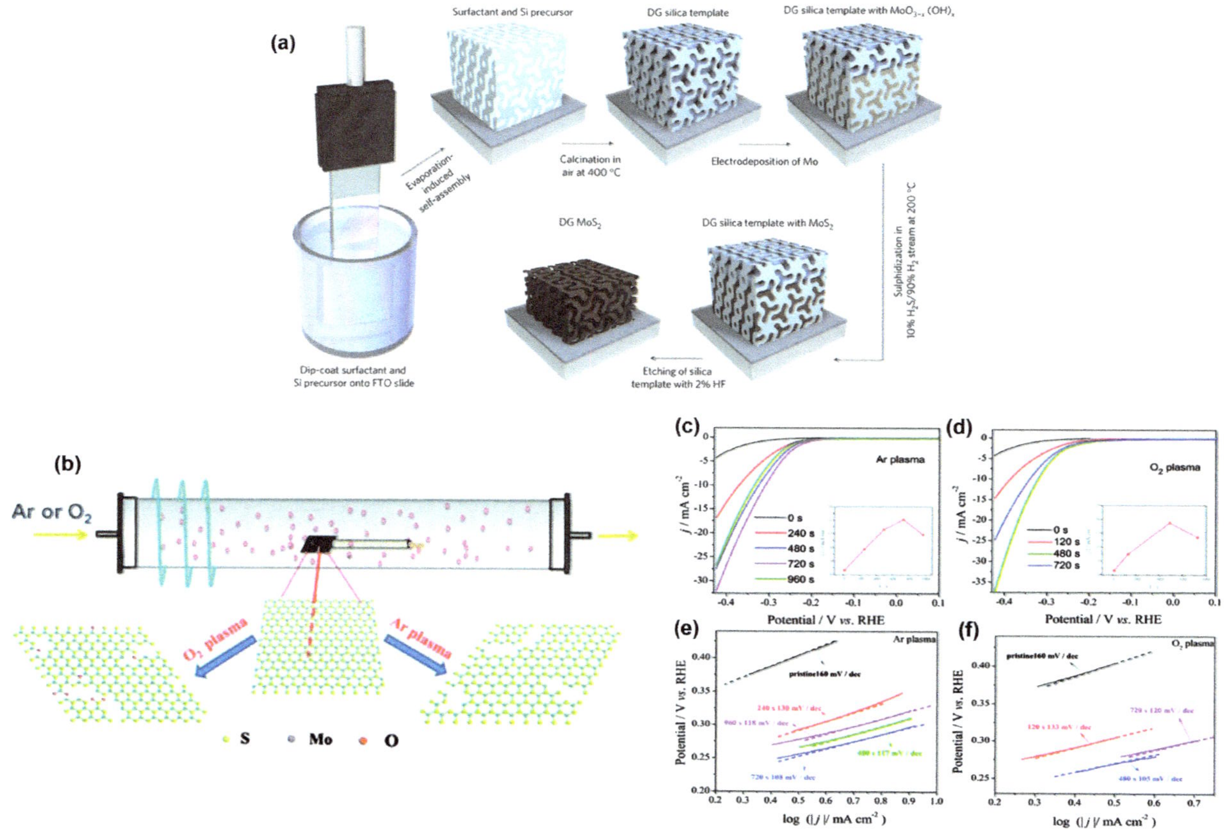

Figure 9.5 (a) Synthesis procedure and a structural model for mesoporous MoS_2 with double-gyroid morphology. (b) Schematic of plasma treatment for MoS_2 nanosheets. (c) Polarization curves of Ar plasma-treated MoS_2. (d) Polarization curves of O_2 plasma-treated MoS_2. (e) Tafel plots of an Ar plasma-sample and (f) Tafel plots of an O_2 plasma-treated sample.
(a) Reproduced from ref. 57 with permission from Springer Nature, Copyright 2012. (b–f) Reproduced from ref. 107 with permission from the Royal Society of Chemistry.

involved in activating the catalytically inert basal planes of MoS_2 nanosheets. For example, Tao *et al.*[107] described a common plasma approach for making active edge sites of MoS_2 nanosheets, which is beneficial to greatly enhance the electrocatalytic performance for the HER. Moreover, MoS_2 thin films on Ti foil are synthesized by a typical chemical vapor deposition technique using the flexible extents of Ar or O_2 plasma for the formation of defect centers, which thereby alters the electronic properties. By controlling the plasma irradiation time, the surface of the MoS_2 structure might be suitably controlled with improved electrocatalytic activity for the HER (Figure 9.5(b–f)).

9.5.1.3 Engineering Strain

The strain in MoS_2 nanosheets alters the electronic property and thereby optimizes electrocatalytic performance.[3,110,111] Strain engineering is likely to have a particularly strong effect on MoS_2 since it undergoes a strain-induced semiconducting to metallic transition with considerable charge redistribution. Li *et al.*[66] described effective tuning of the basal planes in a 2H-MoS_2 monolayer by producing and straining the S-vacancies (Figure 9.6). DFT calculations indicate that the S-vacancies introduced gap states permitting favorable hydrogen adsorption. The number of S-vacancy sites increases the strength of hydrogen adsorption, allowing the simultaneous operation of the hydrogen adsorption free energy (ΔG_H) and the active site density. Straining the S-vacancy sites shifts the gap states nearer to the Fermi level resulting in an optimal ΔG_H. Experimentally, the optimized strained MoS_2 monolayers with S-vacancies acting as active sites display high performance intrinsic HER activity.

9.5.1.4 Doping Engineering

Heteroatom doped MoS_2 nanosheets can disturb the basal plane and also expand the interlayer spacing and thereby alter their d-band electronic properties, which can facilitate high electrocatalytic performance.[53,112] Heteroatom doping engineering is usually further classified into metal-doped catalysts and nonmetal-doped catalysts. The design of various kinds of nonmetal-doped MoS_2 nanosheets with unique electronic properties gives an improved electrocatalytic performance. Xie *et al.*[43] incorporated an oxygen atom in MoS_2 nanosheets using controllable disorder engineering and their studies show that more unsaturated sulfur atoms are generated as active sites as the degree of disorder increases, and oxygen atom incorporation can efficiently reduce the band gap, which improves the intrinsic conductivity of the MoS_2 electrocatalyst (Figure 9.7(a–d)). Therefore, by controlling the degrees of disorder, the optimized oxygen incorporated MoS_2 electrocatalyst with an average degree of disorder was determined, which holds both the increased active sites and the high intrinsic conductivity, revealing higher electrocatalytic performance for the HER. Consequently,

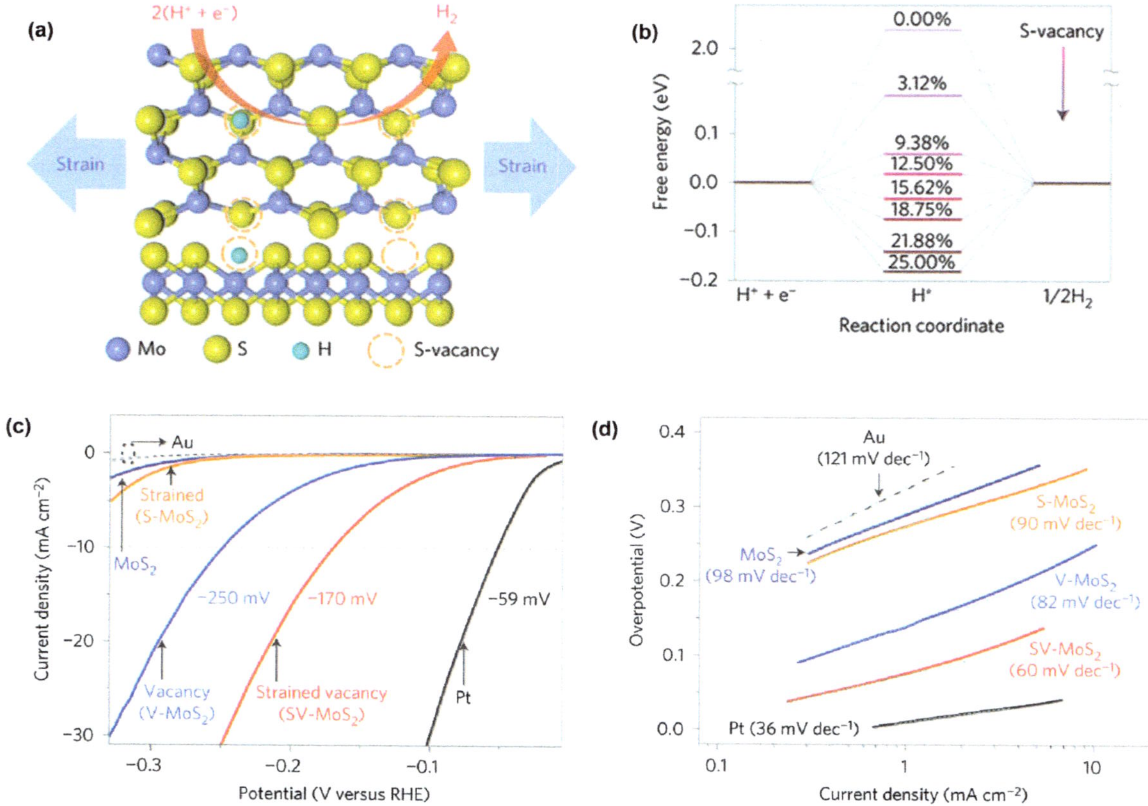

Figure 9.6 (a) Schematic of the top and side views of MoS_2 with strained S-vacancies on the basal plane, where S-vacancies serve as active sites for hydrogen evolution and applied strain further tunes the HER activity. (b) Free energy *versus* the reaction coordinate of HER for the S-vacancy range of 0–25%. (c) LSV curves for the Au substrate, Pt electrode, as-transferred MoS_2, strained MoS_2 without S-vacancies, unstrained MoS_2 with S-vacancies, and strained MoS_2 with S-vacancies. (d) Corresponding Tafel plots of the LSV curves in c.
Reproduced from ref. 66 with permission from Springer Nature, Copyright 2015.

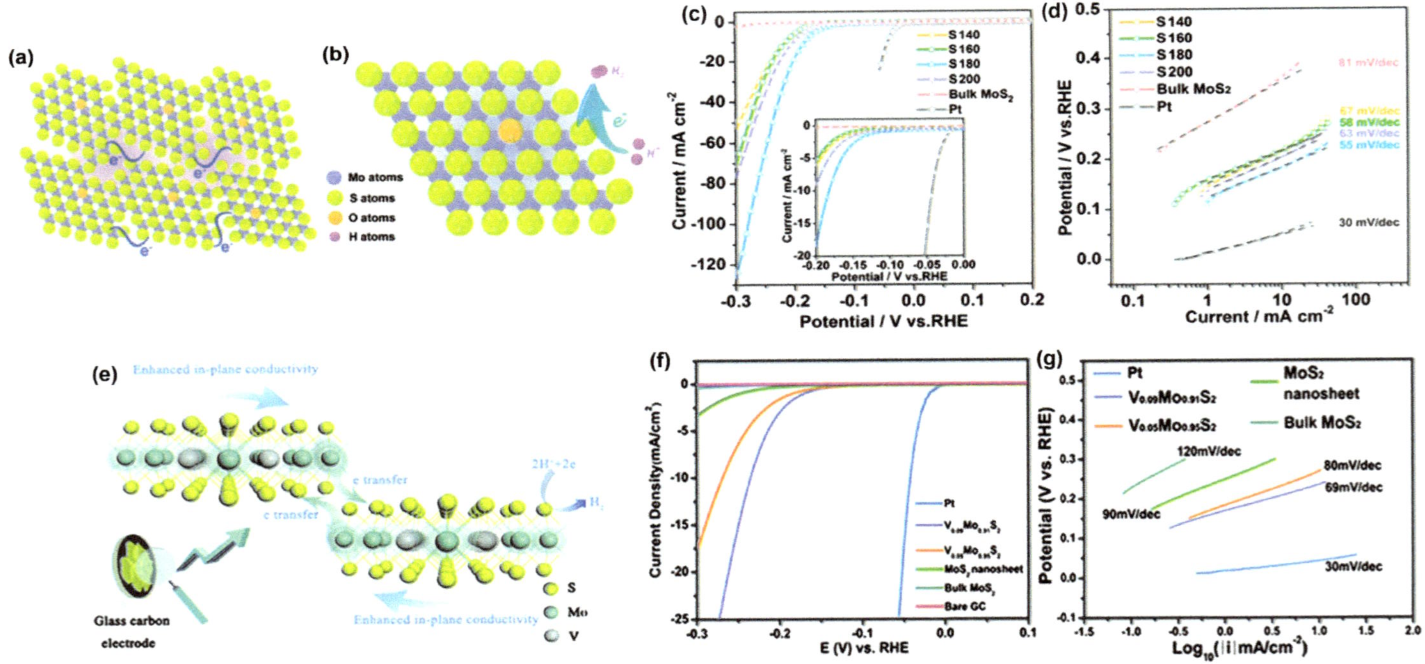

Figure 9.7 (a) Schematic illustration of the disordered structure in oxygen incorporated MoS_2 nanosheets. (b) The model illustrates the HER process of an individual oxygen-incorporated MoS_2. (c) Polarization curves of the oxygen incorporated MoS_2 nanosheets. Inset: enlargement of the region near the onset potential, and (d) corresponding Tafel plots. (e) Schematic illustration of the enhanced catalytic activity of the vanadium-doped MoS_2 nanosheet catalysts due to the enhanced in-plane conductivity *via* vanadium doping and the improved electron transfer between the overlapped nanosheet edges triggered by their semimetallic nature based on the calculation results. (f) Polarization curves recorded on a GC electrode consisting of various vanadium-doped MoS_2 nanosheet catalysts and commercial MoS_2. (g) The corresponding Tafel plots derived from (f). (a–d) Reproduced from ref. 43 with permission from American Chemical Society, Copyright 2013. (e–g) Reproduced from ref. 78 with permission from the Royal Society of Chemistry.

the optimized electrocatalyst shows a low onset overpotential of 120 mV, a small Tafel slope of 55 mV dec^{-1}, as well as the attainment of excellent long-term stability, indicating that the optimized oxygen-incorporated MoS_2 nanosheets could be a promising substitute for Pt-based HER electrocatalysts.

The incorporation of metal atoms into the MoS_2 nanosheets as co-catalysts has been utilized, in addition to the nonmetal-doped MoS_2 catalysts, to create high-performance electrocatalytic activity. Sun *et al.*[78] described a novel form of intralayer vanadium-doped MoS_2 nanosheets with a fewer number of S–(V, Mo)–S layers that possess an interesting semimetallic nature in a wide range of temperatures. The vanadium-doped MoS_2 electrocatalysts exhibited significantly improved in-plane conductivity and more carrier concentration than those of pure MoS_2 nanosheets, respectively. Benefiting from the improved electrical properties and the 2D in-plane conducting paths that avoid the inferior inner 3D electron transfer, the vanadium-doped MoS_2 nanosheets possess an improved catalytic performance with an overpotential of 0.13 V and a smaller Tafel slope, exhibiting significantly improved electrocatalytic HER performance compared with that of a pure MoS_2 system (Figure 9.7(e–g)). Intralayer doping in MoS_2 nanosheets indicates a new way to design advanced 2D electrocatalysts for high-performance energy storage.

9.5.1.5 Interlayer Engineering

The interlayer expansion of MoS_2 nanosheets can decrease the diffusion energy barrier for ion intercalation, high intralayer conductivity, and provide greater accessibility of electrolyte ions, thus benefiting electrochemical energy storage systems.[113,114] The interlayer expansion of MoS_2 nanosheets can also significantly enhance the diffusion kinetics of ions by reducing their diffusion barriers, especially ions with large sizes and multiple valences, which are always involved in other engineering approaches.[113] The accessible interlayer space of the MoS_2 nanosheets provides rich electrochemical active sites and enhances their electrocatalytic performance.[114,115] Therefore, increasing the interlayer spacing of MoS_2 nanosheets by means of intercalation is beneficial as an effective tool for the interlayer engineering approach and also for enhancing electrochemical properties. Chatti *et al.*[116] reported a facile microwave synthesis of interlayer expanded MoS_2 nanosheets vertically aligned on well-conducting reduced graphene (rGO) as a supporting material. The synthesized interlayer expanded the MoS_2-rGO nanocomposite exhibiting an outstanding HER performance in acidic solution with good stability (Figure 9.8(a–c)). Due to the synergy of the interlayer expanded spacing of MoS_2 ($d_{002} = 10.2$ Å), the very small ultrathin MoS_2 sheets and the preferential vertical alignment of MoS_2 sheets on the rGO ensure higher concentration and accessibility of the edge sites and improved electrical conductivity.

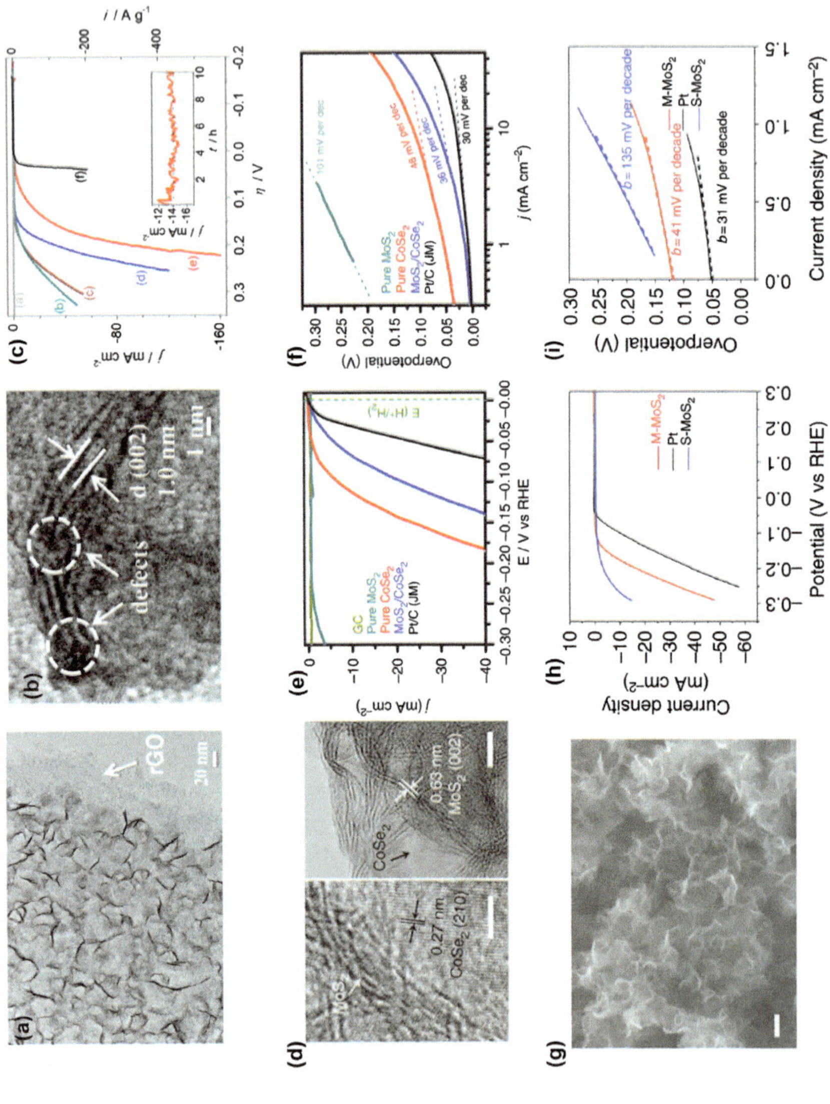
(a)
rGO
20 nm
(b)
defects
d (002)
1.0 nm
1 nm
(c)
(d)
MoS2
0.27 nm
CoSe2 (210)
CoSe2
0.63 nm
MoS2 (002)
(e)
GC
Pure MoS2
Pure CoSe2
MoS2/CoSe2
Pt/C (JM)
E / V vs RHE
j (mA cm−2)
(f)
101 mV per dec
48 mV per dec
36 mV per dec
30 mV per dec
Overpotential (V)
j (mA cm−2)
(g)
(h)
M-MoS2
Pt
S-MoS2
Potential (V vs RHE)
Current density (mA cm−2)
(i)
b = 135 mV per decade
b = 41 mV per decade
b = 31 mV per decade
Overpotential (V)
Current density (mA cm−2)

9.5.1.6 Interface Engineering

Interface engineering is a significant strategy for constructing high-performance electrocatalysts with unique properties. The electronic and chemical properties of interface engineered MoS_2 materials can be tuned by the interaction between two interfaces with divergent materials, which can alter the electrocatalytic performance.[111] The interface engineering approach can be further categorized into two types, namely; engineering heterostructure and synergistic interaction. Heterostructure engineering generally contains chemical bonding between two diverse components, while synergistic interaction involves weak physical interaction between two components with the confined transport of electrons, which are always involved with each other.[111] The poor conductivity of MoS_2 can also be increased by designing several heterostructure and synergetic composites.[96] The interfacial construction of MoS_2 nanosheets is a combination of supportive materials such as graphene, carbon nanotubes, and also weak hydrogen binding on the Mo-edges.[111] Thus, interface engineering through the hybridization of MoS_2-based materials could be utilized to alter their electronic and chemical properties. The MoS_2 layers are arranged in parallel on conductive materials, which limits the electron transport from the conductive material to the active sites due to a lack of appropriate conductive pathways.[53,117] Thus, the combination of conducting materials and MoS_2 layers is an alternative and efficient way to prepare unique composite materials, which can prevent MoS_2 from aggregation and restacking and thus greatly improve the electrocatalytic activity of MoS_2 materials as a consequence of the synergistic effect between the MoS_2 nanosheets and the supportive materials.[29] Hence, several heterostructure or interfacial materials have been constructed for rapid electron transport during the electrocatalytic process attaining improved electrocatalytic performance.

Figure 9.8 (a) TEM and (b) HRTEM images of the microwave synthesized MoS_2-rGO composite. (c) Polarization curves of the microwave synthesized MoS_2-rGO composite compared with different modified electrodes for the HER; the inset shows a long-term chronoamperogram measured for MoS_2-rGO. (d) HRTEM images of a MoS_2/$CoSe_2$ hybrid structure (Scale bars, 5 nm). (e) Polarization curves for HER on a bare GC electrode and modified electrodes containing MoS_2/$CoSe_2$ hybrid, pure MoS_2, pure $CoSe_2$ and commercial Pt/C catalyst. (f) Tafel plot for the various catalysts derived from e. (g) SEM image of metallic MoS_2 nanosheets (Scale bar, 100 nm). (h) Polarization curves of the metallic MoS_2 and semiconducting MoS_2 nanosheets. (i) Corresponding Tafel plots obtained from the polarization curves of h.
(a–c) Reproduced from ref. 116 with permission from American Chemical Society, Copyright 2017. (d–f) Reproduced from ref. 80, https://doi.org/10.1038/ncomms6982, under the terms of the CC BY 4.0 license, https://creativecommons.org/licenses/by/4.0/. (g–i) Reproduced from ref. 91, https://doi.org/10.1038/ncomms10672, under the terms of the CC BY 4.0 license, https://creativecommons.org/licenses/by/4.0/.

In addition, engineering the MoS_2 electrocatalysts on porous supportive materials can afford additional electron transport pathways over porous tunnels.[53]

For example, Gao *et al.*[80] prepared a heterojunction-like structure of MoS_2/$CoSe_2$. The hybrid interface showed improved performance in electrocatalytic activity compared to pure MoS_2 and $CoSe_2$ systems. The MoS_2/$CoSe_2$ electrocatalyst showed an excellent HER electrocatalytic performance with a low onset potential of −11 mV, a small Tafel slope of 36 mV dec^{-1} and a high exchange current density (Figure 9.8(d–f)) in the acidic electrolyte. This is the first non-noble metal catalyst to approach the performance of a state-of-the-art Pt/C catalyst. Similarly, Zheng *et al.*[39] reported a solvent-evaporation-aided intercalation, followed by a solvothermal treatment technique for the synthesis of size-controlled MoS_2 nanosheets/reduced graphene oxide (rGO) composites. The rGO sheets not only give the oxygen-containing functional groups for MoS_2 precursor attachment, but also control the size of the MoS_2 nanosheets owing to the space confinement effect on the rGO layers. Significantly, the resulting MoS_2/rGO composite displays an improved electrocatalytic performance for the HER with a low onset overpotential of 140 mV, a high cathodic current density, and small Tafel slope of 41 mV dec^{-1}. The excellent electrocatalytic performance of the MoS_2/rGO composite could be attributed to the collective effect of more exposed active sites and high intrinsic electrical conductivity.

9.5.1.7 *Phase Engineering*

Tuning of the basal planes by the phase transition from 2H- to 1T-MoS_2 nanosheets alters their band gaps (*i.e.*, semiconducting to metallic behavior) and improves their electron transfer kinetics, which induces high electrocatalytic performance.[1,111] The alteration of one of the S layers in 2H phase MoS_2 produces a 1T phase MoS_2 crystal structure possessing an octahedral coordination of metal atoms corresponding to the trigonal symmetry. The electrocatalytic performances of these two phases of MoS_2 are completely different. The catalytically active sites in 2H-MoS_2 are created only from the unsaturated atoms and dangling bonds present at the edges and the corners, while in 1T-MoS_2, the basal planes are also catalytically active in addition to the unsaturated atoms and dangling bonds, due to the presence of metallic behavior. Thus, due to the presence of metallic behavior in the 1T-MoS_2 nanosheets, rapid electrode kinetics and high intrinsic electrical conductivity were exhibited for high-performance electrocatalytic applications.[118,119] Geng *et al.*[91] reported a hydrothermal process for preparing stable 2D metallic MoS_2 nanosheets in water. The stability of the metallic MoS_2 corresponds with the adsorption of a monolayer of water molecules on both sides of the nanosheets, which reduces restacking and also prevents aggregation in the water. The metallic MoS_2 nanosheets show excellent stability in water and improved electrocatalytic activity for the HER, with a low potential of −175 mV at a current density of 10 mA cm^{-2} and a Tafel slope of

41 mV dec^{-1}. In contrast, semiconducting MoS_2 exhibited poor HER activity with a potential of −274 mV at 10 mA cm^{-2} and a Tafel slope of 135 mV dec^{-1} (Figure 9.8(g–i)). Thus, the electron transfer property in metallic 1T-MoS_2 is an important factor for improving the electrocatalytic HER performance.

9.5.2 New Outcomes for Tuning the Catalytic Activity of MoS_2-based Materials

More interestingly, Yu *et al.*[101] reported the layer-dependent electrocatalysis of MoS_2 materials and described the layer-dependent interlayer hopping of the electron mechanism. Their studies indicate that the atoms in both the edge site and basal plane could be catalytically active sites, and the edge sites displayed improved electrocatalytic performance, because they provide an easier path for the hopping of electrons than the atoms in the basal plane (Figure 9.9(a–d)). The result shows that increasing the hopping efficiency of electrons is a significant development for the rational design of MoS_2 materials with optimum electrocatalytic activities. The efficiency of hopping

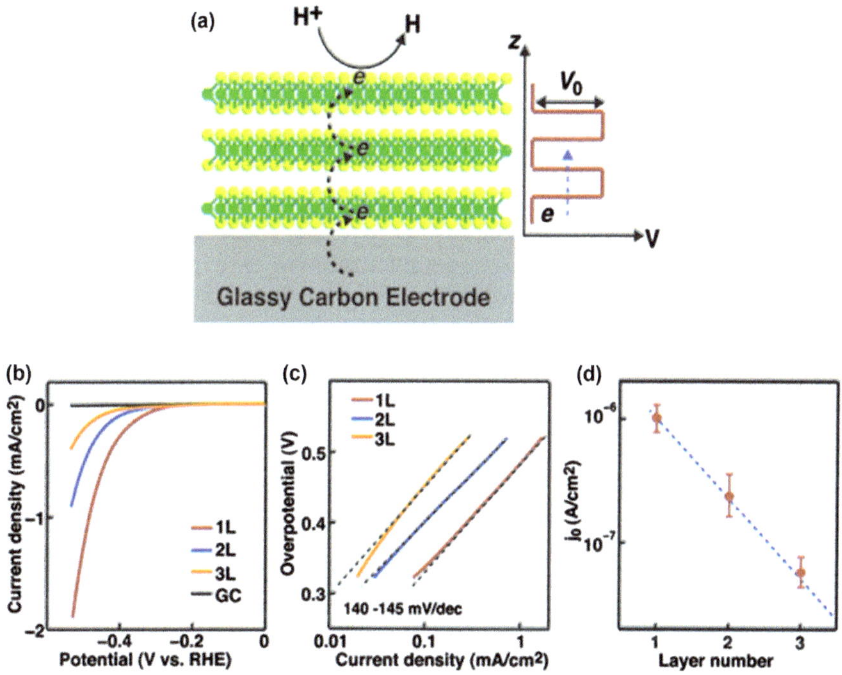

Figure 9.9 (a) Interlayer hopping of electrons in the vertically upward direction of MoS_2 layers. (b) Polarization curves of the synthesized monolayer (1L), bilayer (2L), trilayer (3L) MoS_2 films and bare GC substrates. (c) Tafel plots of the MoS_2 films. (d) The exchange current density of the MoS_2 film *versus* the number of layers.
Reproduced from ref. 101 with permission from American Chemical Society, Copyright 2014.

is controlled by the interlayer coupling of the electron orbitals. Thus, approaches that can increase interlayer coupling, such as intercalation of metal ions or atoms, are expected to improve the electrocatalytic activity of MoS_2 materials.

9.6 Conclusions and Outlook

In this chapter, we have outlined current progress in the improvement of 2D layered MoS_2 nanosheets and their composite materials for achieving high electrocatalytic performance for the HER. The discussion was aimed at the design and optimization of MoS_2-based materials and the emerging approaches for enhancing their electrocatalytic activity. Tweaking of the active sites and the charge transport conductivities are recognized as important concerns in MoS_2-based electrocatalysts for achieving high electrocatalytic performance. The electrocatalytic performance of MoS_2-based materials can be significantly improved with the aid of different types of approaches such as engineering edges and engineering defects for increasing the active sites, and strain engineering, doping engineering, interlayer engineering, interface engineering, and phase engineering for improving the charge transport property. Moreover, the advantages of both increased active sites and improved charge transport properties can be united by suitable structural engineering, which is significant for researchers for high electrocatalytic performance in the emerging 2D layered MoS_2-based electrocatalysts. Furthermore, 2D layered MoS_2 nanosheets and their composite materials have been found to be efficient electrocatalysts for high electrocatalytic HER performance.

Despite the significant attempts demonstrated to date in the research area of MoS_2 nanosheets and their composite materials as electrocatalysts, there are still some significant tasks in interpreting these attempts into practical applications for making devices, and also in both the development of in-depth knowledge and enhancement in the electrocatalytic performance. The preparation of MoS_2-based materials on a large scale, with high quality and unique properties is essential for practical applications. The doping and design of heterostructures of MoS_2 with active materials will lead to tailoring the electrical and chemical properties for high electrocatalytic performance. One of the significant challenges is to engineer high electrocatalytic performance for the HER by the interfacial design and optimization of MoS_2 nanosheets and their composite materials in a highly controllable fashion. Nevertheless, at the current stage, most researchers have been involved in the basic combination/hybridization of MoS_2 nanosheets with other conducting materials, without an in-depth analysis on their structural arrangement, the interaction between each component, optimization of components and surface exposure. In addition, increasing the interlayer distance of MoS_2 nanosheets has been given less consideration to improve the HER activity, although increasing the interlayer is an effective approach to tune intralayer conductivity and gives the electrolyte ions greater accessibility. Moreover, researchers should focus more on the effective preparation of metallic 1T-MoS_2 nanosheets for high electrocatalytic performance due to the active basal planes and high intrinsic electrical conductivity. A future direction lies in the current emerging design strategies that will continue to be used for enhancing the electrocatalytic activity of MoS_2 nanosheets and their composite materials. Interface engineering of TMD materials such as $MoSe_2$, WS_2, VS_2, *etc.*, with various support materials to provide a novel emerging material with tunable electrical and chemical properties is one of the most promising prospects in this research, which will definitely provide a new door for the exploration of potential applications in catalysis, electronic devices and energy storage.

Acknowledgements

This work was supported by the University of Science and Technology Beijing. S.J. acknowledges the China Postdoctoral Science Foundation for funding research grant (No. 2018M631334).

References

1. M. Chhowalla, H. S. Shin, G. Eda, L.-J. Li, K. P. Loh and H. Zhang, *Nat. Chem.*, 2013, **5**, 263–275.
2. H. Zhang, *ACS Nano*, 2015, **9**, 9451–9469.
3. S. Jayabal, G. Saranya, J. Wu, Y. Liu, D. Geng and X. Meng, *J. Mater. Chem. A*, 2017, **5**, 24540–24563.
4. M. Chhowalla, Z. Liu and H. Zhang, *Chem. Soc. Rev.*, 2015, **44**, 2584–2586.
5. Y. Huang, J. Guo, Y. Kang, Y. Ai and C. M. Li, *Nanoscale*, 2015, **7**, 19358–19376.
6. S. Manzeli, D. Ovchinnikov, D. Pasquier, O. V. Yazyev and A. Kis, *Nat. Rev. Mater.*, 2017, **2**, 17033.
7. G. Zhang, H. Liu, J. Qua and J. Li, *Energy Environ. Sci.*, 2016, **9**, 1190–1209.
8. V. Mlinar, *Phys. Chem. Chem. Phys.*, 2017, **19**, 15891–15902.
9. H. Tao, Y. Gao, N. Talreja, F. Guo, J. Texter, C. Yan and Z. Sun, *J. Mater. Chem. A*, 2017, **5**, 7257–7284.
10. Q. Fu and X. Bao, *Chem. Soc. Rev.*, 2017, **46**, 1842–1874.
11. X. Huang, Z. Zeng and H. Zhang, *Chem. Soc. Rev.*, 2013, **42**, 1934–1946.
12. X. Duan, C. Wang, A. Pan, R. Yu and X. Duan, *Chem. Soc. Rev.*, 2015, **44**, 8859–8876.
13. Q. H. Wang, K. K. Zadeh, A. Kis, J. N. Coleman and M. S. Strano, *Nat. Nanotech.*, 2012, **7**, 699–712.
14. W. Zhang, P. Zhang, Z. Su and G. Wei, *Nanoscale*, 2015, **7**, 18364–18378.
15. B. L. Li, J. Wang, H. L. Zou, S. Garaj, C. T. Lim, J. Xie, N. B. Li and D. T. Leong, *Adv. Funct. Mater.*, 2016, **26**, 7034–7056.

16. M.-R. Gao, Y.-F. Xu, J. Jiang and S.-H. Yu, *Chem. Soc. Rev.*, 2013, **42**, 2986–3017.
17. X. Zhang, Z. Lai, C. Tan and H. Zhang, *Angew. Chem., Int. Ed.*, 2016, **55**, 8816–8838.
18. J. Ping, Z. Fan, M. Sindoro, Y. Ying and H. Zhang, *Adv. Funct. Mater.*, 2017, **27**, 1605817.
19. C. Tan and H. Zhang, *Chem. Soc. Rev.*, 2015, **44**, 2713–2731.
20. W. T. Hong, M. Risch, K. A. Stoerzinger, A. Grimaud, J. Suntivich and Y. S. Horn, *Energy Environ. Sci.*, 2015, **8**, 1404–1427.
21. J. Xie and Y. Xie, *Chem. – Eur. J.*, 2016, **22**, 3588–3598.
22. A. Splendiani, L. Sun, Y. Zhang, T. Li, J. Kim, C.-Y. Chim, G. Galli and F. Wang, *Nano Lett.*, 2010, **10**, 1271–1275.
23. D. Xiao, G.-B. Liu, W. Feng, X. Xu and W. Yao, *Phys. Rev. Lett.*, 2012, **108**, 196802.
24. I. Song, C. Park and H. C. Choi, *RSC Adv.*, 2015, **5**, 7495–7514.
25. A. N. Enyashin, L. Yadgarov, L. Houben, I. Popov, M. Weidenbach, R. Tenne, M. B. Sadan and G. Seifert, *J. Phys. Chem. C*, 2011, **115**, 24586–24591.
26. M. Acerce, D. Voiry and M. Chhowalla, *Nat. Nanotechnol.*, 2015, **10**, 313–318.
27. G. Eda, T. Fujita, H. Yamaguchi, D. Voiry, M. Chen and M. Chhowalla, *ACS Nano*, 2012, **6**, 7311–7317.
28. M. Kertesz and R. Hoffmann, *J. Am. Chem. Soc.*, 1984, **106**, 3453–3460.
29. X. Cao, C. Tan, X. Zhang, W. Zhao and H. Zhang, *Adv. Mater.*, 2016, **28**, 6167–6196.
30. J. Dolinska, A. Chidambaram, W. Adamkiewicz, M. Estili, W. Lisowski, M. Iwan, B. Palys, E. J. R. Sudholter, F. Marken, M. Opallo and L. Rassaei, *J. Mater. Chem. B*, 2016, **4**, 1448–1457.
31. G. Yang, C. Zhu, D. Du, J. Zhu and Y. Lin, *Nanoscale*, 2015, **7**, 14217–14231.
32. X. Huang, Z. Zeng, S. Bao, M. Wang, X. Qi, Z. Fan and H. Zhang, *Nat. Commun.*, 2013, **4**, 1444.
33. J. Kim, S. Byun, A. J. Smith, J. Yu and J. Huang, *J. Phys. Chem. Lett.*, 2013, **4**, 1227–1232.
34. Y. Zhang, Q. Ji, G.-F. Han, J. Ju, J. Shi, D. Ma, J. Sun, Y. Zhang, M. Li, X.-Y. Lang, Y. Zhang and Z. Liu, *ACS Nano*, 2014, **8**, 8617–8624.
35. X. Xu, Z. Y. Fan, S. J. Ding, D. M. Yu and Y. P. Du, *Nanoscale*, 2014, **6**, 5245–5250.
36. T. Liu, C. Wang, X. Gu, H. Gong, L. Cheng, X. Shi, L. Feng, B. Sun and Z. Liu, *Adv. Mater.*, 2014, **26**, 3433–3440.
37. S. K. Kim, J. J. Wie, Q. Mahmooda and H. S. Park, *Nanoscale*, 2014, **6**, 7430–7435.
38. L. Ma, Y. Hu, G. Zhu, R. Chen, T. Chen, H. Lu, Y. Wang, J. Liang, H. Liu, C. Yan, Z. Tie, Z. Jin and J. Liu, *Chem. Mater.*, 2016, **28**, 5733–5742.
39. X. Zheng, J. Xu, K. Yan, H. Wang, Z. Wang and S. Yang, *Chem. Mater.*, 2014, **26**, 2344–2353.

40. K. Zhou, J. Liu, Y. Shi, S. Jiang, D. Wang, Y. Hu and Z. Gui, *ACS Appl. Mater. Interfaces*, 2015, 7, 6070–6081.
41. X. Zhang, A. Selkirk, S. Zhang, J. Huang, Y. Li, Y. Xie, N. Dong, Y. Cui, L. Zhang, W. J. Blau and J. Wang, *Chem. – Eur. J.*, 2017, **23**, 3321–3327.
42. X. Huang, M. Leng, W. Xiao, M. Li, J. Ding, T. L. Tan, W. S. V. L. Lee and J. Xue, *Adv. Funct. Mater.*, 2017, **27**, 1604943.
43. J. Xie, J. Zhang, S. Li, F. Grote, X. Zhang, H. Zhang, R. Wang, Y. Lei, B. Pan and Y. Xie, *J. Am. Chem. Soc.*, 2013, **135**, 17881–17888.
44. D. Gopalakrishnan, D. Damien and M. M. Shaijumon, *ACS Nano*, 2014, **8**, 5297–5303.
45. D. O. Dumcenco, H. Kobayashi, Z. Liu, Y. S. Huang and K. Suenaga, *Nat. Commun.*, 2013, **4**, 1351.
46. K. Pramoda, M. Kaur, U. Gupta and C. N. R. Rao, *Dalton Trans.*, 2016, **45**, 13810–13816.
47. X. Huang, B. Zheng, Z. Liu, C. Tan, J. Liu, B. Chen, H. Li, J. Chen, X. Zhang, Z. Fan, W. Zhang, Z. Guo, F. Huo, Y. Yang, L.-H. Xie, W. Huang and H. Zhang, *ACS Nano*, 2014, **8**, 8695–8701.
48. C. L. Tan, X. Y. Qi, X. Huang, J. Yang, B. Zheng, Z. F. An, R. F. Chen, J. Wei, B. Z. Tang, W. Huang and H. Zhang, *Adv. Mater.*, 2014, **26**, 1735–1739.
49. L. Yan, H. Shi, X. Sui, Z. Deng and L. Gao, *RSC Adv.*, 2017, 7, 23573–23582.
50. Z. Gu, P. D. Luna, Z. Yang and R. Zhou, *Phys. Chem. Chem. Phys.*, 2017, **19**, 3039–3045.
51. D. Voiry, A. Goswami, R. Kappera, C. e Silva Cde, D. Kaplan, T. Fujita, M. Chen, T. Asefa and M. Chhowalla, *Nat. Chem.*, 2015, **7**, 45–49.
52. Y. Wang, B. Chen, D. H. Seo, Z. J. Han, J. I. Wong, K. Ostrikov, H. Zhang and H. Y. Yang, *NPG Asia Mater.*, 2016, **8**, e268.
53. Y. Chen, K. Yang, B. Jiang, J. Li, M. Zeng and L. Fu, *J. Mater. Chem. A*, 2017, **5**, 8187–8208.
54. J. V. Lauritsen, J. Kibsgaard, S. Helveg, H. Topsoe, B. S. Clausen, E. Laegsgaard and F. Besenbacher, *Nat. Nanotechnol.*, 2007, **2**, 53–58.
55. T. Liao, Z. Sun, C. Sun, S. X. Dou and D. J. Searles, *Sci. Rep.*, 2014, **4**, 6256.
56. M. V. Bollinger, K. W. Jacobsen and J. K. Norskov, *Phys. Rev. B*, 2003, **67**, 085410.
57. J. Kibsgaard, Z. Chen, B. N. Reinecke and T. F. Jaramillo, *Nat. Mater.*, 2012, **11**, 963–969.
58. X. Chia, A. Y. S. Eng, A. Ambrosi, S. M. Tan and M. Pumera, *Chem. Rev.*, 2015, **115**, 11941–11966.
59. M. S. Dresselhaus and I. L. Thomas, *Nature*, 2001, **414**, 332–337.
60. X. Zou and Y. Zhang, *Chem. Soc. Rev.*, 2015, **44**, 5148–5180.
61. K. Ojha, S. Saha, P. Dagar and A. K. Ganguli, *Phys. Chem. Chem. Phys.*, 2018, **20**, 6777–6799.
62. D. Merki and X. Hu, *Energy Environ. Sci.*, 2011, **4**, 3878–3888.
63. M. Zeng and Y. Li, *J. Mater. Chem. A*, 2015, **3**, 14942–14962.

64. S. J. R. Neale, D. A. C. Brownson, G. C. Smith, D. A. G. Sawtell, P. J. Kellya and C. E. Banks, *Nanoscale*, 2015, 7, 18152–18168.
65. E. Skulason, V. Tripkovic, M. E. Bjorketun, S. Gudmundsdottir, G. Karlberg, J. Rossmeisl, T. Bligaard, H. Jonsson and J. K. Nørskov, *J. Phys. Chem. C*, 2010, **114**, 18182–18197.
66. H. Li, C. Tsai, A. L. Koh, L. Cai, A. Contryman, A. H. Fragapane, J. Zhao, H. S. Han, H. C. Manoharan, F. A. Pedersen, J. K. Nørskov and X. Xiaolin Zheng, *Nat. Mater.*, 2016, **15**, 48–53.
67. Z. Lv, N. Mahmood, M. Tahir, L. Pan, X. Zhang and J. Zou, *Nanoscale*, 2016, **8**, 18250–18269.
68. G. W. Shim, W. Hong, S. Y. Yang and S.-Y. Choi, *J. Mater. Chem. A*, 2017, **5**, 14950–14968.
69. R. Subbaraman, D. Tripkovic, D. Strmcnik, K. C. Chang, M. Uchimura, A. P. Paulikas, V. Stamenkovic and N. M. Markovic, *Science*, 2011, **334**, 1256–1260.
70. Z. Xing, X. Yang, A. M. Asiri and X. Sun, *ACS Appl. Mater. Interfaces*, 2016, **8**, 14521–14526.
71. N. Danilovic, R. Subbaraman, D. Strmcnik, K. C. Chang, A. P. Paulikas, V. R. Stamenkovic and N. M. Markovic, *Angew. Chem., Int. Ed.*, 2012, **51**, 12495–12498.
72. M. A. Abbas and J. H. Bang, *Chem. Mater.*, 2015, **27**, 7218–7235.
73. Y. Huang, R. J. Nielsen, W. A. Goddard and M. P. Soriaga, *J. Am. Chem. Soc.*, 2015, **137**, 6692–6698.
74. Q. Tang and D.-e. Jiang, *ACS Catal.*, 2016, **6**, 4953–4961.
75. Z. Wu, B. Fang, Z. Wang, C. Wang, Z. Liu, F. Liu, W. Wang, A. Alfantazi, D. Wang and D. P. Wilkinson, *ACS Catal.*, 2013, **3**, 2101–2107.
76. Y. Yin, J. Han, Y. Zhang, X. Zhang, P. Xu, Q. Yuan, L. Samad, X. Wang, Y. Wang, Z. Zhang, P. Zhang, X. Cao, B. Song and S. Jin, *J. Am. Chem. Soc.*, 2016, **138**, 7965–7972.
77. D. Voiry, R. Fullon, J. Yang, C. e Silva Cde, R. Kappera, I. Bozkurt, D. Kaplan, M. J. Lagos, P. E. Batson, G. Gupta, A. D. Mohite, L. Dong, D. Er, V. B. Shenoy, T. Asefa and M. Chhowalla, *Nat. Mater.*, 2016, **15**, 1003–1009.
78. X. Sun, J. Dai, Y. Guo, C. Wu, F. Hu, J. Zhao, X. Zeng and Y. Xie, *Nanoscale*, 2014, **6**, 8359–8367.
79. J. Ekspong, T. Sharifi, A. Shchukarev, A. Klechikov, T. Wågberg and E. Gracia-Espino, *Adv. Funct. Mater.*, 2016, **26**, 6766–6776.
80. M.-R. Gao, J.-X. Liang, Y.-R. Zheng, Y.-F. Xu, J. Jiang, Q. Gao, J. Li and S.-H. Yu, *Nat. Commun.*, 2015, **6**, 5982.
81. Y. Li, J. Wang, X. Tian, L. Ma, C. Dai, C. Yang and Z. Zhou, *Nanoscale*, 2016, **8**, 1676–1683.
82. Y.-J. Tang, Y. Wang, X.-L. Wang, S.-L. Li, W. Huang, L.-Z. Dong, C.-H. Liu, Y.-F. Li and Y.-Q. Lan, *Adv. Energy Mater.*, 2016, **6**, 1600116.
83. J. Zhang, S. Liu, H. Liang, R. Dong and X. Feng, *Adv. Mater.*, 2015, **27**, 7426–7431.

84. Y. Zhang, L. Zuo, Y. Huang, L. Zhang, F. Lai, W. Fan and T. Liu, *ACS Sustainable Chem. Eng.*, 2015, **3**, 3140–3148.
85. J. Zhang, T. Wang, L. Liu, K. Du, W. Liu, Z. Zhu and M. Li, *J. Mater. Chem. A*, 2017, **5**, 4122–4128.
86. S.-K. Park, D. Y. Chung, D. Ko, Y.-E. Sung and Y. Piao, *J. Mater. Chem. A*, 2016, **4**, 12720–12725.
87. H. Huang, L. Chen, C. Liu, X. Liu, S. Fang, W. Liua and Y. Liu, *J. Mater. Chem. A*, 2016, **4**, 14577–14585.
88. Y. Luo, D. Huang, M. Li, X. Xiao, W. Shi, M. Wang, J. Su and Y. Shen, *Electrochim. Acta*, 2016, **219**, 187–193.
89. K. Ojha, S. Saha, S. Banerjee and A. K. Ganguli, *ACS Appl. Mater. Interfaces*, 2017, **9**, 19455–19461.
90. E. H. Bafrooei and N. S. Shamszadeh, *RSC Adv.*, 2016, **6**, 95979–95986.
91. X. Geng, W. Sun, W. Wu, B. Chen, A. Al-Hilo, M. Benamara, H. Zhu, F. Watanabe, J. Cui and T.-P. Chen, *Nat. Commun.*, 2016, **8**, 10672.
92. L. Cai, W. Cheng, T. Yao, Y. Huang, F. Tang, Q. Liu, W. Liu, Z. Sun, F. Hu, Y. Jiang, W. Yan and S. Wei, *J. Phys. Chem. C*, 2017, **121**, 15071–15077.
93. A. Ejigu, I. A. Kinloch, E. Prestat and R. A. W. Dryfe, *J. Mater. Chem. A*, 2017, **5**, 11316–11330.
94. D. Voiry, M. Salehi, R. Silva, T. Fujita, M. Chen, T. Asefa, V. B. Shenoy, G. Eda and M. Chhowalla, *Nano Lett.*, 2013, **13**, 6222–6227.
95. D. Wang, X. Zhang, S. Bao, Z. Zhang, H. Fei and Z. Wu, *J. Mater. Chem. A*, 2017, **5**, 2681–2688.
96. Y. Guo, X. Zhang, X. Zhang and T. You, *J. Mater. Chem. A*, 2015, **3**, 15927–15934.
97. R. Li, L. Yang, T. Xiong, Y. Wu, L. Cao, D. Yuan and W. Zhou, *J. Power Sources*, 2017, **356**, 133–139.
98. P. Cao, J. Peng, J. Li and M. Zhai, *J. Power Sources*, 2017, **347**, 210–219.
99. L. Yang, W. Zhou, J. Lu, D. Hou, Y. Ke, G. Li, Z. Tang, X. Kang and S. Chen, *Nano Energy*, 2016, **22**, 490–498.
100. W. Zhou, K. Zhou, D. Hou, X. Liu, G. Li, Y. Sang, H. Liu, L. Li and S. Chen, *ACS Appl. Mater. Interfaces*, 2014, **6**, 21534–21540.
101. Y. Yu, S.-Y. Huang, Y. Li, S. N. Steinmann, W. Yang and L. Cao, *Nano Lett.*, 2014, **14**, 553–558.
102. J. Zhang, T. Wang, P. Liu, S. Liu, R. Dong, X. Zhuang, M. Chen and X. Feng, *Energy Environ. Sci.*, 2016, **9**, 2789–2793.
103. J. Zhang, T. Wang, D. Pohl, B. Rellinghaus, R. Dong, S. Liu, X. Zhuang and X. Feng, *Angew. Chem., Int. Ed.*, 2016, **55**, 6702–6707.
104. B. Zhang, J. Liu, J. Wang, Y. Ruan, X. Ji, K. Xu, C. Chen, H. Wan, L. Miao and J. Jiang, *Nano Energy*, 2017, **37**, 74–80.
105. X. Geng, W. Wu, N. Li, W. Sun, J. Armstrong, A. Alhilo, M. Brozak, J. Cui and T. Chen, *Adv. Funct. Mater.*, 2014, **24**, 6123–6129.
106. X. Yin, H. Dong, G. Sun, W. Yang, A. Song, Q. Du, L. Su and G. Shao, *Int. J. Hydrogen Energy*, 2017, **42**, 11262–11269.

107. L. Tao, X. Duan, C. Wang, X. Duan and S. Wang, *Chem. Commun.*, 2015, **51**, 7470–7473.
108. Y. Ouyang, C. Ling, Q. Chen, Z. Wang, L. Shi and J. Wang, *Chem. Mater.*, 2016, **28**, 4390–4396.
109. H. Wang, Q. Zhang, H. Yao, Z. Liang, H.-W. Lee, P.-C. Hsu, G. Zheng and Y. Cui, *Nano Lett.*, 2014, **14**, 7138–7144.
110. J. T.-W. Wang, Z. Wang, S. Pathak, W. Zhang, D. W. deQuilettes, F. Wisnivesky-Rocca-Rivarola, J. Huang, P. K. Nayak, J. B. Patel, H. A. M. Yusof, Y. Vaynzof, R. Zhu, I. Ramirez, J. Zhang, C. Ducati, C. Grovenor, M. B. Johnston, D. S. Ginger, R. J. Nicholas and H. J. Snaith, *Energy Environ. Sci.*, 2016, **9**, 2892–2901.
111. H. Jin, C. Guo, X. Liu, J. Liu, A. Vasileff, Y. Jiao, Y. Zheng and S.-Z. Qiao, *Chem. Rev.*, 2018, **118**, 6337–6408.
112. P. Liu, J. Zhu, J. Zhang, P. Xi, K. Tao, D. Gao and D. Xue, *ACS Energy Lett.*, 2017, **2**, 745–752.
113. K. D. Rasamani, F. Alimohammadi and Y. Sun, *Mater. Today*, 2017, **20**, 83–91.
114. Y. Zhu, L. Peng, Z. Fang, C. Yan, X. Zhang and G. Yu, *Adv. Mater.*, 2018, 1706347.
115. M.-R. Gao, M. K. Y. Chan and Y. Sun, *Nat. Commun.*, 2015, **6**, 7493.
116. M. Chatti, T. Gengenbach, R. King, L. Spiccia and A. N. Simonov, *Chem. Mater.*, 2017, **29**, 3092–3099.
117. B. Qu, X. Yu, Y. Chen, C. Zhu, C. Li, Z. Yin and X. Zhang, *ACS Appl. Mater. Interfaces*, 2015, **7**, 14170–14175.
118. D. Voiry, A. Mohiteb and M. Chhowalla, *Chem. Soc. Rev.*, 2015, **44**, 2702–2712.
119. M. A. Lukowski, A. S. Daniel, F. Meng, A. Forticaux, L. Li and S. Jin, *J. Am. Chem. Soc.*, 2013, **135**, 10274–10277.

Subject Index